58's Are the Greatest

William G. Gessner, Sr.
and
Charles Conklin

HERITAGE BOOKS
2006

HERITAGE BOOKS
AN IMPRINT OF HERITAGE BOOKS, INC.

Books, CDs, and more—Worldwide

For our listing of thousands of titles see our website
at
www.HeritageBooks.com

Published 2006 by
HERITAGE BOOKS, INC.
Publishing Division
65 East Main Street
Westminster, Maryland 21157-5026

Cover design by Charles Conklin

International Standard Book Number: 0-7884-2592-7

ACKNOWLEDGEMENTS

I wish to acknowledge the following friends who were always quick to offer advice and help in the writing of this book. Without that advice and assistance it would have been an impossible chore. Thank you all very much.

I could never list them all, if I have missed a few, I'm sorry and apologize.

The following individuals stand out in my thoughts

John L. Barber
Victor Engel
Louis Behling
Robert Butweiler
George Kass
Richard A. King
George H. Murphy
Dr. Philippe A. Regenstreif
Jack A. Brickell
William Dibbern
Paul Foster
Richard King
Paul Klatt
Frank Sellers
Chuck Suydam
Emile Timmersman
Dean Thomas
James Tillinghast
Otto Witt
William Woodin

I also wish to thank my son William G. Gessner, Jr. for his assistance and proofreading the manuscript in spite of being 1500 miles away.

Last but far from least is my friend who has encouraged and supported all my efforts over the past twelve years in many ways, Mora Rea. Thank you Mora.

DEDICATION

This book is gratefully dedicated to my long-suffering wife Teena who cheerfully shared a pastime of travel, adventure and a lot of typing.

Bill

In dedication to my wonderful family and loving wife Dianne whose encouragement and support made this work possible.

Chas

DEDICATION

This book is gratefully dedicated to my long-suffering wife Teena who cheerfully shared a pastime of travel, adventure and a lot of typing.

Bill

In dedication to my wonderful family and loving wife Dianne whose encouragement and support made this work possible.

Chas

CONTENTS

INTRODUCTION

The reader must not be mislead by the book's title: .58's ARE THE GREATEST. It is not meant to infer that the size of the cartridge is the largest, rather, I am referring to the varied types of ignition systems, case materials, types of weapons and length of years in use.

The low grade, slow to ignite powder in use during the early 1800's required large amounts to enable an average projectile of 6/10ths of an inch to be propelled to sufficient velocity to be of use, hence the great size of the .58 caliber case.

We have taken great care to insure that all of the cartridges displayed in side view on a "grid" page are life size and true to measurements. This should enable someone to place a cartridge on top of an identifying picture with the hope that it will match the size as close as possible if not exact. The headstamps are not displayed as actual size in order to show as much detail as possible. However no guarantee of accuracy is made or implied by the authors, since measurements (particularly case lengths) in the referenced time period can slightly vary from one cartridge to another even amongst two of the same type. These variations of tolerance were sometimes more prevalent from one manufacturer to another due to the relatively primitive manufacturing conditions of the time and are certainly not encountered in modern weapons or munitions.

While it is our hope that this book will inspire interest from a new generation of cartridge collectors, our primary goal was to contribute to this interesting field of study, which has been a continual source of enjoyment for these many years.

I

Centerfire

BERDAN CARTRIDGES

When one thinks of .58 caliber center fire cartridges, the natural reaction is to think of the common Berdan 1 5/8" Musket round; on the surface, a not-so-spectacular, ho-hum sort of a cartridge with an "if you've seen one, you've seen them all" approach.

Nothing could be further from the truth.

Consider these facts. The Berdan cartridge (named after Col. Hiram Berdan, a Union officer) was one of the largest self-contained rifle cartridges ever manufactured in America. It was the basis for a successful primer used extensively throughout Europe. It had a unique case construction containing its own primer anvil. It was made in both copper and brass. It had a life span in excess of fifty years, and was accepted throughout the world as a reliable and dependable combination when used with the various weapons for which it was designed.

The cartridge was developed (notice I did not say invented - more about that later) by Col. Hiram Berdan of Berdan's Sharpshooters fame during the Civil War. In addition to his claim to fame through the organization of this group of sharpshooters, he was a skilled and talented firearms inventor and designer. He resigned his commission in January 1864 to pursue these talents and by November 1864 had submitted a breech-loading rifle for trial. He became associated with Union Metallic Cartridge Company during the mid-sixties and did much of his cartridge development in their plant at Bridgeport, Connecticut.

There are three cartridge patents issued to Col. Berdan that are of major importance in the .58 caliber field.

1. #52818-2-1866- "CAVITY FOR FULMINATE IN BASE OF PIN EXTENDING BACK FROM BULLET".

   This specimen had the primer set flush with the rim and utilized the pin as an anvil to explode the fulminate.

2. #53338-3-20-1866-"OUTSIDE CAP SUNK BELOW THE SURFACE OF THE HEAD OF THE SHELL".

   This is the first type Berdan case with the teat (actually the anvil) thrown up to the side and with a single flash hole. This differed from his later Patent (#82587) which moved the anvil to the center of the case and had primer flush with the rim. There is serious question here if this first style was original with Berdan, or whether he might have again "borrowed" the idea from an earlier English patent. One must keep in mind that these earliest attempts to get away from rim fire ignition to center fire still retained the case construction of the rim fire. The center fire cases were the same as the rim fire except that a reinforcing cup containing the fulminate was inserted and the cup retained by crimping, as with Benet or Martin Bar Anvil methods.

3. #82587-9-1868-"INTERNAL REINFORCE CUP FITS LOOSELY INTO SHELL BUT TIGHTLY AROUND TURNED IN PORTION OF HEAD. PATCHED BULLET. EXTERIOR PRIMING CAP".

This was Berdan's final successful patent (relating to .58s) and a reliable cartridge that would serve U.M.C. in good stead for years to come. The anvil was relocated in the center of the case with flash holes on either side. He had now designed a truly satisfactory re-loadable case and the idea of "patching" the bullet below the case mouth to prevent leading was introduced. The patch would be difficult to identify as the very early rounds had the patch reach only to the case mouth; it did not extend up the projectile as did the .42 Russian Berdan round. The first style primer was nothing more than a flattened percussion cap which required the integral anvil in the case. In 1869, Alfred C. Hobbs was issued a patent for a better designed primer that was highly reliable, and as both Berdan and Hobbs were employed by U.M.C., the two patents were combined into a most successful cartridge, Marcellus Hartley saw the potential of Berdan's cartridge, and lost no time in acquiring the manufacturing rights to both patents #53338 and #82587 for U.M.C.

The principles utilized in these patents were the basis of cartridge construction for many years thereafter.

A paradox is that while the Berdan primer was never accepted in America, it was a tremendous success elsewhere in the world. The Boxer (English) style primer with its removable anvil as adapted to most all American ammunition became an instant success and has continued in popularity to present times.

I mentioned at the beginning of this chapter that Colonel Berdan developed rather than invented the Berdan cartridge. This is the story. During a visit to the Frankford Arsenal, Colonel Berdan (associated at that time with U. M. C.), observed the manufacturing of cartridge cases under the supervision of Colonel Benet, who had designed the inside-primed ignition system. Observing how the primer pocket was formed on the machine by a "bumping" process, Berdan got the idea of how the method could be improved upon. He "acquired" some sample cases, measurements and types and sizes of tools and beat a hasty retreat to Bridgeport, Connecticut.

If you read very carefully Ordnance Memorandum No. 14, "METALLIC CARTRIDGES" Major T. J. Treadwell's 1873 report of activities at the Frankford Arsenal and the remarks under PLATE XXXIX you will find:

> "After seeing the pocket formed of continuous metal, Benets (sic) at this Arsenal 1866, he (Berdan VERY QUIETLY APPLIED (emphasis by author) his teat anvil to it soon however changing the teat or projection from side as at O to the center, Fig. X, forming in connection with the HOBBS primer and an interior gas check cup with brass cover a very reliable cartridge."

To discuss only the Berdan cartridge without reviewing Col. Berdan's accomplishments in converting muzzle-loaders to a cartridge system would not be doing justice to the entire Berdan story.

At the close of the Civil War, the arsenals were filled with thousands of muzzle-loading rifles, which were becoming obsolete with the advancement of technology in the manufacturing of self-contained ammunition. Our army, always noted for being slow to adopt new ideas, and forever trying to find the cheapest way out of a bad situation, began to invite individuals to submit a method to convert this vast store of muzzle-loaders to an economical, but efficient, breech-loading weapon.

In January 1864, Col. Berdan resigned his commission in the U. S. Army to invent such a conversion. Within six months he had devised a system, and he had the rifle ready for testing at the November 12, 1864 trials.

The weapon was judged to be satisfactory, but unfortunately was never accepted by the Board. The Allin conversion appeared to be more desirable and was accepted as the official arm.

The conversion designed by Col. Berdan was described as a double-hinged, cam-locked, breech block; it was the first applied to Springfield rifle muskets. These conversions were capable of using either rim fire or center fire ignition systems depending on the location of the firing pin.

Some years later, when Berdan closely examined the single jointed breech block system in use on the M1866 Springfield rifle, he decided that it was an infringement of his conversion patent. He sued the United States Government in the Court of Claims seeking damages and compensation. However, the wheels of justice turn slowly and the case lingered in the courts for many years; finally, in 1895, the suit was settled for $95,000.

Russia was one of Berdan's first customers and adopted the conversion known as the No. 1 Berdan. This weapon was issued only to an elite corps of Russian marksmen. (sounds like Berdan's Regiment of Sharp-Shooters transplanted in another land!) The special .58 Russian musket cartridge for this weapon is discussed elsewhere in this book.

Remington immediately purchased the rights to manufacture the Berdan rifle (along with his cartridge patents) and soon established customers throughout the world. Although the weapon was accepted by foreign military powers as a means to re-arm with a reliable breech-loader. Neither the American public nor the United States Military Board ever took an interest in the gun. As a civilian venture, it was a failure.

A tribute to the excellence of the Berdan cartridge and its manufacture by the Union Metallic Cartridge Co.., Bridgeport Conn. is paid in the book METALLIC CARTRIDGES, MANUFACTURED AND TESTED AT THE FRANKFORD ARSENAL, PHILADELPHIA, PA. 1860-1873:

> "Of class second, one of the best and most extensively made and experimented with is the Berdan, as made by the Union Metallic Cartridge Company of Bridgeport, Connecticut, in large numbers, for the Russian government, for use in the Berdan breech loading rifle. This cartridge has been most strictly and severely tested during their manufacture, and has proved of great excellence. It is exceedingly ingenious; its re-enforcement simple and effective; its capacity as a reloader fully tested and demonstrated by prolonged and repeated trial, daily, during production of millions of rounds; a number of the shells being reloaded, primed, and fired ten times, and much more extended trials have had for special test of the endurance of the cases in this particular. Its chief distinguishing feature is that its anvil is of the same continuous piece of metal as that of which the case is made. Herein there is no possible displacement of misplacement of the anvil, and it has a fixed position with respect to the primer. The cartridge is singular in this respect, and superior to its rivals that require a separate anvil. In it was a very happy idea hit upon by the inventor of making his anvil by a simple return of the metal of the pocket for the primer. All other anvils are its inferiors in that they have to be handled in assembling the parts of the shell. Another advantage is, it presents a point to the primer inside, rendering it sensitive to the blow of the hammer. The use of the special Hobbs' primer is most excellent in this combination. Other varieties of an excellent re-enforcement may be referred to, as exhibited in the drawings under this class."

THE .58 MUSKET:

.58 Musket
.58 Transformed Musket
.58 Berdan N. M. (New Model)
.58 Caliber N. M. (New Model)

It was described as being adapted to the Needham, Remington, Peabody Transformed, Converted Springfield and Remington Rifles. No doubt foreign countries adapted it to their own weapons, with slight variations; here we are concerned only with American made specimens.

The cartridge has the following minimum and maximum dimensions:

CASE LENGTH	RIM	O.A.	TOTAL GRAINS
1.64" - 1.66"	.712" - .741"	2.07" 2.18"	777 - 910

Generally speaking, the cartridge case is brass, 1 5/8" in length, has a powder charge of eighty grains and a bullet with a diameter of .580", weighing approximately 530 grains. The earliest cases were made of copper and had a bullet very similar to the Minie bullet. Later specimens have a pronounced rounded shape and a relatively flat nose.

The majority of the specimens are unmarked, but had the distinctive ringed head common to U.M.C. manufacture. Winchester also made the round, which was usually not head-stamped. However, Winchester did make a head-stamped round. Their 1891 catalog shows only the .58 Carbine round as being head-stamped, yet I have never heard of such a case in any collection. Strangely enough, the .58 Musket round was illustrated without a head-stamp in the same edition and it was not until Winchester printed their 1913 ammunition catalog does one find the round marked: W.R.A. Co. 58 Musket. I doubt if any United States ammunition manufacturer other than U.M.C., W.R.A. Co., And the United States Cartridge Company made the ammunition. Certainly no government arsenal would have made it because the gun was never an official weapon of our Army.

In addition to the types listed on the following pages, there had to be countless numbers of similar rounds from foreign countries. They would comprise a collection in themselves if they were available. Berdan conversions were sold to Russia, Spain, Japan, Mexico, Puerto Rico, Chile, China, Egypt, Sweden and Denmark, and while U.M.C. supplied the bulk of their ammunition needs, there had to be rounds produced within each country. An example is the Spanish order to U.M.C. of .58 caliber rounds with a case length of 1.63"; was this just a variation of the Russian Berdan Improved No. 1, or a totally different cartridge?

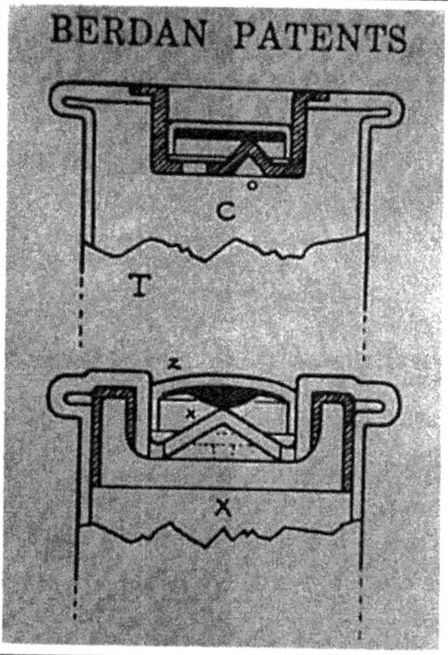

H. Berdan.

Cartridge.

No 53388. Patented Mar. 20. 1866.

Fig: 1

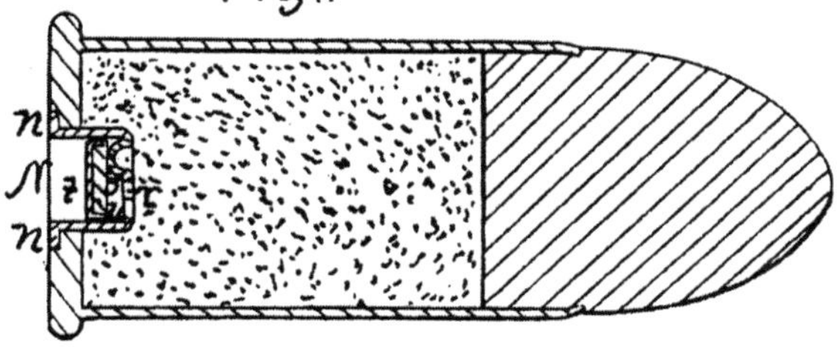

Fig: 2

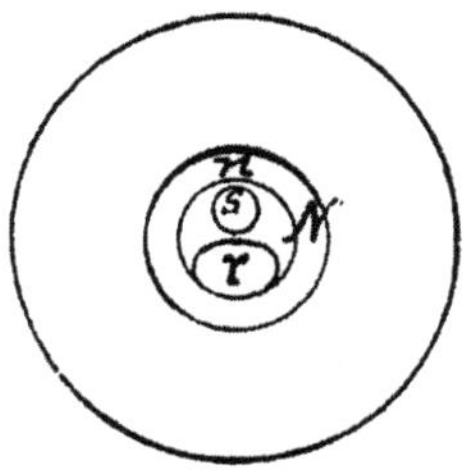

Witnesses:

J. W. Coombs

Henry T. Brown

Inventor:

H Berdan

"One advantage of this cartridge is that it may be loaded before the application of the priming, thus not only rendering its manufacture less dangerous than that of other primed cartridges, but permitting the cartridges to be stored in an unprimed condition, and the priming to be applied just before they are required to be used, thereby not only rendering the storage safer, but preventing the deterioration to which the fulminate priming is subject in cartridges when kept for a long time.

What I claim as my invention, and desire to secure by Letters Patent, is the fixed permanent teat like projection [S] at the bottom of the cup [N] in the head of a cartridge, in combination with the pellet or other priming inserted into and protected within the said cup, substantially as herein set forth."

TITLE: .58 BERDAN MUSKET

HEADSTAMP: Plain

DIMENSIONS:

Rim:	.730"
Head:	.642"
Neck:	.626"
Bullet:	.599"
Case:	1.100"

OTHER DETAILS:

Bullet Grains:	530
Powder Grains:	90
Bullet:	Short, Rounded Nose
Case Material:	Brass
Total Grains:	788

OVERALL: 1.445"

This has to be one of the very early specimens of Berdan.

The round has the typical ringed head peculiar to Union Metallic Cartridge Company. The overall length is shorter than any of the other Berdan Muskets and its weight is much less.

Positive identification of these early rounds is difficult without the original box or container. The general appearance is unlike the many other musket cartridges examined.

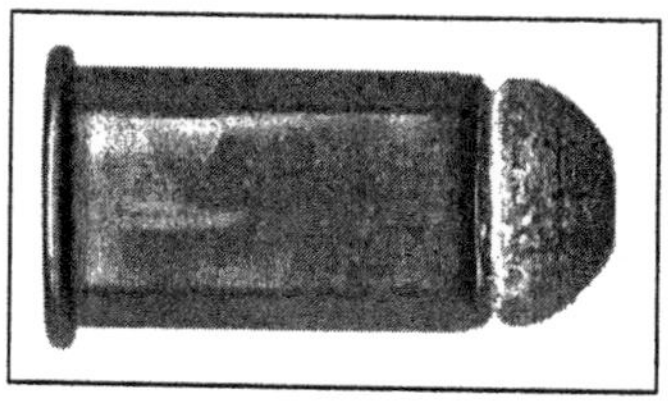

TITLE: .58 BERDAN CARBINE – BOARD DUMMY

HEADSTAMP: Plain

DIMENSIONS:

Rim:	.724"
Head:	.644"
Neck:	.631"
Bullet:	.603"
Case:	1.111"

OTHER DETAILS:

Bullet Grains:	530
Powder Grains:	----
Bullet:	Lead
Case Material:	Brass
Total Grains:	670

OVERALL: 1.605"

It was the custom of ammunition manufacturers to place a colorful board containing dummy rounds of all their products in hardware stores as an advertising gimmick. These rounds can be identified by two large holes in the case for running the wire through the case to attach the specimen to the board.

Assembler's or functioning dummies differ in that they usually have only one hole in the case. Many times a wooden rod is inserted to keep the bullet from being forced down inside the case when being used.

This cartridge is not head stamped, but does have the raised wide ring circling the head which is indicative of Winchester ammunition of this era.

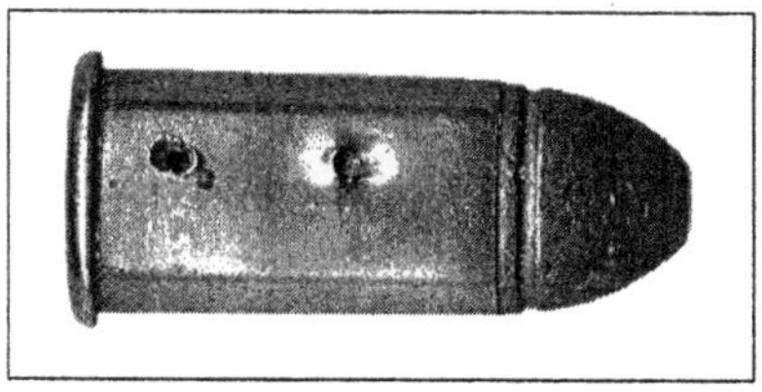

TITLE: 15 mm RUSSIAN CARBINE

HEADSTAMP: 2 G 78 Mx

DIMENSIONS:		OTHER DETAILS:	
Rim:	.684"	Bullet Grains:	556
Head:	.619"	Powder Grains:	40
Neck:	.607"	Bullet:	Tapered Nose
Bullet:	.608"	Case Material:	Brass
Case:	1.119"	Total Grains:	636

OVERALL: 1.672"

Russia adapted the Berdan cartridge system and applied it to a variety of their own muskets and carbines. It would appear logical that while these dimensions vary from the American Union Metallic Cartridge Company round, it was based on the one and one-quarter inch Berdan carbine case.

Although similar to the Berdan round, notice that this specimen is much lighter. This "A" based round has about ten percent less total weight than the average Berdan Carbine cartridge.

It is the author's opinion that the Mx head stamp indicates that the manufacturer was Gevelot & Gaupillat of Issy-Les-Moulineau, France, who on contract made military ammunition from 1870 – 1885 land not of Bulgarian origin as stated in White and Munhall's book, "Cartridge Head-Stamp Guide".

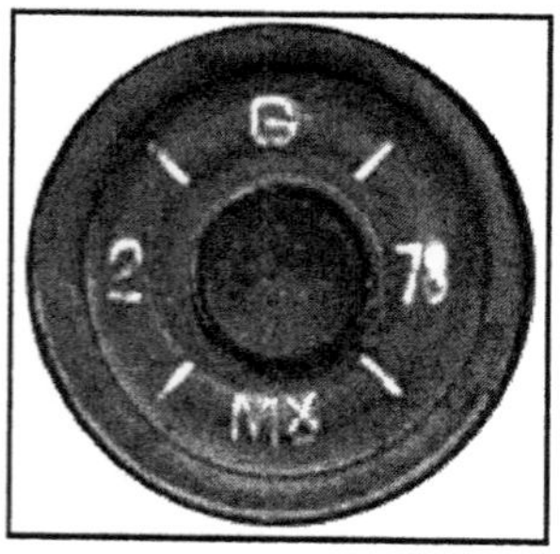

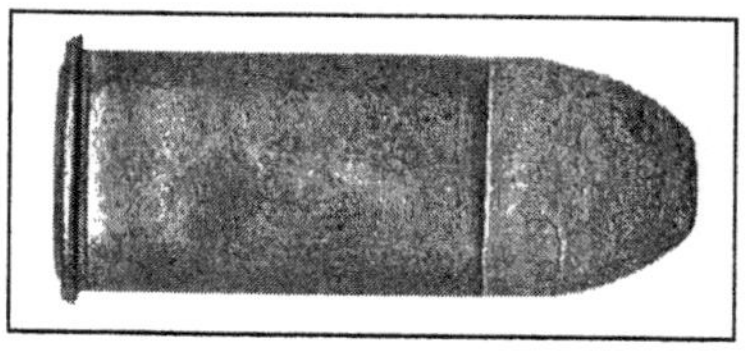

TITLE: .58 BERDAN CARBINE-BLANK

HEADSTAMP: Plain

DIMENSIONS:

Rim:	.730"
Head:	.641"
Neck:	.631"
Bullet:	----
Case:	1.125"

OTHER DETAILS:

Bullet Grains:	----
Powder Grains:	40
Bullet:	----
Case Material:	Brass
Total Grains:	233

OVERALL: 1.125"

This is the blank round to fit the standard Berdan carbine.

The ringed head would indicate manufacture by the Union Metallic Cartridge Company

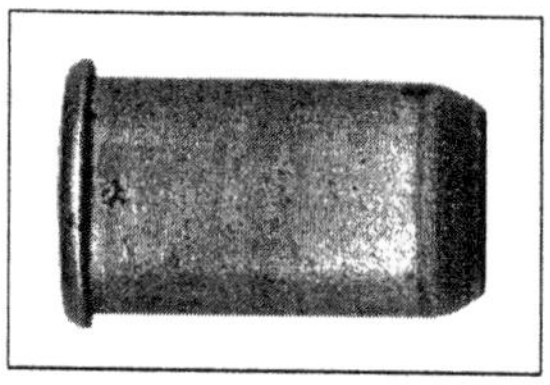

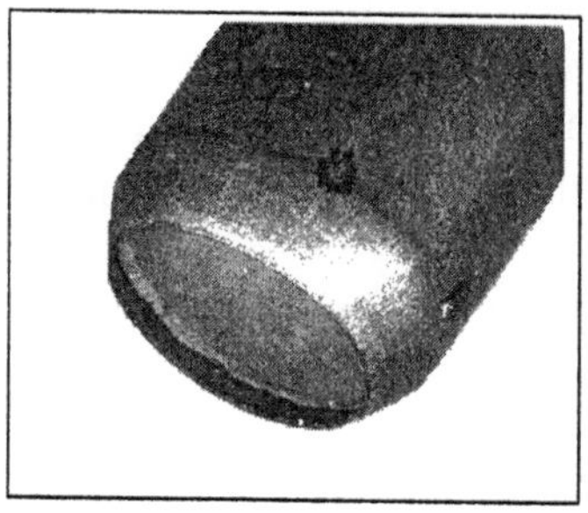

TITLE: .58 BERDAN CARBINE

HEADSTAMP: Plain

DIMENSIONS:		OTHER DETAILS:	
Rim:	.744"	Bullet Grains:	530
Head:	.641"	Powder Grains:	40
Neck:	.617"	Bullet:	Rounded, Flat Nose
Bullet:	.598"	Case Material:	Brass
Case:	1.125"	Total Grains:	732

OVERALL: 1.625"

This short cased version of the Berdan musket cartridge was used in the Remington carbine as well as the Berdan carbine, and will usually be found loaded by the Union Metallic Cartridge Company.

U.M.C. rounds are unmarked, but have the typical ringed head to identify them. The W.R.A. Co. catalog of 1891 pictures their carbine round with a head stamp, while the common Berdan Musket cartridge is not illustrated with a head stamp until the 1913 catalog. I have yet to learn of any carbine so marked.

In 1870, Cuba placed an order for 89,000 carbines with Remington. A very large order for ammunition for these carbines was also given to U.M.C. Co.

The early boxes of carbine ammunition lists the following patents under which the ammunition was made:

Berdan's Patent of 3/20/1866
Berdan's Patent of 9/29/1868
S. W. Woods Patent of 4/1/1862
S. W. Woods Patent of 4/3/1872

The fact that these "shells" can be reloaded many times was also stressed on the green box label.

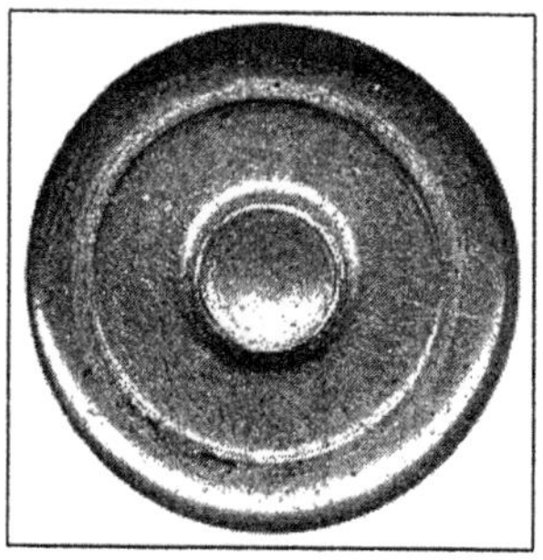

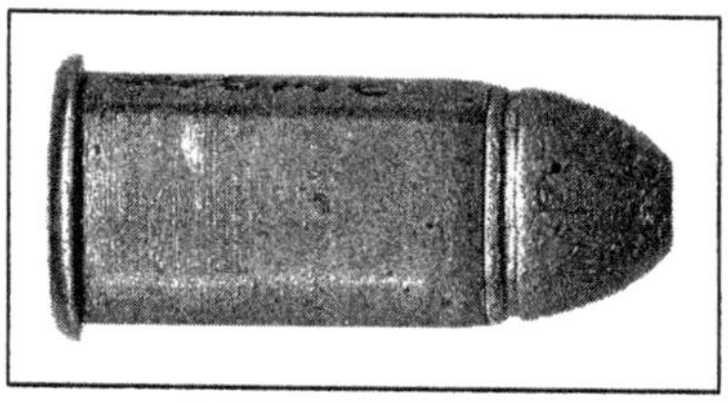

TITLE: .58 FRANKFORD ARSENAL ORIGINAL BERDAN

HEADSTAMP: Plain

DIMENSIONS:

Rim:	.733"
Head:	.647"
Neck:	.635"
Bullet:	.558"
Case:	1.330"

OTHER DETAILS:

Bullet Grains:	
Powder Grains:	
Bullet:	Lead, Flat Nose
Case Material:	Brass
Total Grains:	610

OVERALL: 1.688"

The War Department instructed Frankford Arsenal to celebrate the one-hundredth anniversary of the independence of the United States, at the international Exposition in Philadelphia in 1876.

Great strides had been made in America immediately after the Civil War in developing firearms and ammunition as well as improved technology in the manufacturing of these weapons. This enabled the United States to lead other countries of the world in equipping her army. It was decided by the planning committee that it would be fitting to create an exhibit showing how arms and ammunition manufacturing had been automated in America. The War Department instructed Frankford Arsenal and Springfield Armory to build such a display. Part of the items on display depicted the evolution of cartridges – paper through center fire.

The cartridge described here is Specimen #107 in Section II of this display titled "Miscellaneous Center Primed Musket and Carbine Cartridges". The full description of the round is as follows"

> #107 Brass case, Berdan type, caliber .57, with shallow compressed primer pocket. Burr of vent forms the anvil. Flat nose bullet, flat .308
> O-1.704, C=1.340, R-.730, H-.648, B-.605"

Close examination of Hiram Berdan's Patent #53338-3-1866describes this exact specimen. The text reveals that the reason for recessing the primer below the base of the cartridge was intended as a safety feature. With a recessed primer, pointed bullets would be prevented from striking the primer. Another feature claimed in the patent papers was that the cartridges could be loaded and stored unprimed. When the ammunition was ready to be issued they could then have the primers attached. The purpose to ensure fresh primers rather than having primers deteriorate while in storage.

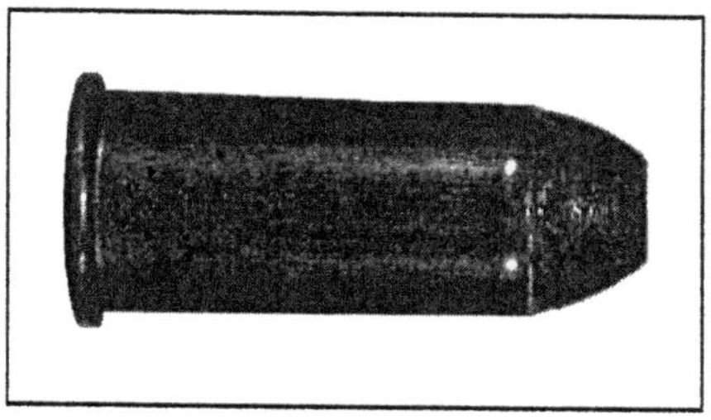

TITLE: .60 RUSSIAN KRNKA RIFLE

HEADSTAMP: RUSSIAN MARKINGS B III O 75 (r)

DIMENSIONS:

Rim:	.777"
Head:	.671"
Neck:	.617"
Bullet:	.601"
Case:	1.470"

OTHER DETAILS:

Bullet Grains:	----
Powder Grains:	----
Bullet:	Lead
Case Material:	Brass
Total Grains:	----

OVERALL: 2.070"

One of the earliest Union Metallic Cartridge Co. cartridge boards displayed this round and apparently erroneously labeled it "60 Russian Berdan".

There has never been a .60 caliber Berdan rifle. Perhaps the reference was to the construction of the Berdan cartridge case with the raised anvil in the primer pocket.

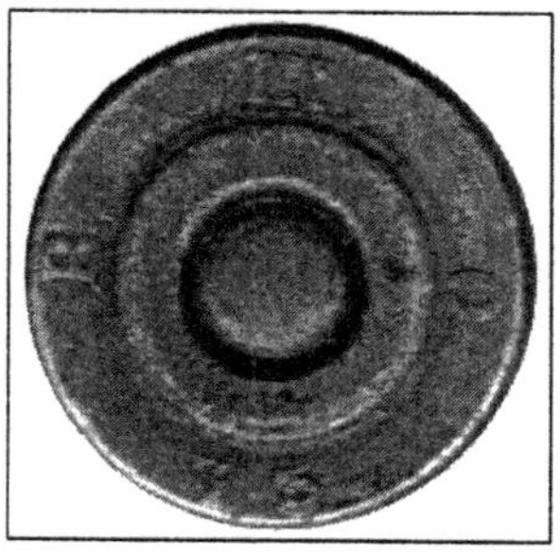

TITLE: .58 RUSSIAN-CARBINE

HEADSTAMP: Plain

DIMENSIONS:		OTHER DETAILS:	
Rim:	.737"	Bullet Grains:	500
Head:	.658"	Powder Grains:	80-85
Neck:	.593"	Bullet:	Round Nose
Bullet:	.538"	Case Material:	Brass
Case:	1.570"	Total Grains:	700

OVERALL: 2.035"

Remarks here must naturally follow the same pattern as the specimen with the 1.562" case length. Similarities between critical measurements are obvious and the total grains weight are within ten grains of each other.

This specimen came from the White & Munhall collection and is coded 8469 on the side of the case. Unfortunately, in their reference file they simply titled it "Snider". The only distinctive features are the tiny copper primer and very slightly raised rib circling the head of the cartridge.

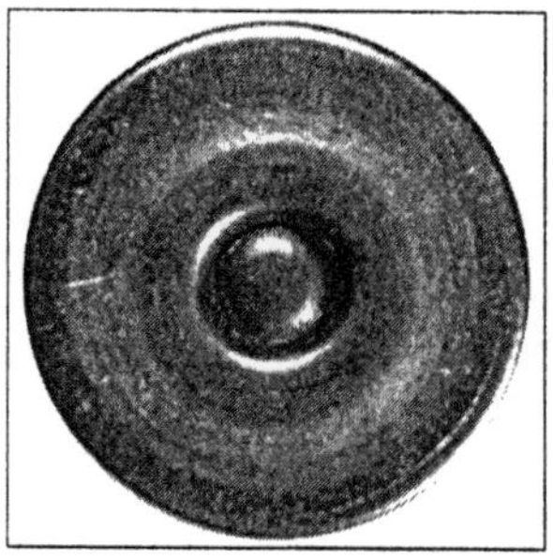

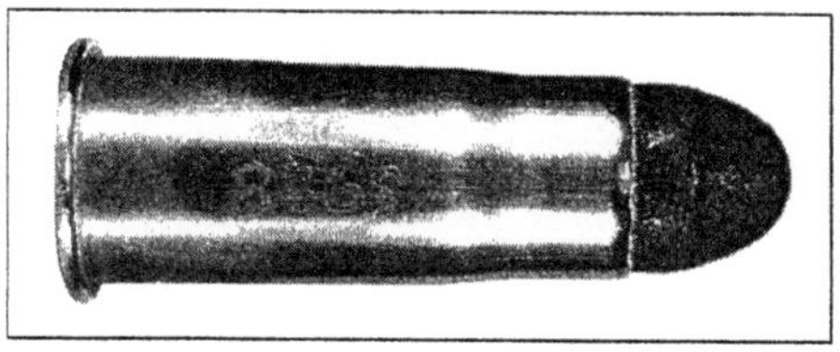

TITLE: 15.24 x 40 Rmm RUSSIAN KRNKA M. 1869.
TRANSFORMED

HEADSTAMP: B.R.O. 1871 (Raised)

DIMENSIONS:

Rim:	.804"
Head:	.672"
Neck:	.652"
Bullet:	.595"
Case:	1.575"

OTHER DETAILS:

Bullet Grains:	355
Powder Grains:	78
Bullet:	Lead, Conical
Case Material:	Brass
Total Grains:	833

OVERALL: 2.880"

This early self-contained Russian round was Russia's first conversion from a muzzle-loader to a breech-loading rifle. It is shorter than the so-called No. 1 improved, and, it had a thick folded head type case instead of the U.M.C. No. 1 cartridge that was of the solid head construction variety.

Two other head-stamps are known; B.R.O. 1872 and B.R.O. 1875, all with the same type raised lettering. The name and location of the Russian Arsenal is unknown.

Like other countries, Russia wished to convert their vast stock of muzzle-loading rifles over to a breech-loader system. The logical choice at this time (1860s) was the Berdan conversion. The No 1 Berdan rifles were only issued to their sharpshooter battalions. The rifle was known as the Model 1866.

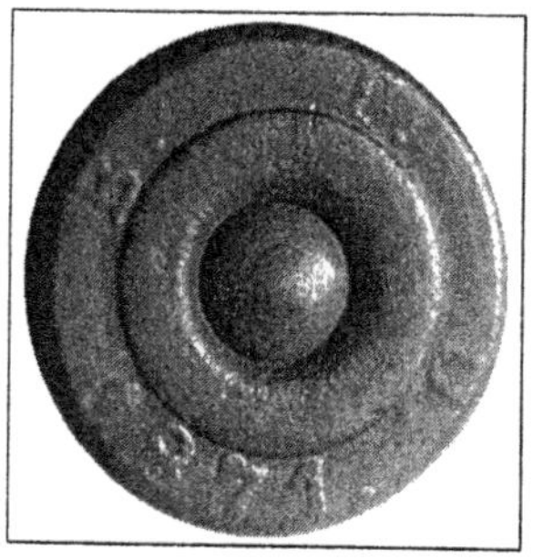

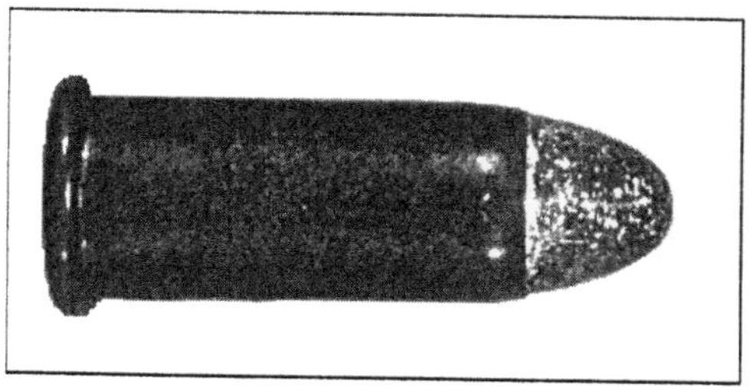

Volume One of the I.C.C.A. has an article by Col. Berk Lewis describing this and other early Russian rifle cartridges)
Titled: THE RUSSIAN NUMBER 1 AND NUMBER 2 BERDAN CARTRIDGES

TITLE: .58 BERDAN MUSKET-BLANKS

HEADSTAMP: Plain

DIMENSIONS:		OTHER DETAILS:	
Rim:	.739	Bullet Grains:	
Head:	.645	Powder Grains:	77
Neck:	.630	Bullet:	Blank
Bullet:		Case Material:	Brass
Case:	1.585" – 1.646"	Total Grains:	Various

OVERALL: 1.585" – 1.646"

To list all the dimensions, shapes and sizes of the many blanks would be quite a task; it seems each manufacturer had its own idea as to what the case legnth, case mouth and style of crimp should be.

Total weight of the various types ranged from 262 grains to 308 grains.

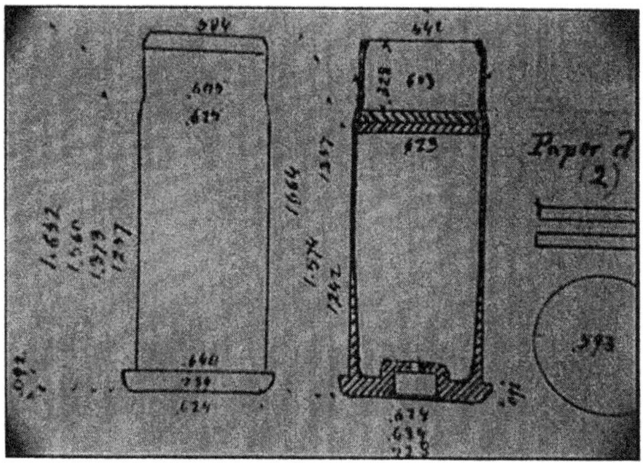

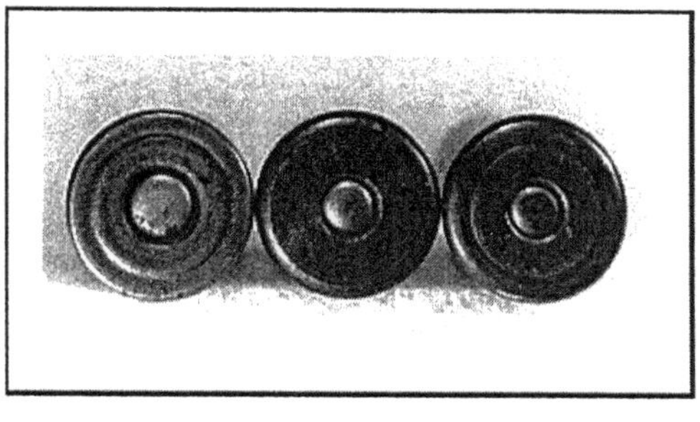

TITLE: MADRID BERDAN RIFLE MUSKET

HEADSTAMP: Plain

DIMENSIONS:

Rim:	.750"
Head:	.657"
Neck:	.619"
Bullet:	.586"
Case:	1.621"

OTHER DETAILS:

Bullet Grains:	NA
Powder Grains:	NA
Bullet:	Lead Smooth
Case Material:	Brass
Total Grains:	NA

OVERALL: 2.138"

A very scarce rifle specimen for the 15 x 41 R Madrid Berdan 1859 converted Spanish muskets.

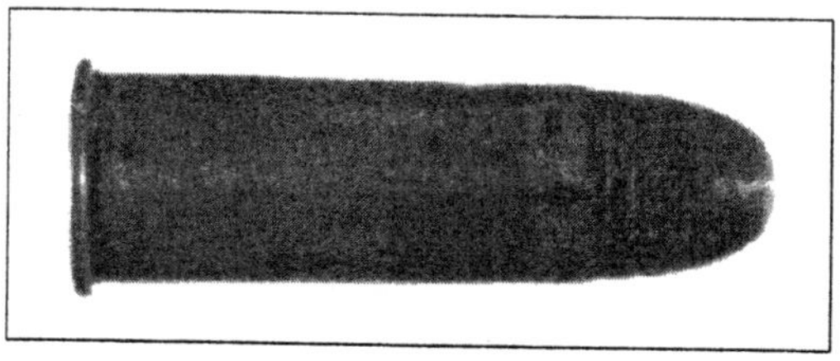

TITLE: .58 MADRID/JAPANESE

HEADSTAMP: Plain

DIMENSIONS:

	MADRID	JAPANESE
Rim:	.774	.774"
Head:	.656"	.672"
Neck:	.619"	.630"
Bullet:	?	?
Case:	1.627"	1.638"

OTHER DETAILS:

Bullet Grains:	----
Powder Grains:	----
Bullet:	----
Case Material:	Brass
Total Grains:	----

OVERALL: 2.09" (Both)

These dimensions were taken from a specimen attached to an 1875 Union Metallic Cartridge Company board which had been made up for the Frankford Arsenal.

The round was titled "FOR MADRID/JAPANESE".

No specimen is known or available

TITLE: .58 BERDAN MUSKET COPPER CASE

HEADSTAMP: Plain

DIMENSIONS:

Rim:	.720"
Head:	.654"
Neck:	.636"
Bullet:	.659"
Case:	1.638"

OTHER DETAILS:

Bullet Grains:	530
Powder Grains:	80
Bullet:	Lead
Case Material:	Copper
Total Grains:	780

OVERALL: 2.146"

A scarce variant of the musket cartridge in a copper case instead of the common brass cased round.

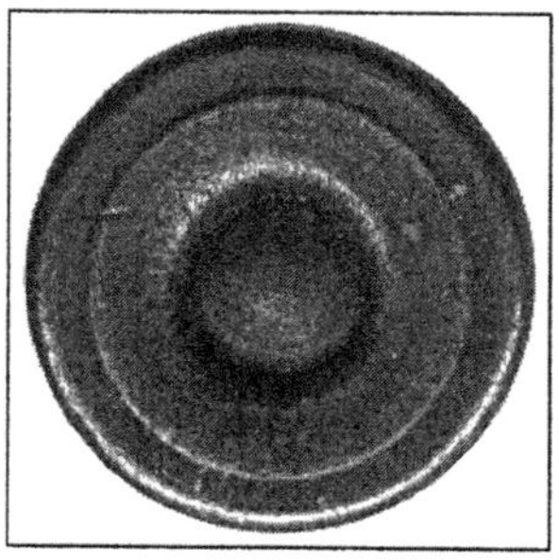

TITLE: .58 RUSSIAN BERDAN IMPROVED NO. 1

HEADSTAMP: Plain

DIMENSIONS:		OTHER DETAILS:	
Rim:	.729"	Bullet Grains:	?
Head:	.639"	Powder Grains:	80-85
Neck:	.615"	Bullet:	Pointed Nose
Bullet:	.585"	Case Material:	Brass
Case:	1.640"	Total Grains:	900

OVERALL: 2.315"

If the reader will refer to case-length 1.600", he will find remarks on the first self-contained Russian Berdan Musket cartridge. This slightly longer cased round soon replaced the shorter one and it became known as the Improved No. 1. The Improved No. 1 was used in the Russian Model 1866 Military Rifle.

Both the original Berdan No. 1 and the Improved No. 1 assume the old muzzle-loading weapons were for the less skillful soldier.

A very excellent article on the Russian Berdan rifles and ammunition was written by the late Colonel B. R. Lewis in Vol. 1 of the International Cartridge Collectors Association annual publication.

TITLE: .58 MUSKET-PHILLIPS CONVERSION

HEADSTAMP: Plain

DIMENSIONS:

Rim:	.731"
Head:	.642"
Neck:	623"
Bullet:	.578"
Case:	1.641"

OTHER DETAILS:

Bullet Grains:	530
Powder Grains:	80
Bullet:	Flat Nose, Grooved
Case Material:	Brass
Total Grains:	795

OVERALL: 2.075"

Specific mention is made on the label that this Berdan Musket cartridge is for a Philips Conversion Musket. Nothing about the dimensions would indicate anything special about the round. It would appear to be the regular Berdan Musket cartridge.

The projectile has the configuration of those bullets loaded in magnetic based Berdan cartridges. This is not an iron base specimen.

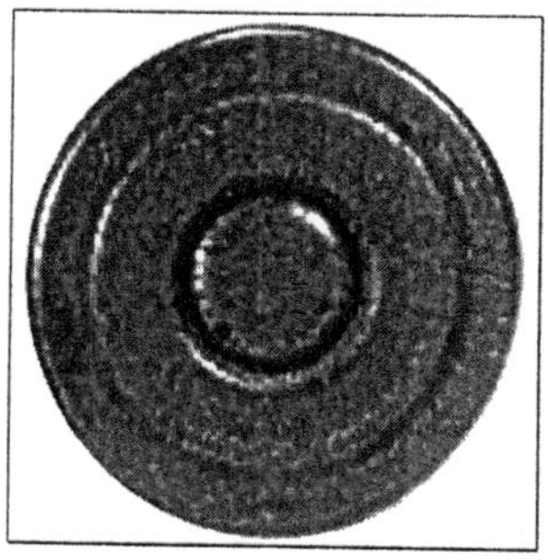

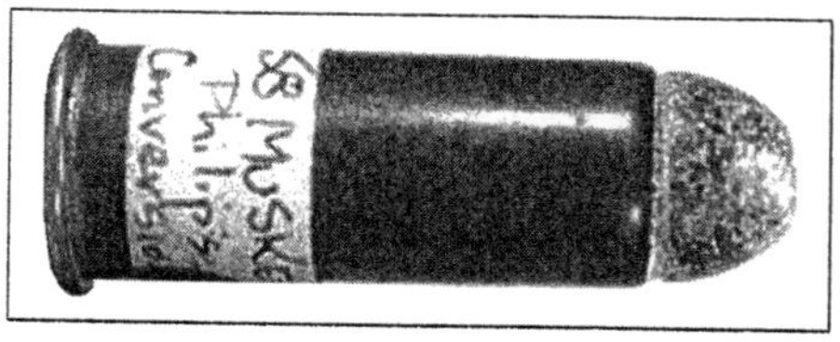

TITLE: .58 BERDAN MUSKET-MAGNETIC

HEADSTAMP: Plain

DIMENSIONS:

Rim:	.741"
Head:	.644"
Neck:	.625"
Bullet:	.591"
Case:	1.642"

OTHER DETAILS:

Bullet Grains:	530
Powder Grains:	80
Bullet:	Smooth
Case Material:	Brass
Total Grains:	828

OVERALL: 2.065"

It has only been in the past few years that this variation of the Berdan Musket has come to light. The primer-reinforcing ring is made of steel or iron and is of the earliest manufacture; later on it was made with a brass ring.

In other calibers, many attempts to use steel or iron for reinforcing the base of the cases failed due to the reaction of the black powder upon the metal of the case head.

Two ways to aid in identification of these iron-cupped rounds is that the grease groove seldom appears above the case mouth, as it does in the later rounds, and the cartridge base is magnetic.

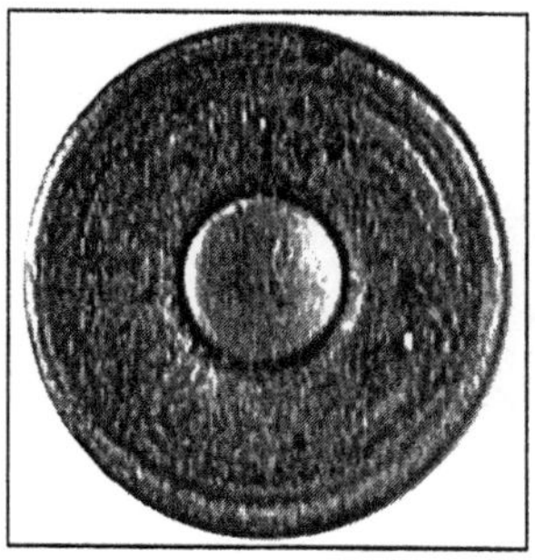

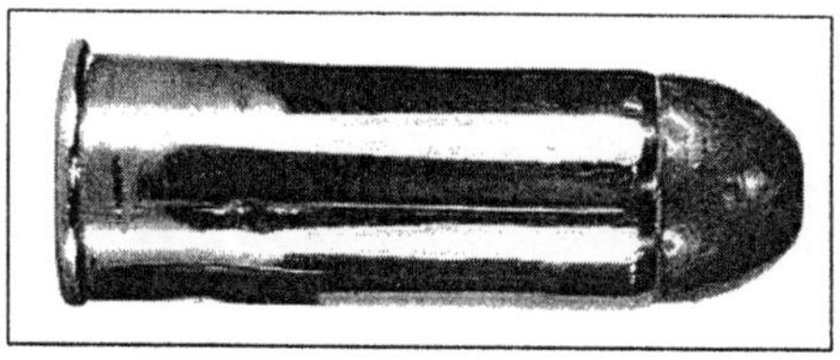

TITLE: .58 BERDAN MUSKET-BOARD DUMMY

HEADSTAMP: Plain

DIMENSIONS:

Rim:	.730"
Head:	.640"
Neck:	.630"
Bullet:	.590"
Case:	1.644"

OTHER DETAILS:

Bullet Grains:	530
Powder Grains:	None
Bullet:	One Groove
Case Material:	Brass
Total Grains:	710

OVERALL: 2.005"

The cartridge board dummy usually has two or more holes punched into the case so that a thin wire can be inserted, providing a means of attaching the specimen to a cartridge board. These boards contained specimens of the company's products and were given to retail stores to display as advertising material.

The salesman's sample was simply an unloaded round with a hole drilled in the case and/or a pierced primer.

It has only been since the 1960s that active interest has been shown in collecting dummy cartridges. It stands to reason that a dummy round is much scarcer, in any caliber, than its loaded counterpart. Only a limited number would be made, especially cartridge board specimens, as compared with the number of loaded rounds manufactured.

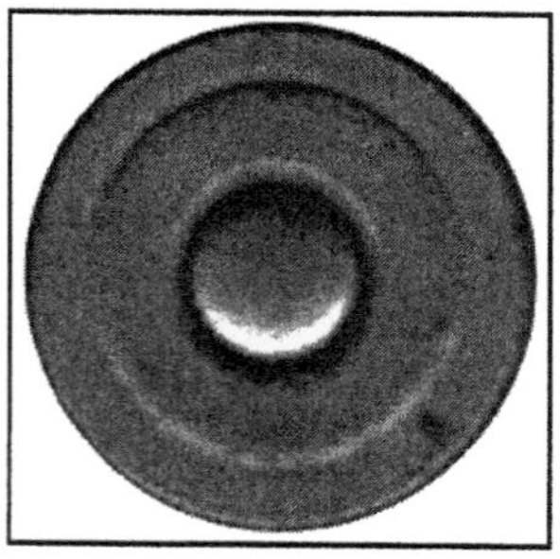

TITLE: .58 BERDAN MUSKET

HEADSTAMP: Plain

DIMENSIONS:

Rim: .732"
Head: .655"
Neck: .625"
Bullet: .595"
Case: 1.652"

OTHER DETAILS:

Bullet Grains: 530
Powder Grains: 80
Bullet: Tapered, One Groove
Case Material: Brass
Total Grains: 766

OVERALL: 2.118"

As pointed out elsewhere in the book, cartridge specimens from White & Munhall were indexed and cataloged but unfortunately were not very explicit in their description.

The case on this specimen is marked 10347 and the index simply stated "Musket". The round is primed with a very small copper primer that may be the Orcutt Primer.

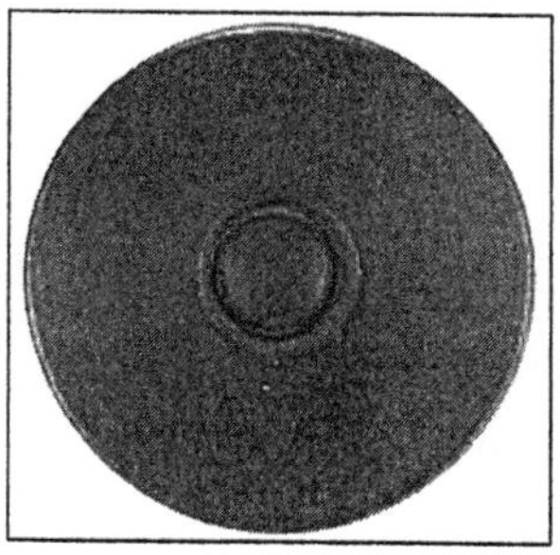

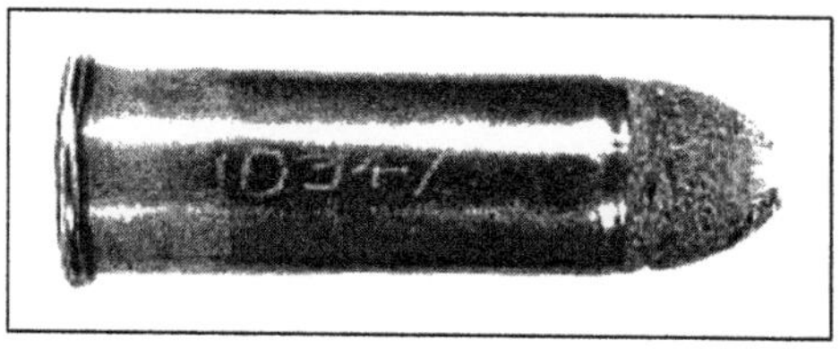

TITLE: .58 BERDAN MUSKET-IRON HEAD

HEADSTAMP: Plain

DIMENSIONS:		OTHER DETAILS:	
Rim:	.727"	Bullet Grains:	530
Head:	.643"	Powder Grains:	80
Neck:	.626"	Bullet:	Tapered 1 Groove
Bullet:	.605"	Case Material:	Brass
Case:	1.656"	Total Grains:	788

OVERALL: 2.125"

An extremely desirable and rare specimen of an experimental case construction, under Milbank's patent #131017 of September 3, 1872.

The head bears a strong resemblance to the iron head, coiled, Boxer cases for the British Snider, except the washer around the copper primer is flatter and wider on this specimen.

These cases are seldom in good condition, as rust, corrosion and the ravages of time take their toll on the iron washers.

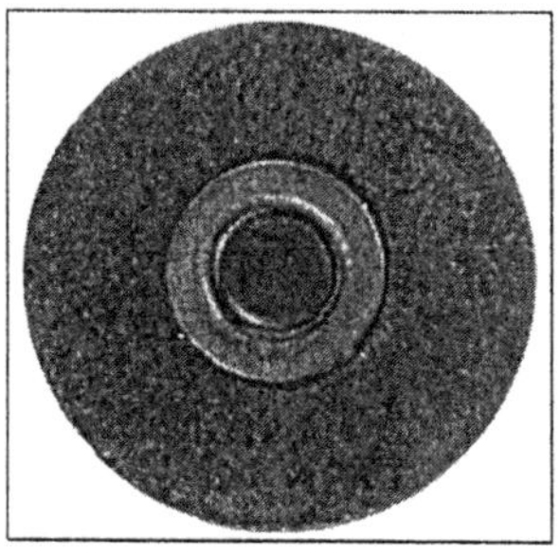

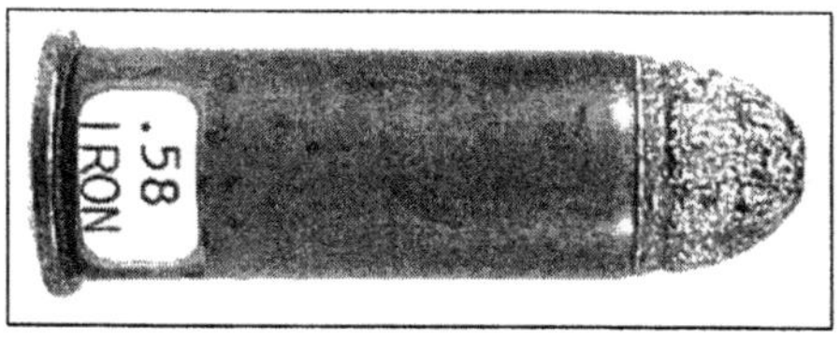

TITLE: .58 BERDAN MUSKET-MILBANK PRIMER

HEADSTAMP: Plain

DIMENSIONS:

Rim:	.739"
Head:	.640"
Neck:	.628"
Bullet:	.597"
Case:	1.656"

OTHER DETAILS:

Bullet Grains:	530
Powder Grains:	80
Bullet:	One Groove
Case Material:	Brass
Total Grains:	788

OVERALL: 2.125"

This Milbank primed specimen is a desirable Berdan Musket priming variation. It is easily identified by the primer, which at first appears to be a misfire.

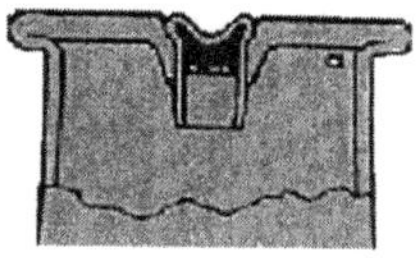

In May 1870, a patent was granted to Isaac Milbank for this ignition system and assigned to Winchester Repeating Arms Co. for manufacture. A batch of the .58 Milbanks (they were also made in a number of other calibers) were forwarded to Frankford Arsenal for testing. Although labeled as central fire reloading cartridges and of solid head construction, they had in fact the old style folded head type case.

The primer is similar to a small rim fire case with a recess in the center holding the priming composition; some priming mixture contained in the folded rim. The priming element is covered by a paper wad; in order to aid in placing the primer in the primer pocket, the open end was slightly closed.

The Milbank Primer's primary "claim to fame" was waterproof, ease of reloading, sureness of fire. Despite these qualities, variances in metal thickness and poor construction resulted in chambering problems and eventually the idea was scrapped. These cartridges weigh approximately fifty grains less than the average Berdan Musket round.

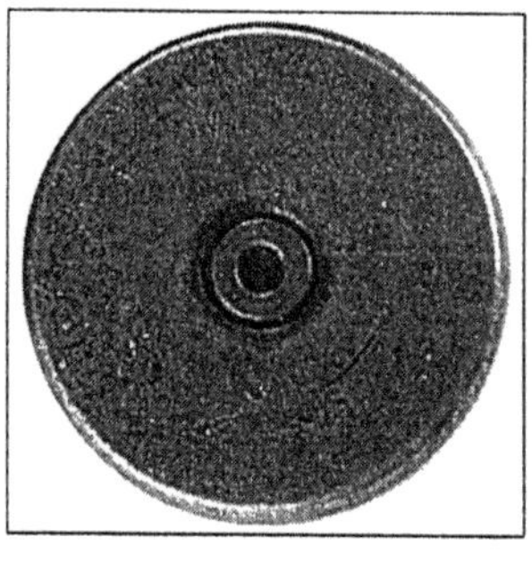

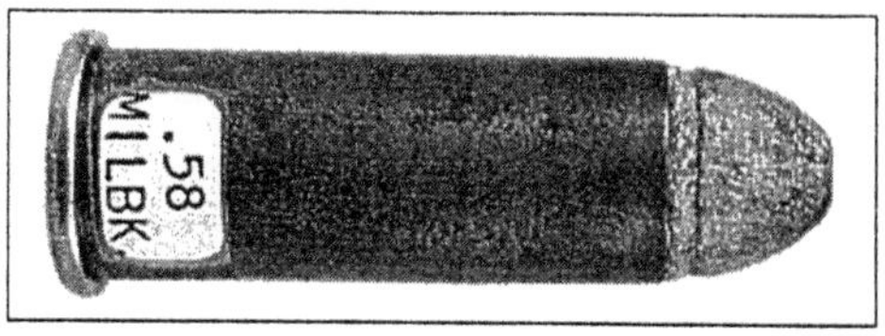

TITLE: .58 BERDAN MUSKET-PATCHED

HEADSTAMP: Plain

DIMENSIONS:

Rim:	.729"
Head:	.650"
Neck:	.622"
Bullet:	.591"
Case:	1.657"

OTHER DETAILS:

Bullet Grains:	?
Powder Grains:	80-85
Bullet:	Pointed, Patched
Case Material:	Brass
Total Grains:	777

OVERALL: 2.306"

An extremely scarce specimen of the .58 Berdan Musket.

There is a good possibility of other specimens having a paper patch, but they would be hidden from view being below the case mouth and not visible as this one is.

Another confusing feature is that the round gives the appearance of being quite heavy due to its long pointed bullet. Actually the round weighs less than other representative Berdan cartridges.

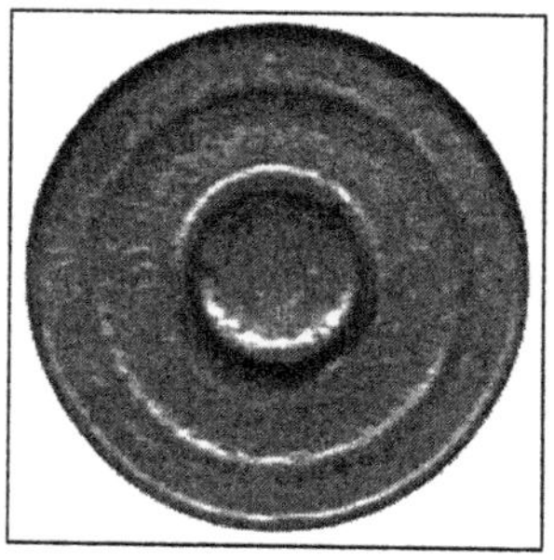

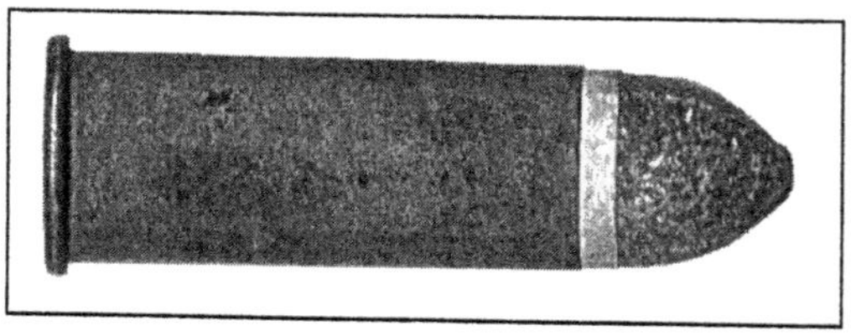

TITLE: .58 BERDAN MUSKET

HEADSTAMP: Plain

DIMENSIONS:

Rim:	.730"
Head:	.642"
Neck:	.626"
Bullet:	.599"
Case:	1.658"

OTHER DETAILS:

Bullet Grains:	530
Powder Grains:	90
Bullet:	Short, Rounded Nose
Case Material:	Brass
Total Grains:	788

OVERALL: 2.004"

This has to be one of the very early Berdan specimens.

The round has the typical ringed head peculiar to Union Metallic Cartridge Company. The overall length is shorter than most other Berdan musket rounds.

Positive identification of these early rounds is difficult without the original box or container. The general appearance is unlike the many other musket cartridges examined.

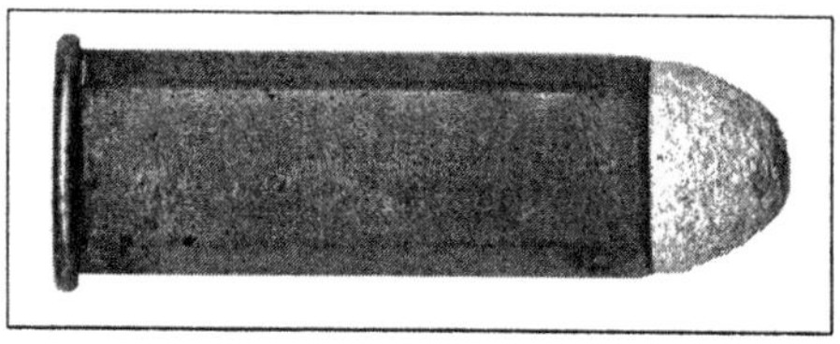

TITLE: .58 BERDAN – ASSEMBLER'S FUNCTIONING DUMMY

HEADSTAMP: Plain

DIMENSIONS:

Rim:	.736"
Head:	.658"
Neck:	.630"
Bullet:	.595"
Case:	1.658"

OTHER DETAILS:

Bullet Grains:	530
Powder Grains:	None
Bullet:	Grooved, Flat Nose
Case Material:	Brass
Total Grains:	710

OVERALL: 2.080"

Three types of dummy rounds will be encountered: one is the assembler's functioning round to be used in testing the mechanism and operating the action. These can be identified by a hole in the side of the case. The usual practice was to insert a small wooden rod inside the case from the base of the bullet to the primer pocket. This rod prevented the bullet from being pushed down into the case when it was repeatedly used by the gunsmith.

Recognizable by the small hole in the side of the case, this is an assembler's functioning dummy.

This specimen is a product of the Union Metallic Cartridge Company.

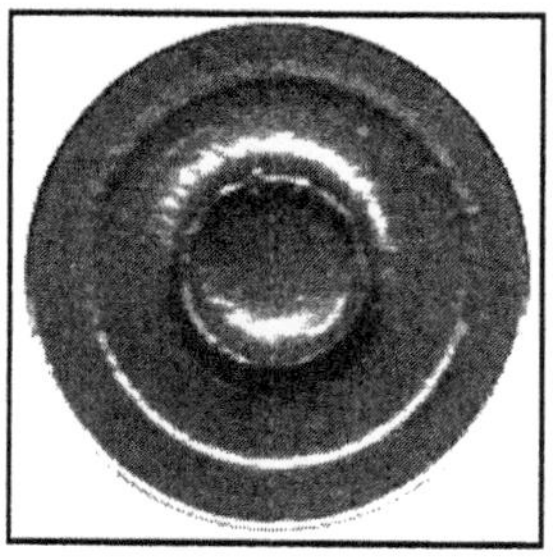

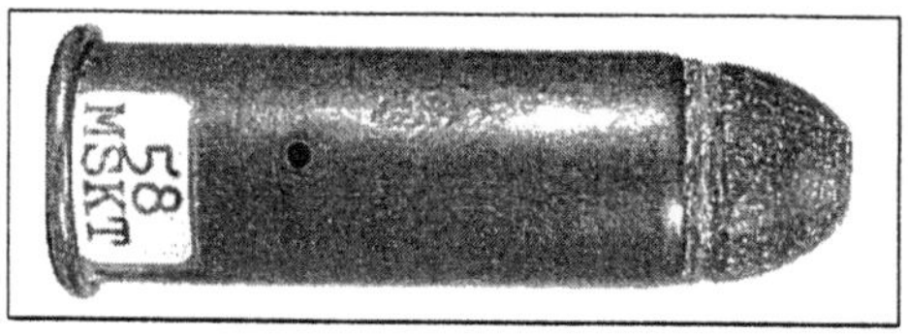

TITLE: .58 BERDAN MUSKET

HEADSTAMP: Plain

DIMENSIONS:

Rim: .735"
Head: .649"
Neck: .626"
Bullet: .596"
Case: 1.661"

OTHER DETAILS:

Bullet Grains: 530
Powder Grains: 80-85
Bullet: Flat Nose-One Groove
Case Material: Brass
Total Grains: 780

OVERALL: 2.011"

A very scarce specimen of the Berdan musket, manufactured by the United States Cartridge Company of Lowell, Massachusetts.

A distinguishing feature of this cartridge is the very tiny copper primer .

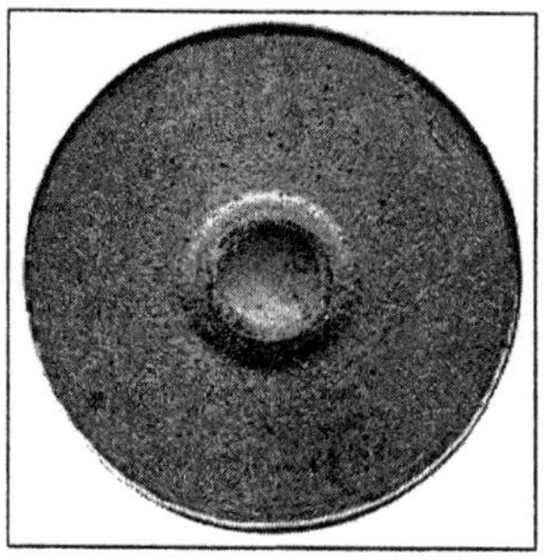

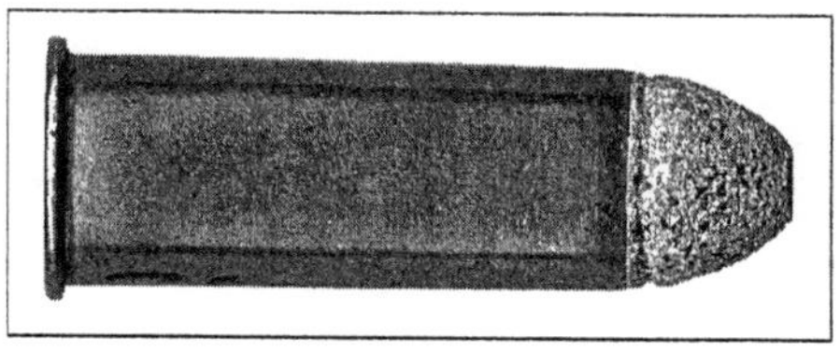

TITLE: 58 MUSKET ORCUTT PRIMER

HEADSTAMP: Plain

DIMENSIONS:

Rim:	.715”
Head:	.648”
Neck:	.624”
Bullet:	.584”
Case:	1.662”

OTHER DETAILS:

Bullet Grains:	
Powder Grains:	
Bullet:	Lead
Case Material:	Brass
Total Grains:	

OVERALL: 2.900”

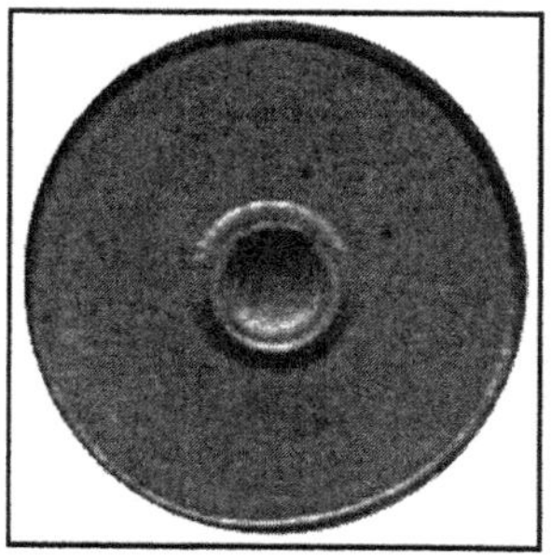

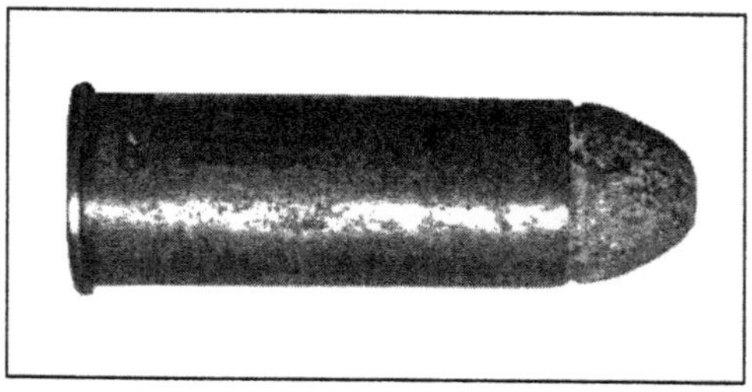

TITLE: .58 BERDAN MUSKET - W.R.A. CO.

HEADSTAMP: W.R.A. CO. .58 Musket

DIMENSIONS:

Rim: .735"
Head: .640"
Neck: .631"
Bullet: .592"
Case: 1.664"

OTHER DETAILS:

Bullet Grains: 530
Powder Grains: 85
Bullet: Pointed, One Groove
Case Material: Brass
Total Grains: 789

OVERALL: 2.18"

This is one of the more desirable musket rounds because it is head stamped. In fact, it is the only musket round that is head stamped.

It is strange that Winchester pictures this head stamped round for the first time in their 1913 catalog, yet the carbine round is shown with a head stamp in their 1891 catalog.

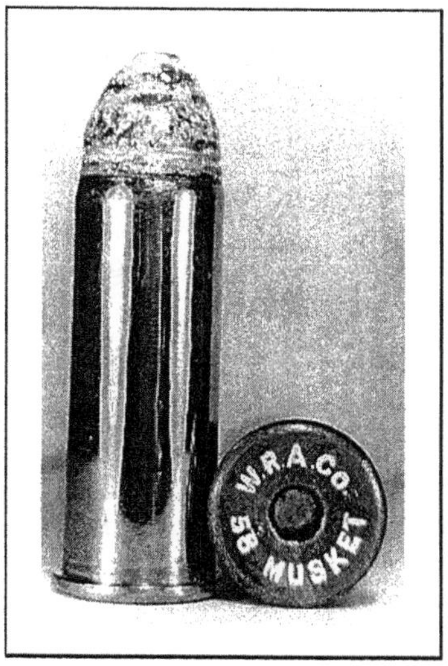

BRITISH AMMUNITION

During the early 1860s, Britain reluctantly began to accept the fact that their army would have to be equipped with a breech-loading weapon. The mainstay weapon was the "ENFIELD RIFLE MUSKET-PATTERN 1853" a muzzle loading gun which had served its purpose well. With constant improvement in gunpowder, steel, manufacturing know-how, and countless new weapons being designed throughout the world, it was inevitable that a change had to be made.

The British were impressed with the results of the use of the Dreyse needle gun that was in the hands of the Prussians during the Danish-Prussian War of 1864. The weapon used a self-contained cartridge and its rate of fire and ease of use soon convinced Britain and the rest of the world, that the end of the muzzle-loader was near.

In August 1864, an Ordnance Selection Committee advertised for a conversion system to be applied to the vast numbers of M1853 Enfield Rifles along with 1,000 rounds of ammunition. Apparently, it was not the intention of the Committee to ask for a completely new breech-loading weapon at this time, but a means of converting the numerous Enfields over to a breech-loading, self-contained cartridge system was needed.

Immediately upon announcement of the Committee's intention of acquiring a self-contained cartridge system, large numbers voiced their opposition to the plan. Many comments were expressed as to the dangers involved, the gun would blow up and kill the soldier, the ammunition would explode either when stored or carried by the soldier. Others expressed concern about the cost of both weapon and the wasting of ammunition, while others felt that the old muzzle-loader was a much more accurate piece.

In spite of the opposition, fifty different systems of conversion were submitted to the Committee for consideration. Of these, forty-two were rejected, leaving only eight worthy of evaluation. Five of the finalists still maintained the outside capping system while the Snider system was the only non-capping variety. Six newly tested Enfields were issued to each of the six inventors for conversion and these were to be returned to the Committee when completed for consideration and testing.

During 1865 after a thorough examination and the passing of a variety of tests proposed by the ordnance Selection Committee, the conversion and cartridge submitted by a Philadelphia wine merchant by the name of Jacob Snider was accepted.

The system was officially adopted on September 18, 1866.

Within less than a year's time, 200,000 Enfield rifles had been converted and thirty million rounds of Boxer ammunition had been fabricated. This was a most note-worthy achievement that was recognized by the Cornhill magazine in September 1866.

This quote is from that magazine:

> "We should not care to recall the ignorant and unjust clamour which was raised on the introduction of this arm and ammunition, the alarming prophecies of failure, and the manner in which the slightest and most unimportant difficulties were magnified into grave defects, condemnatory of the system, were it not that it would be impossible otherwise to do justice to one of the most satisfactory features of the year's progress, viz the complete success which has attended the introduction of the Snider system; and the confidence with which the arm and ammunition are now regarded by the whole Army."

BASIC .577 CALIBER CARTRIDGES
Enfield Pattern 1853 - .577 and .55 Caliber

Our primary consideration of British cartridges will start with those rounds that fit the Enfield Pattern 1853 Rifle of .577 Caliber.

The basic English round has an official military nomenclature of:

"CALIBRE .577 ENFIELD RIFLE-MODEL 1853"

The pictured specimen has an over-all length of 2.876", a total grain weight of 645 grains and a bullet diameter of .568". Powder charges varied from 60-70 grains. The majority of projectiles weighed 530 grains and came in four variations; hollow base, iron plug, clay plug and the last was of boxwood.

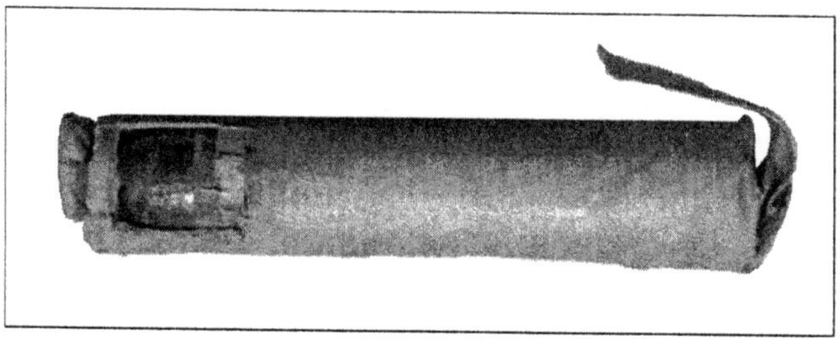

A discussion of the bullet types is in order here as without an understanding of the problem, one cannot begin to realize the need for the variety of bullet styles encountered.

One of the greatest difficulties encountered with a muzzle-loading weapon is the constant problem with powder residue build-up in the rifle barrel. In order to ensure accuracy, it is necessary to have a tight fit between rifle bore and the bullet. The tight fit worked fine for a few rounds, but soon it became extremely difficult to force the projectile down the bore. The resulting "pounding home the shot" only resulted in a deformed projectile that greatly reduced its accuracy. To resort to an easily loaded bullet meant that while it could easily be rammed down the barrel, its looseness in the bore when traveling out the gun meant erratic shots; no two ever hitting the mark in close proximity to one another.

It was only through the talents and efforts of two English gunsmiths, Messrs. Metford and Pritchett, that a more satisfactorily designed bullet was developed. Mr. Metford suggested to Mr. Pritchett the shape and dimension of the proper projectile for the 1853 Enfield Rifle and it became known as the Pritchett Bullet.

This bullet had a length nearly twice the diameter of the bullet and had an unplugged base. The French Minie bullet had the hollow base filled with a metal plug inserted so that upon firing, the force of the explosion would push against the plug, expand the base and as it traveled down the barrel, the bullet tightly gripped the grooves and rifling while scrapping out some of the powder residue along with it. The reader will be reminded that the original Pritchett lacked a plug of any kind. The reason for this being that recent testing by a gunsmith by the name of Boucher had done extensive testing with projectiles and reconfirmed Delvigne's tests of some twelve years previous (1840), that a "leaden projectile will expand at the impact of powder gases and that all will expand, from the simple cylindrical solid plug to the most elaborate hollow bullet, and the expansion will be more or less according as the bullets are longer or shorter".

Notwithstanding all these facts, the base plug was restored as the powder used by the Army was of an inferior and unreliable nature and could not be depended upon to effectively expand the bullet.

Another interesting fact regarding the shape of bullets was the result of the experiment by General Jacob to develop and design an exploding rifle bullet. Tests proved that the extra length needed in the nose to contain the charge aided in stabilizing the flight of the projectile. Later on, the Boxer cartridge had a hollow tip not unlike that of an exploding round.

The conclusion of all this resulted in the design of a bullet which had a diameter of .568" while the bore of the Enfield barrel measured .577"; thereby permitting the shooter to easily load the bullet knowing that the base would expand upon firing and give a tight bore fit.

The author would be derelict in his reporting if he failed to mention a most interesting sidelight attributable to the Enfield cartridge. For clarity the following is quoted from remarks made in "Rifles and Ammunition" by Ommundsen & Robinson, pages 78 and 79:

> "The Enfield rifle cartridge resembled the Minie very closely, and is only mentioned here because it is of some historical importance. It is said that the use of a particular kind of tallow as a lubricant on the Enfield cartridge was the cause of the excuse for the disturbance that began the Indian (Sepoy) Mutiny. It is certain that in 1857, a native worker at the ordnance works at Dum-Dum, some six miles northwest of Calcutta, spread a rumour that the new lubricating tallow was a mixture of hog's fat and cow's fat. If the rumour was designed with more devilish ingenuity, for such a mixture offended the religious feelings of both Hindus and Moslems. A deputation of native soldiers waited on the head of the Dum-Dum works, and suggested that the bullets should be lubricated with beeswax. No notice was taken of them at the time, but it is significant that two years later, the lubricant of the Enfield cartridge was changed to pure beeswax"

Constant complaints of the difficulty encountered while trying to load the Enfield rifle led to switching over to five parts beeswax and one part tallow during August 1857. Lead oxidation caused by the fatty substance used in the lubricant was the reason the guns were so hard to load. Finally, in April 1859, an entire revamping of the ammunition used in the Enfield was ordered. The lubricant was to be entirely beeswax, the bullet diameter was reduced to .55 inches and 1.09 inches in length, and, in order to facilitate the biting off of the end of the cartridge, the outer envelope containing the bullet was attached to the inner envelope containing the powder by a strip of gummed paper. Previously, the two were twisted together beyond the cylinder of the powder container.

The .55 caliber Enfield pictured here is 2.968" overall, weighs a total of 477 grains, and the bullet diameter measures .521".

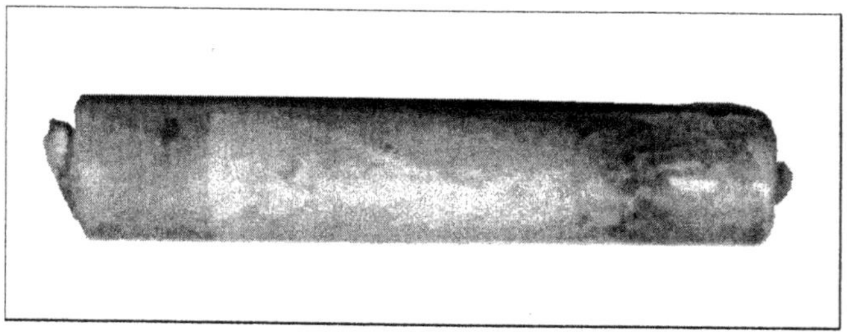

During the American Civil war from 1861-1864, large numbers of Enfield rifles were either traded or sold to the Confederate army. It was unthinkable that the South could depend upon a reliable and constant supply of ammunition from England, so substitutes had to be found and manufactured in their own factories.

The common round was of regular paper construction that had to have the paper envelope base bitten off to expose the powder. There were also combustible types which were manufactured under the patents of Hazard, Johnston and Dow and yet another by Bartholow. While these were of a fragile nature, they were reasonably waterproof and they did permit the rifleman to load his weapon much more quickly.

This photo depicts the Johnston & Dow cartridge. It weighs a total of 542 grains, has a .584" diameter bullet, the case measures 1.19" long and the cartridge has a total over-all length of 1.944 inches.

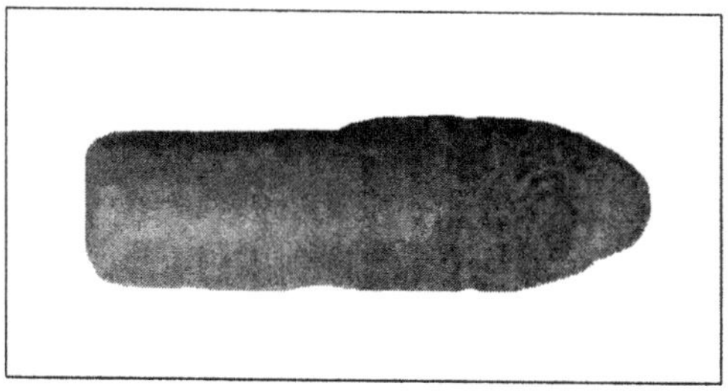

The photo below depicts another Johnston & Dow combustible for the Enfield, but with varying measurements. This one has a total weight of 584 grains, a case length of 1.232", the bullet measures .575" and the over-all length is 2.07".

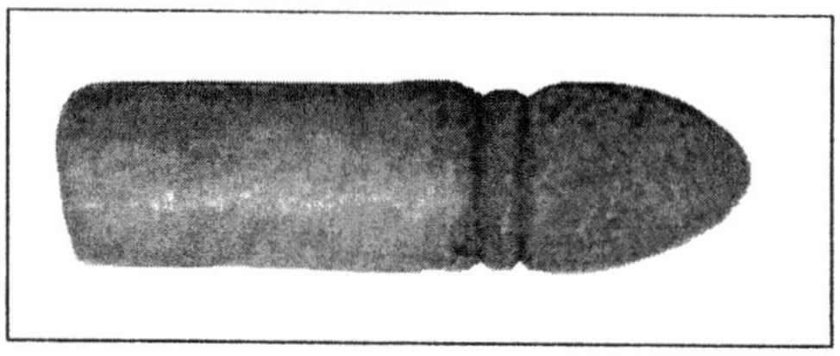

At this point in time, we take leave of the Enfield cartridges and go on to the era of the Boxer breech-loading rifle and the great variety of types (Marks of ammunition used in the gun.)

DETAILS	CARTRIDGE BODY DIA.	BL'T DIA.	O.A.	TOTAL GRS.
.577 BORE				
BRITISH ENFIELD				
Circa '53-'54 No Plug in bullet - no base reinforcement wrap	.561"	.568"	2.876"	645
Circa '55-'58 Wood Plug - no base reinforcement wrap	.567"	.568"	2.876"	630
Wood Plug and wide reinforcement wrap 1 1/2 times around base	.570"	.608"	2.820"	627
CANADIAN ENFIELD				
Circa 1860 Wood Plug - narrow reinforcement wrap Note slit around bullet	.591" .591"	.575" .584"	2.890" 2.972"	625 623
.55 CALIBER				
White paper, diagonal seam with pull tab	.523"	.520"	2.968"	477
Salmon paper, pull tab & paper patched bullet	.524"	.510"	2.968"	475

The conversion system and related ammunition, which was accepted by the Ordnance Selection Committee in 1866, was the invention of Jacob Snider, a Philadelphia wine merchant. The cartridge was a combination of the French Potet and Schneider ammunition which were made for and used by sportsmen and were not unlike a modern day shotgun shell. The paper case of the Schneider, while suitable for sporting purposes under generally protected controlled conditions, was not to the liking of the military. They felt, and rightly so, that the paper case would swell when damp or wet and prevent the rounds from chambering, plus the fact that the case would not withstand the hard use to which they would be subjected while under military usage.

Before we get to the Boxer cartridge, a description of the Daw cartridge, one of, if not, the first center fire cartridge to be introduced in England. It is appropriate to discuss it here, because of its influence on the Ordnance Committee and the resulting controversy between Messrs. Daw and Boxer over patent infringements.

George Henry Daw, a gun maker of Threadneedle Street, London, during the early 1860s, was considered to have been the first person to introduce the principle of the center fire cartridge in England.

The Daw paper cased cartridge was used by sportsmen in shotguns and was similar to a present day shotgun shell. It was felt that Daw had designed his round after the French ammunition originated by Clement Potet and Francois Eugent Shneider. Mr. Daw traveled extensively throughout Europe, and apparently after observing the uniqueness of the two French cartridges, decided to introduce them to the shooting fraternity, and later on, to the military in England.

A description of the Daw cartridge as taken from an old British publication is as follows:

> "This cartridge is composed of four parts, two of which are compound. The bullet, as in the Boxer, has the wooden picket through half its longer axis, and the clay plug in the base for expansion. The shell is also of brass, and in three parts, retaining the copper cap for the fulminate in the centre of the base. In almost all other respects this cartridge differs materially from the Boxer. It is much shorter than the present service cartridge, and the inventor consequently claims for it greater facility of extraction from the regulation arm. The cylindrical portion of the case is composed of a little over one fold of thin metal, which, being united is perfectly gas-tight and, from the slightness of this one fold, of little or no weight, and of great flexibility and toughness. The latter is a marked advantage over stouter metal, as the case, on the ignition of the charge, cannot fail to take the form of the breech, or chamber, in which it is enclosed, and cannot impede, in the least, the extraction of the shell after the explosion, as the sudden alteration of the temperature after the gas leaves the barrel, causes a slight contraction of the metallic shell. It is waterproof as well as gas-tight, requiring no paper covering or lubrication, as, in addition to the shell being joined by cement or solder, it grips the bullet closely above the cannelures, or grooves, and thus the projectile, with the powder and fulminate, are held together in one compact body by the slight shell. The weight of sixty rounds of this cartridge is five pounds eleven ounces, the weight of bullet 405 grains, that of shell and fulminate 150 grains and the charge of gunpowder is 65 grains."
>
> The result of the practical experiments under orders of the British War Department, as regards TO THE MERITS OF FILLING CARTRIDGES, IS AGAINST THE SAME, AS PER REPORT OF Capt. Majendie, Royal Volunteers, who states that the only advantage which is considered likely to result from refilling is economy. A cartridge which admits of refilling, first sight may seem to save the value of the case on each time it is refilled; but it must be borne in mind
>
> 1st. That in active service the cases could not be collected at all.
>
> 2nd. It seems doubtful if they could all be collected even at practise.
>
> 3rd. If the cartridges used for refilling are more expensive than those which are not refilled, from the apparent saving must be deducted the original cost
>
> 4th. Under any circumstances credit must be taken for the value of the metal in the cases of the cartridges which are only fired once.
>
> It will thus be seen that the saving is very far from what might be supposed at first sight. The Captain further states that he is strongly opposed to the idea of refilling at all, the cartridge cases returned to him have been in so bad a condition from dirt, verdigris, etc., that the expense of cleaning and preparing them for refilling would be very considerable, also not keeping free in use or carrying."

It was this type round that was submitted by Jacob Snider to be used in the rifle submitted for the trials in 1864. It is generally accepted that while not specifically mentioned, the military service Enfield bullet with plugged base was loaded in this ammunition.

Immediate dissatisfaction was expressed over the performance of the rifle's accuracy and the problem was traced to the inferior ammunition used in the tests. The Select Committee rejected the cartridge. Inadequate as the original ammunition was, it did prove to the Committee that the principle of the self-contained cartridge was far superior to the muzzle-loading system then in service.

The Committee set about to find a substitute for the first type ammunition. They commissioned Col. Edward Mounier Boxer, Supt. of the Royal Laboratory at Woolwich Arsenal to develop a superior cartridge.

The composite pattern (Daw's design) was the basis for Col. Boxer's redesigned cartridge. He still used the base cup and primer pocket with some changes in the form and shape of the anvil and pocket, but the major construction variation was the change from a paper case to the use of a rolled sheet brass tubular case. On January 15, 1866, a patent was issued to Col. Boxer covering these changes. These first cartridges were known as the Pattern I. Tests with this first model cartridge indicated that now, at least, the combined Snider conversion system with the newly designed ammunition was a step in the right direction, although further improvements were needed.

Much of Col. Boxer's problems were attributable to the use of the Enfield bullet, Pattern of 1859. The projectile weighed 530 grains, was 1.09 inches in length and had a diameter of .550". Col. Boxer redesigned the bullet. He increased the diameter to 573", used four cannellures filled with beeswax instead of the old paper patched bullet, increased the length to 1 1/2inches, retained the pressed clay base plug, but added a hole in the nose of the bullet which was filled with a Sycamore wood plug. The bullet now weighed 525 grains which was only five grains less than the original Enfield. Although the bullet was lengthened, the hole in the nose helped reduce the total weight. Boxer was aware that previously Col. Jacobs had performed extensive experimentation with bullet weights, shapes and lengths of projectiles while developing the explosive round attributable to him. During these tests, it was found that the hole in the nose of the lengthened projectile stabilized the bullet in flight and greatly improved the length of its flight and added to the accuracy of the gun.

As we have mentioned, the Snider conversion using the newly designed bullets, now equaled or slightly improved its accuracy over the old Enfield muzzle-loading rifle.

Upon learning of Col. Boxer's improvements in his cartridge, Daw set about to make changes in his own ammunition. The major change was Daw's substitution of sheet brass rolled one and a quarter turns and soldered at the joint. This replaced the troublesome paper case of his original round. Comparison of the Boxer and Daw rounds indicated that they were nearly identical. Ironically, the real originators of this type self-contained cartridge, Messrs. Potet and Schneider, were almost forgotten in history, while Boxer and Daw continued to squabble over which of them really should lay claim to being the originator of the first successful self-contained cartridge in the British Military Service.

As previously mentioned, the 1866 Committee's purpose was to select a suitable conversion of the M1853 Enfields to get the breech-loading principle started as soon as possible and then at some convenient point in time, decide upon a repeating breech-loader. The next step to achieve these goals was to set up a prize competition to pick the best breech-loader available and next, the best ammunition to use in it.

On October 29, 1867, it was announced that a prize of L400 was to be awarded for the best service cartridge submitted. The present Boxer service round was to be used as the standard either to be equaled or bettered in order to be successful. In the hopes that a better round would come forth, the Committee agreed that the Boxer round would not be eligible for the prize.

The conditions of the contest that would determine the winner would be:

1. Improved accuracy
2. Weight
3. Freedom from deterioration
4. Economy
5. Able to withstand rough usage and serviceability

During the new trials, under the rules of which the Boxer cartridge was not eligible to compete, Boxer worked to improve his anvil design and primer pocket. His efforts resulted in the so-called "Safety Anvil". The new anvil had projections or "shoulders" which rested upon the edges of the primer cup in order to keep its point above the priming compound rather than down deep into the cup where the slightest blow could explode the round. This change was included in Boxer's second patent #2653 of October 13, 1866.

The Pattern II cartridge using this new "safety anvil" was approved by the Small Arms Committee on August 14, 1867.

The Daw came in first and had a few things that were superior to the Boxer ammunition, which was the standard for the competition.

The Daw was cheaper to manufacture, equal in ability to withstand rough usage, it had an edge in resisting varying temperatures, and equal to serviceability and was under, by one ounce, the amount of weight established by the Committee. Ironically, there were no tests for accuracy since the cartridges submitted by Daw would not chamber in the Snider Rifle! One defect noted in the Daw was its tendency to foul the rifle barrel to a greater extent than the Boxer ammunition.

All facts considered the Daw was adjudged the winner and was awarded the prize of L400. Remember that the Boxer cartridge was not eligible to compete.

Upon reporting the results of the contest on February 12, 1868, the Select Committee reported that of the forty-nine entries, only the Daw was superior. However, the Committee did agree that the Daw was less than equal to the present Boxer cartridge. Because the Daw ammunition did comply with the requirements of the Committee's conditions to be eligible for the prize, the award was given to him.

One defect of the early cartridges was the anvil in both the Daw and Boxer rounds. It was not properly shaped and resulted in the possibility of premature explosion if dropped or carelessly handled. The danger lay from the anvil resting upon the primer compound in the primer.

Col. Boxer was extremely upset over the awarding of the prize to his rival, Daw, but the Committee was adamant that Boxer was not eligible as stated in the conditions of the rules. Boxer contended that the Daw cartridge was an infringement of his patent and that his Pattern II Boxer Cartridge was superior to the Daw and should have been permitted to enter the contest. Still, the Committee would not budge in their opinion and informed Boxer that inasmuch as there was no intention to adopt the Daw cartridge, he should drop the entire matter.

In turn, Daw felt no less disturbed by the action of the Committee in not accepting his cartridge for military usage and that Col. Boxer's cartridge was in fact an infringement of his (Daw's) patent issued on March 8, 1867.

All discussions to the contrary, it was of no avail and gradually the Daw cartridge became a part of English military history and was forgotten. These Daw rounds are extremely rare now and are seldom seen except in a few collections throughout the world.

With the final approval of Col. Boxer's Pattern II cartridge on July 8, 1867, we are ready to understand the construction of such an odd looking round as compared to our modern solid drawn cased ammunition. The reader must keep in mind that during the 1860s, the machinery and know-how to form solid drawn cases was not available or just in the earliest stages of development. While the wrinkled and flimsy dented body of the Boxer composite round looked inferior and weak by today's standards, it was in reality a reasonably good and efficient cartridge. The rolled case, while purposely undersized, would effectively expand upon firing and seal the chamber against the escape of gas rearward into the face of the shooter. After firing, the brass foil would contract and permit easy extraction of the spent case. This is not to imply that the ammunition was without its problems as many did develop. Poor accuracy which was corrected by improvement in bullet design, split cases, rims pulled off or cracked, were problems that all contributed to the woes of Col. Boxer. Gradually the problems were solved and over a period of years, the Boxer cartridge served the British Army well in all corners of the earth. The Boxer style rounds continued to be made for a number of years, either as the original .577 caliber or in later years when the bore size was reduced to .45 inches.

Head stamps on the British military rounds are almost unheard of except for two small arrows appearing on the flange of the primer pocket indicating their manufacture at the Royal Laboratory, Woolwich Arsenal. Rounds manufactured at the Dum-Dum Arsenal had a "D" and arrow stamped onto the body sheet near the base of the cartridge. Those rounds made at the Kirkee Arsenal in India were stamped with a "K". Some commercial rounds were head-stamped and are pictured later on in the text.

The following information has been extracted from an earlier British military publication that described the construction of the Boxer cartridge. A list of the various types or Marks are also given.

SMALL ARMS AMMUNITION - BOXER -SNIDER

"The same description of cartridge is used for all the B.L. small arms with Snider action in the service, the same cartridge will fit an Enfield rifle or carbine, or a Lancaster carbine. The diameter of the bore is .577", the diameter of the bullet is .573", small enough to drop through the bore, and depending on expansion for its fit, the length of bullet is about 1" (1.04" present pattern), and the length of the cartridge is a very little over 2 1/2". The charge is 70 grs. (very nearly 2 1/2drs.) of R.F.G. Each packet of ten rounds weighs about 1 lb.

Many patterns of the Boxer cartridge have been made, they are all serviceable with the exception of Mark I., which may be recognized by having a Potet base, formed by pressing out the rim of the case cup so as to form a beading round the edge instead of having a brass or iron base disc as the other marks have. Mark V, is a weaker cartridge than the others, it may be recognized by its being the only brown paper covered cartridge without a distinguishing ring or rings marked on it, the brass sheet from which this cartridge is made is .003" thick, instead of .005" like the later patterns, and its base is not so strong as the earlier ones, the cartridge is not to be issued for foreign service, but is to be expended at home stations.

The bullet is made from pure lead, weight 480 grs., the hollow in the head is closed by having the lead spun over it, the hollow parts are necessary in order to get the bullet of a sufficient length for good shooting, and to get its centre of gravity in the proper place, the hollow in the base is also used to give the expansive action to the bullets, the plug, made of clay, and soaked in beeswax, closes the rear cavity, and on firing expands the bullet which has three cannelures, the sides of the bullet as far as the front cannelure are coated with beeswax, the cannelures holding a sufficient supply of the lubricant in their recesses; by the expansion of the bullet the lubricant is squeezed out, and the bore is thoroughly cleaned out by the bullets passing through it. As many as 4,000 rounds have been fired without fouling sufficiently to injure the shooter.

The case is formed of sheet brass covered with brown paper. It is lined with shellac and thin white paper to prevent corrosion from the powder. The case overlaps by about 1/4", and is cemented together with shellac and glue. It is attached to the bullet by being choked into the rear cannelure; the base is strengthened by two cups, and the bottom closed by an iron disc; inside, a paper pellet is pressed against the bottom of the cartridge, a copper cap chamber pierced with a fire-hole passes through the base and rivets the bottom of the cartridge to the case, the top being bulged out over the paper pellet, and the base of the cap chamber holds a brass anvil, on the shoulders of which rests the copper cap, which is primed with cap composition (fulminate of mercury, sulphide of antimony, and chlorate of potash) and varnished. A little wool is placed over the powder to keep it from touching the bullet.

On firing, the cap composition is forced against the point of the anvil, and the flash reaches the charge through the hole in the chamber; the case unwinds, and being pressed firmly against the sides of the bore prevents the gas from escaping in the direction of the breech; after firing, the cartridge contracts to its original proportions, making extraction easy.

The base of the cartridge is very strong owing to its base cups and disc, and it is at the base that strength is essential, indeed if the base is perfect, a very faulty cartridge may be fired without throwing any undue strain on the breech.

The cartridge is found to stand rough usage and wet well, what tries it most is a moist hot climate, it is almost impossible to prevent the moist air from penetrating when sided by great variations of temperature. Experiments have shown that it is impossible to explode these cartridges in a mass, thus firing 1/4 lb. Of powder along with a number of cartridges in an iron cylinder hardly exploded any rounds.

The first four marks had the brass cylinder covered with white paper.

Mark	I	has a Potet base, before described
Mark	II	has a brass disc
Mark	111	has also a brass disc. The bullet is shorter, the distance therefore from choke to top of case is less, but it is hard to distinguish between II and III
Mark	IV	is the only cartridge which has white paper and iron disc
Mark	V	is the only brown paper cover cartridge without a distinguishing ring

Mark	VI	has one black ring
Mark VII	VII	has one black ring and lead spun over head instead of wooden plug
Mark	VIII	has two black rings
Mark	IX	has one red ring, it only differs from VIII in having a bullet of the same size as pattern IV which was supposed to give better shooting at long ranges

Blank cartridges are issued in packets of 10 rounds made up in purple paper.

There are two patterns now issued; one is a condemned ball cartridge of Patterns VIII or IX cut down to the required length and called Mark I, the other has a case made of brown paper and is called Mark III. Both contain pellets made from blank F.G. or L.G. 2 1/2 drs.

Blank Mark I without the numeral II after "rifles" is unsuitable for arms marked II. A new pattern blank, Mark IV is not made. It is filled with loose powder instead of a pressed pellet."

BOXER BUCKSHOT CARTRIDGES

The Boxer Buckshot cartridge is an interesting and very scarce round.

The Buckshot cartridge was introduced in January 1868 about the time of the Fenian uprising. The round was issued to prison guards and for use in riot control. Its purpose then, as now, was to prevent injury to the public who might be in the immediate vicinity of trouble. A very potent round up to seventy yards with it dropping off dramatically beyond that distance.

It appears the primary intention for developing the shot round was riot control. To quote from an article in "The Engineer" of June 16, 1871:

> "the effect of the buckshot-being probably limited to about 150 yards at most-would be confined to the locality of the disturbance, and the lives of individuals half a mile away would not be endangered. Again, as concerning the rioters themselves, a very wholesome punishment might be administered at about 100 yards with comparatively little risk of life being sacrificed, although blood might be drawn, cries elicited, and very great moral effect produced. Lastly, should determined rioters press the troops hard, and really endanger their lives, the effect of buckshot delivered close would be much greater than that of bullets."

The Mark I Buckshot round was similar to the Mark V ball round but different in some respects. The overall case length was shorter as well as the primer pocket and base wad. The reduction was necessary in order to increase the internal capacity in its lower section due to the construction of the mock bullet that contained the buckshot. The "bullet" was longer than the usual lead bullet and it extended further into the powder chamber. By shortening the primer pocket the anvil was shorter and had no shoulder thereby reverting the ignition system back to the non-safety type first used in the earlier Boxer ammunition. Sixteen balls of approximately .27 caliber were contained in a paper container shaped like a bullet. The case set on a cotton wool wad seated on top of a charge of 61 grains of pressed pellets of black powder. Plaster-of-Paris surrounded the buckshot and shellac was placed into the case to hold it in position.

The Mark II Buckshot round made its appearance in 1874. The construction of the case was simplified. Instead of the Mark I special case, the Mark II used the Mark IX case, a regular powder charge, a paper disk and felt wad lubricated with beeswax, was placed on top of the charge. The charge contained thirteen balls packed in bonedust and was placed in a purple colored paper bag and the entire unit was shellacked in place within the case neck. The bag was choked at the mouth and tied with string. The Mark II was issued for use in the Ashantee War of 1873-74 as well as for use by prison guards.

The Special Snider Shot Cartridge, Mark I made its appearance circa 1899. The case was of rolled sheet brass, loose powder charge with mill-board wad on top; the shot cartridge contained twenty-eight balls, a felt wad was placed on top of the shot and the case mouth crimped over to secure the cartridge.

The round was issued to Her Majesty's Prison's Service to replace previous buckshot cartridges then in use. The round was easier and cheaper to produce than previous Marks. In spite of the fact that this buckshot round was made as late as 1899, the round is seldom encountered by collectors.

Shot Cartridges-Paper Sabots

	-1-	-2-	-3-	-4-
Rim	.736"	.736"	.746"	.745"
Head	.660"	.659"	.660"	.660"
Neck	.648"	.628"	.622"	.660"
Case Length	1.915"	1.966"	1.966"	1.928"
Paper Bullet	White	Blue	Yellow	Yellow
Over-all	2.350"	2.515"	2.485"	2.2600"
Powder	61	54		
Case Material	Brown Paper	Brass & Paper	Brass & Paper red stripe	Brass & Paper 2 red stripes
Total Grains	521	657	458	654

The five general classifications of Snider cartridges are as follows:

BOXER CONSTRUCTION

a. Composed of brass sheeting rolled into a tube with reinforcing brass cups and a base disk of brass or iron.

b. A case made from thick brown paper with an attached iron base disk.

SOLID METAL CASE

a. Drawn brass or copper tube approximately two inches long with brass or iron disk riveted to the base.

b. A commercial or sporting round with a 1.6" case length

c. Solid drawn brass case slightly less than two inches in length

Average dimensions of all types are a rim diameter of .745" and a head diameter of .660". Total grain weight of a ball round averaged 727 grains for the Boxer case and 780 grains for a solid brass cased round.

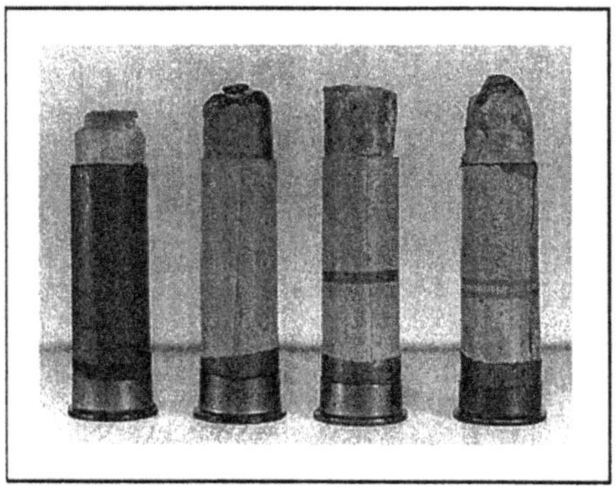

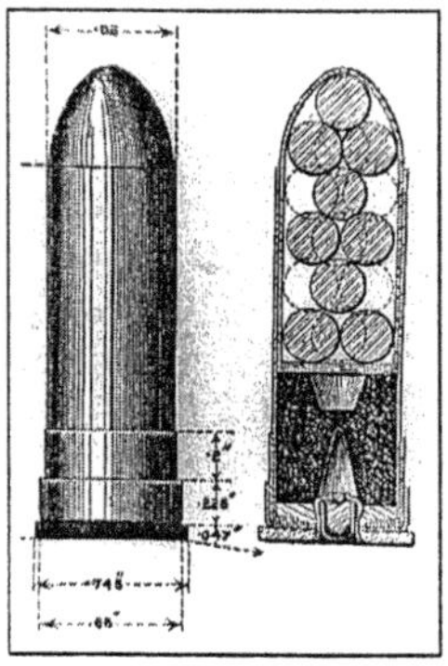

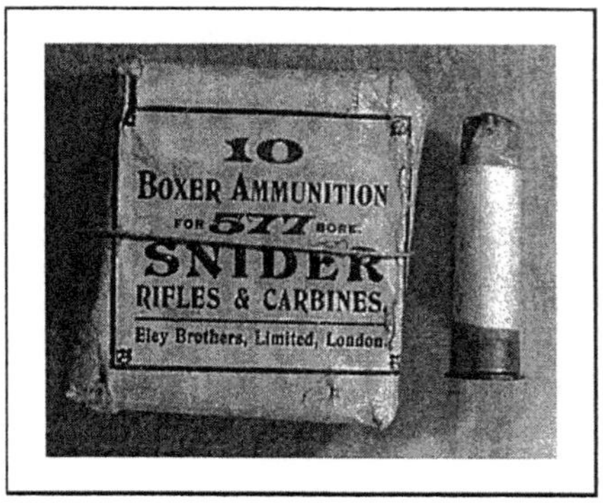

TITLE: .577 SNIDER BLANK-MARK I

HEADSTAMP: Plain

DIMENSIONS:

Rim: .733"
Head: .664"
Neck: .612"
Bullet: -0-
Case: 1.294

OTHER DETAILS:

Bullet Grains: -0-
Powder Grains: ?
Bullet: -0-
Case Material: Lacquered Brass
Total Grains: 220

OVERALL: 1.294"

A very old blank round which has a brass case but for identification has been lacquered.

The applied ringed head or brass cup contains a copper primer. The primer is formed so as to act as a rivet to hold the applied cup to the main section of the case.

This round originated in Australia.

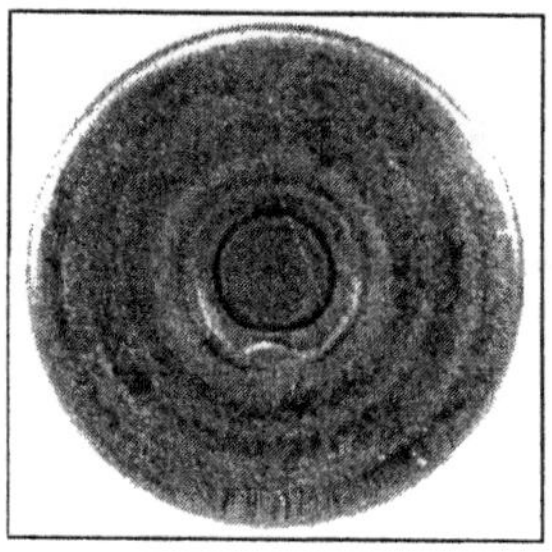

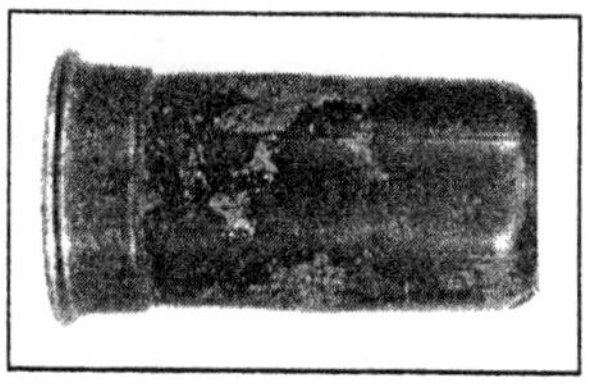

TITLE: SNIDER BOXER BLANK

HEADSTAMP: Plain

DIMENSIONS:

Rim:	.745"
Head:	.656"
Neck:	.652"
Bullet:	Blank
Case:	1.430"

OTHER DETAILS:

Bullet Grains:	N.A.
Powder Grains:	Not known
Bullet:	Blank
Case Material:	Heavy cardboard
Total Grains:	

OVERALL: 1.430"

A Boxer style case of reddish color, heavy cardboard. Brass reinforcing strip at the base near the iron head has very faint crimp lines not usually encountered.

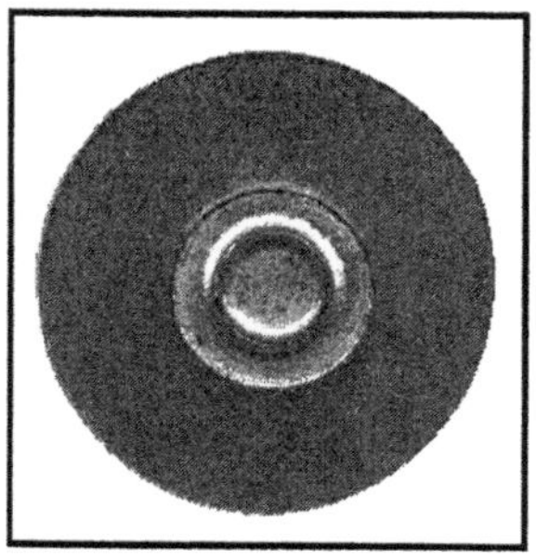

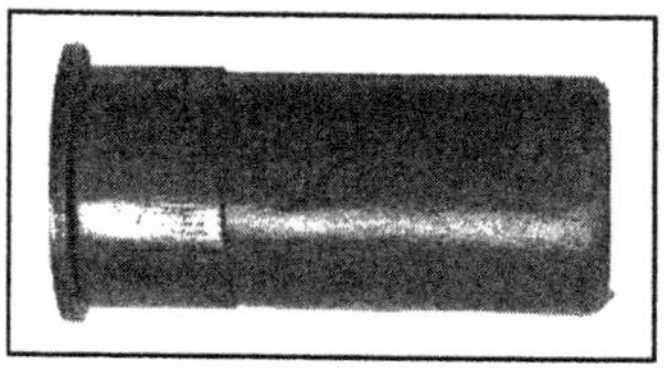

TITLE: SNIDER MARK III BLANK

HEADSTAMP: Plain

DIMENSIONS:	
Rim:	.746"
Head:	.658"
Neck:	.626"
Bullet:	Blank
Case:	1.455"

OTHER DETAILS:	
Bullet Grains:	N.A.
Powder Grains:	40
Bullet:	Blank
Case Material:-Heavy cardboard	
Total Grains:	170

OVERALL: 1.455"

Originally regulations required first class cartridge case materials even in blank loads, but it was soon discovered that the cost was prohibitive. It became more economical to simply take fired empty cases and reload these rounds. Any available cases were converted into blanks at first, then in 1868 a Letter of Change was issued to manufacture the cases out of heavy brown cardboard. These were reliable rounds and remained in service until the rifles were altered to use a .450 caliber barrel. The case is held closed by a heavy roll crimp.

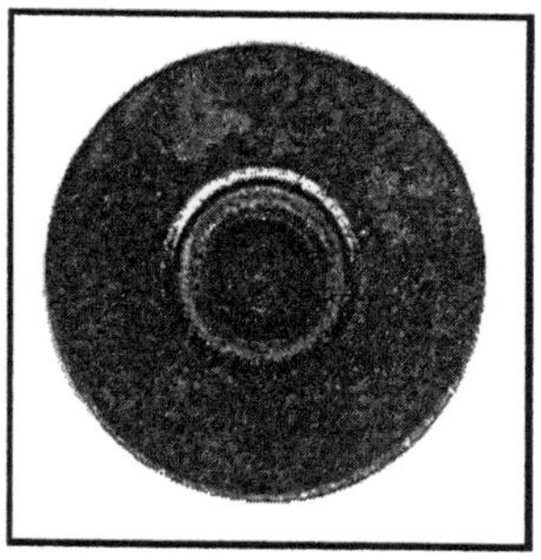

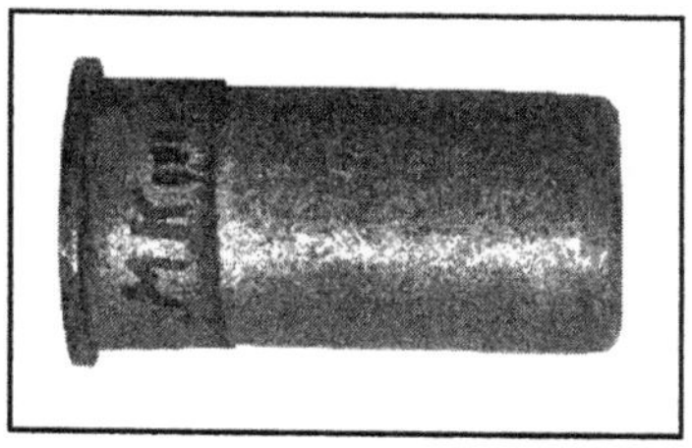

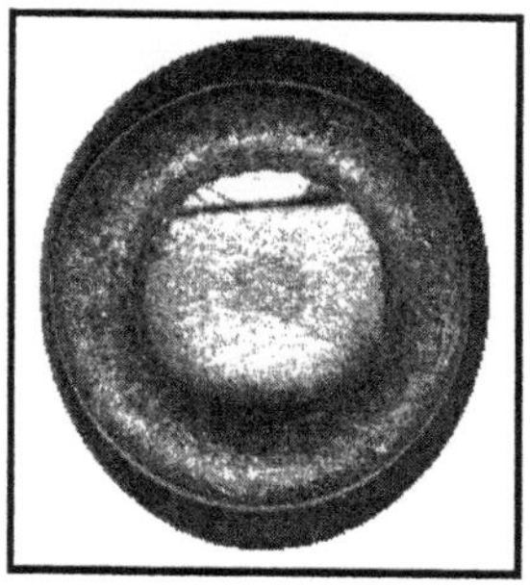

TITLE: SNIDER-MARK II BLANK

HEADSTAMP: Plain

DIMENSIONS:

Rim:	.745"
Head:	.665"
Neck:	.635"
Bullet:	None
Case:	1.456"

OTHER DETAILS:

Bullet Grains:	None
Powder Grains:	2 1/2 Drams FG
Bullet:	None
Case Material:	Brass
Total Grains:	256

OVERALL: 1.456"

Another form of the Mark II blank with deep-set petal crimp over paper wad. No paper covering over case body. Approved April 4, 1867.

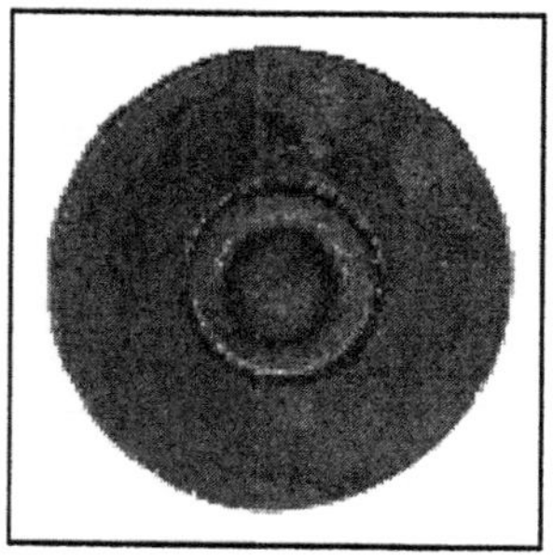

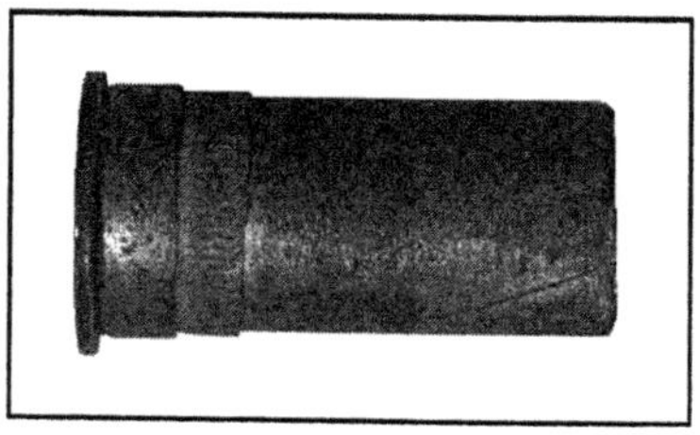

TITLE: SNIDER-MARK II
CARBINE AND RIFLE BLANK

HEADSTAMP: Plain

DIMENSIONS:

Rim:	.752"
Head:	.664"
Neck:	.663"
Bullet:	-0-
Case:	1.500"

OTHER DETAILS:

Bullet Grains:	None
Powder Grains:	2 1//2 Drams FG
Bullet:	None
Case Material:	Brass
Total Grains:	186

OVERALL: 1.500"

The Mark II has a rolled case but without a paper cover, a single brass base cup and an iron base disk. The charge is contained in a paper bag twisted close and the case mouth rose crimped to secure it. Circa 1867.

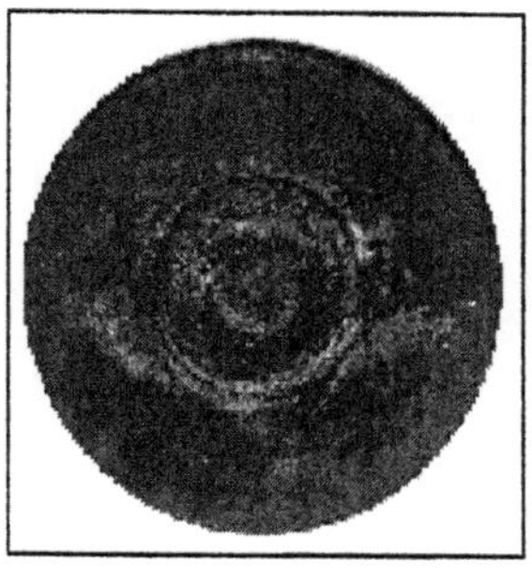

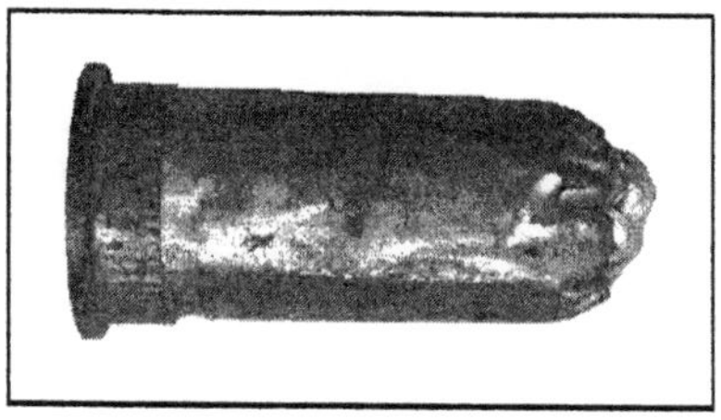

TITLE: .577 SNIDER BLANK

HEADSTAMP: Plain

DIMENSIONS:

Rim: .747"
Head: .664"
Neck: .633"
Bullet: Blank
Case: 1.524"

OTHER DETAILS:

Bullet Grains: N.A.
Powder Grains: Not known
Bullet: Blank
Case Material: Heavy cardboard
Total Grains: 179

OVERALL: 1.524"

A very early blank with brass base. The original Boxer cases were made of brass then later a switch was made to iron for economical reasons. The cartridge is held closed by a heavy roll crimp.

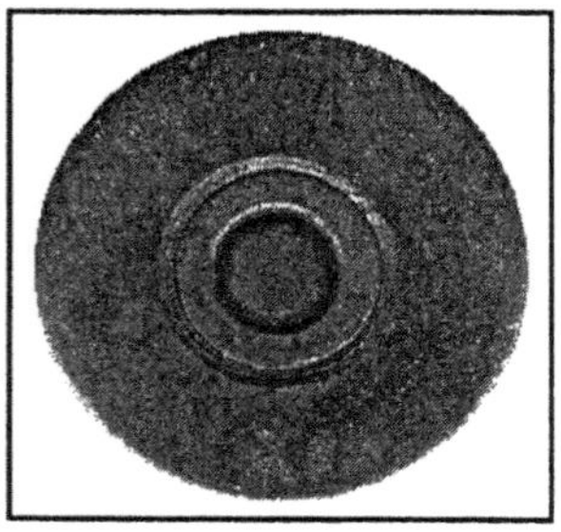

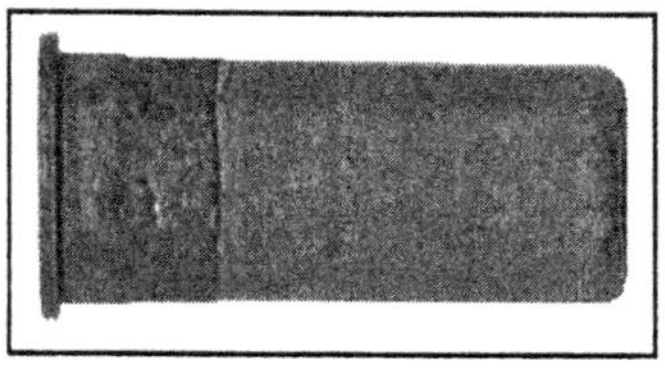

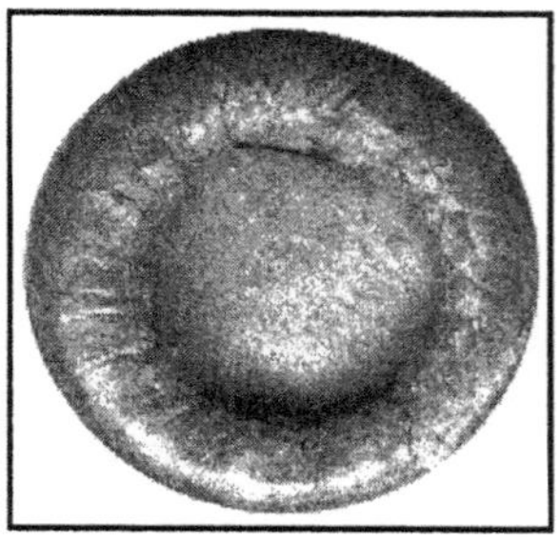

TITLE: TURKISH SNIDER-CARBINE

HEADSTAMP: Plain

DIMENSIONS:

Rim: .736"
Head: .665"
Neck: .590"
Bullet: .556"
Case: 1.553"

OTHER DETAILS:

Bullet Grains: 480
Powder Grains: 80
Bullet: Tapered Point
Case Material: Brass
Total Grains: 603

OVERALL: 2.082"

Both rifle and carbine ammunition was manufactured for the Turkish weapons, but the carbine ammunition is seldom found. This cartridge is not easily recognizable as a carbine round because the case length is approximately 1.60" (40mm) instead of the longer length 1.671" (42mm) made for the rifle round. With manufacturing standards being what they were in the 70s and 80s, such small differences were probably not noticed.

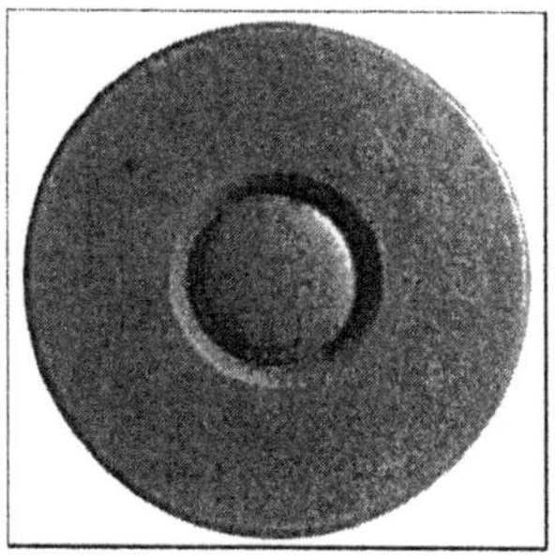

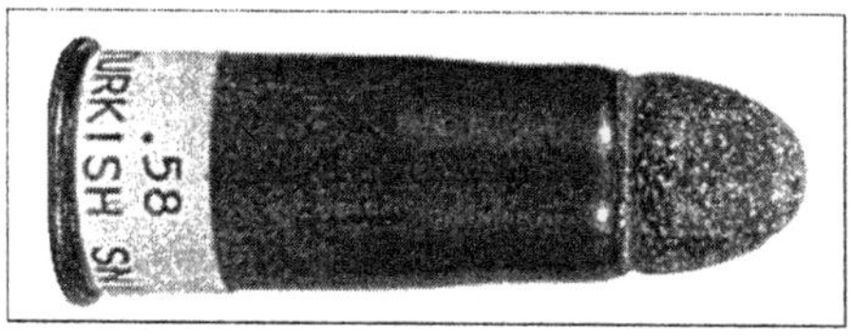

While examining a selection of these Turkish Snider rounds, it was found that this cartridge was approximately one hundred grains less than other Snider rifle specimens.

The ribbed bullet also differs in configuration than the other Turkish Snider rounds.

TITLE: .58 RUSSIAN SNIDER

HEADSTAMP: Plain

DIMENSIONS:

Rim:	.737"
Head:	.654"
Neck:	.586"
Bullet:	.556"
Case:	1.562"

OTHER DETAILS:

Bullet Grains:	500
Powder Grains:	80-85
Bullet:	Lead-Ribbed
Case Material:	Brass
Total Grains:	690

OVERALL: 2.078"

The British Snider cartridge proved to be so reliable and popular that many European countries chose to convert their own muskets to the Snider round.

Both Turkey and Russia made such conversions and the differences between the two rounds were very slight.

The Russian Snider cartridge is slightly longer over all and has a heavier total grain weight. Generally these differences are only recognizable when both cartridges are examined side by side.

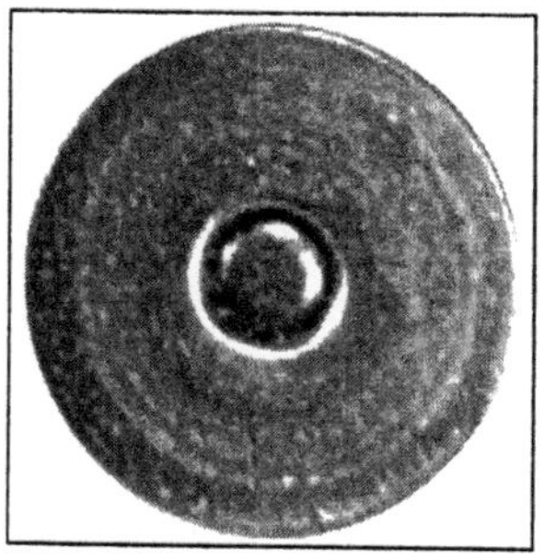

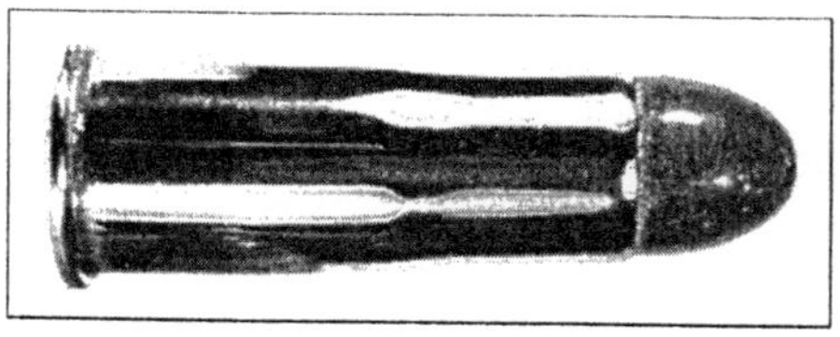

TITLE: .58 TURKISH SNIDER

HEADSTAMP: Plain

DIMENSIONS:

Rim:	.751"
Head:	.655"
Neck:	.588"
Bullet:	.532"
Case:	1.565"

OTHER DETAILS:

Bullet Grains:	480
Powder Grains:	80-85
Bullet:	Round Nose
Case Material:	Brass
Total Grains:	672

OVERALL: 2.020"

This cartridge was loaded by U.M.C and W.R.A. Co in considerable numbers during the mid 1870s. An order for fifteen million rounds from Turkey by U.M.C. in 1874, and in the same year, an order for eighty million, one hundred thousand rounds was placed with W.R.A. Co.

It was large orders such as these, and the resulting profits from such ammunition sales that kept both U.M.C. and W.R.A. Co in good financial condition. The local demands for arms after the Civil War was practically nil. It was this international business that kept both companies from going into bankruptcy.

TITLE: .577 SNIDER BLANK

HEADSTAMP: None

DIMENSIONS:

Rim:	.746"
Head:	.670"
Neck:	.600"
Bullet:	-0-
Case:	1.572"

OTHER DETAILS:

Bullet Grains:	-0-
Powder Grains:	?
Bullet:	-0-
Case Material:	Brass
Total Grains:	179

OVERALL: 1.572"

Close examination of this early blank would indicate that the small copper primer might be a musket cap with the ends flared over to form a rivet to hold the brass head to the base of the tube. The black lacquer still clinging to the brass tube was used for identification and/or protection against corrosion.

The total grains of 179, is really meaningless because the paper container, which extends completely into the case was broken; hence the round holds no charge.

This is a very interesting old round found in Australia.

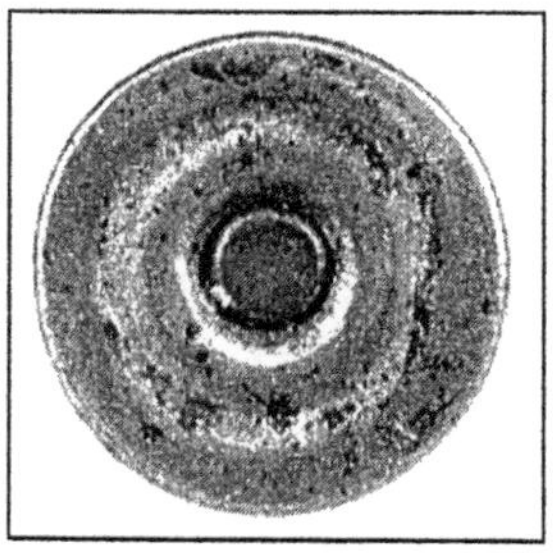

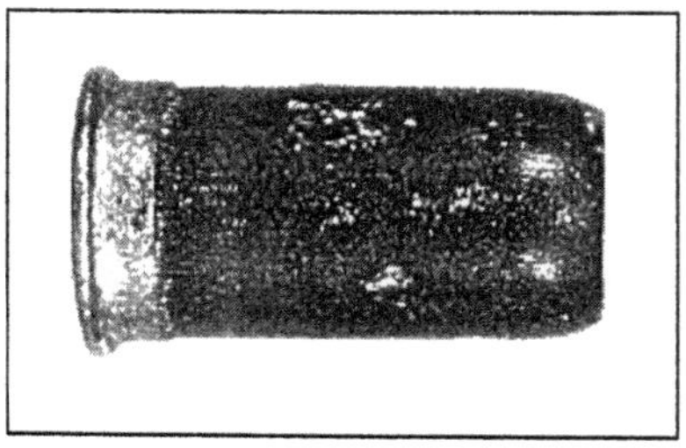

TITLE: .577 SNIDER AND MARTINI HENRY BLANK

HEADSTAMP: Plain

DIMENSIONS:

Rim: .744"
Head: .665"
Neck: .610"
Bullet: None
Case: 1.590"

OTHER DETAILS:

Bullet Grains: -0-
Powder Grains: 68
Bullet: None
Case Material: Coiled Brass
Total Grains: 244

OVERALL: 1.590"

The official nomenclature of this blank is: Cartridge, Small Arm Blank, Martini-Henry and Snider Arms (Mark IV, Converted).

Originally, the thinking was that the construction of blank ammunition should be of first class materials and treated like rifle rounds, to be reloaded and used again. However the cost of picking up, cleaning and storing, and reloading proved to be too costly and troublesome. The final solution was to make the blank case as cheaply as possible; use rejected cases and then after firing simply dispose of the case without giving thought to reloading. This specimen is a reused rifle case evidenced by the extra cups on the base. Two small arrows impressed on the primer cup indicates manufacture by the Woolrich Arsenal.

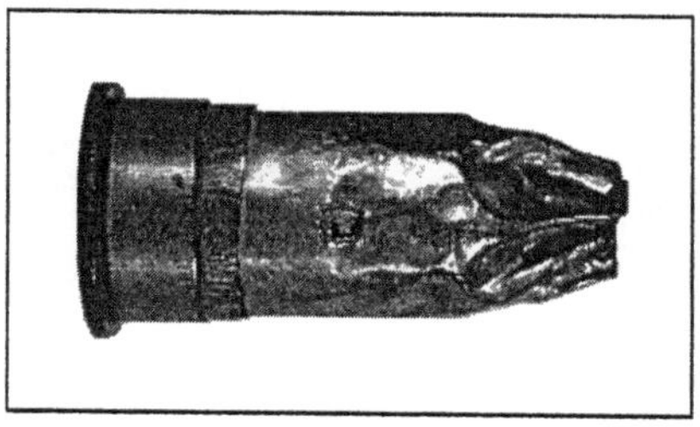

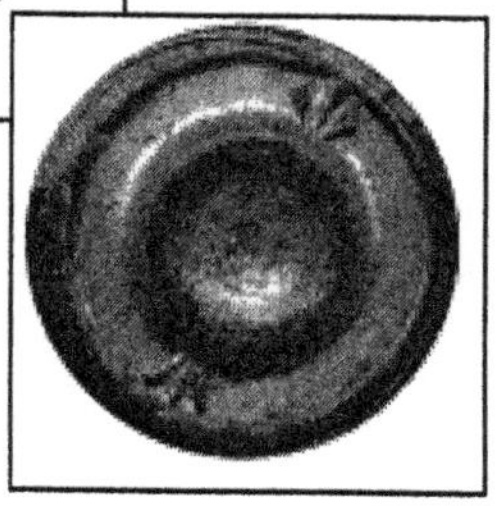

TITLE: .577 SNIDER-SPORTING

HEADSTAMP: 57 SNIDER F.W.L. & CO.

DIMENSIONS:		OTHER DETAILS:	
Rim:	.739"	Bullet Grains:	480
Head:	.659"	Powder Grains:	70
Neck:	.601"	Bullet:	Round Nose
Bullet:	.564"	Case Material:	Brass
Case:	1.609"	Total Grains:	732

OVERALL: 2.125"

In order to maintain the shooting interest of the returning soldiers after each war, it was common practice for arms manufacturers to develop or otherwise alter the military rifle so it could be used for sporting purposes.

The British Snider Rifle was no exception. Consequently the Dominion Cartridge Co. manufactured a sporting version of the British Army Snider cartridge for the F. W. Lamplough Co.

The F.W.L. & Co. was a wholesale hardware and sporting goods store located in Montreal, Canada. It was in business from approximately 1904 to 1914.

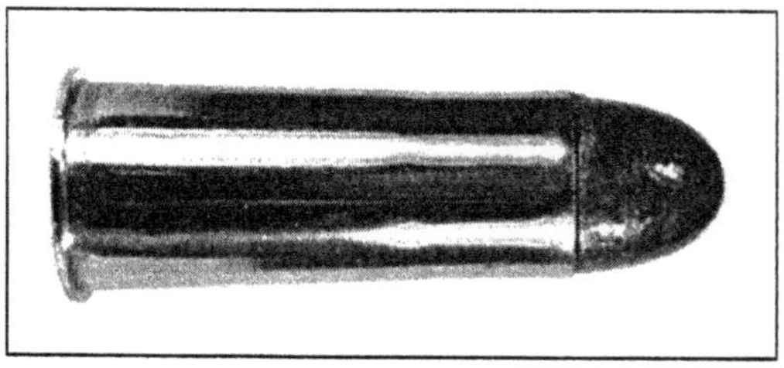

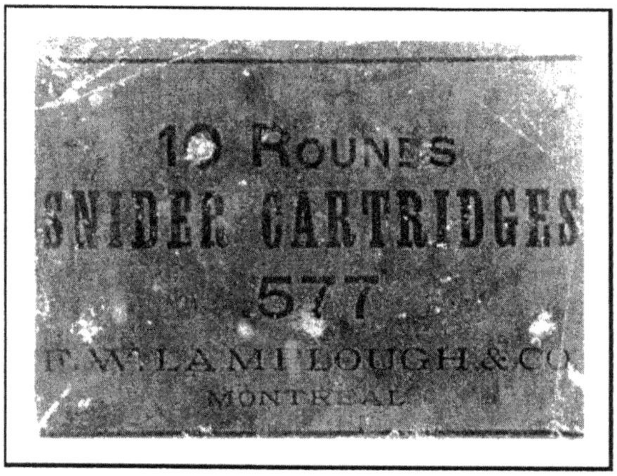

TITLE: .57 SNIDER SPORTING

HEADSTAMP: Plain

DIMENSIONS:

Rim:	.755"
Head:	.655"
Neck:	.606"
Bullet:	.588"
Case:	1.611"

OTHER DETAILS:

Bullet Grains:	480
Powder Grains:	50-60
Bullet:	One Groove
Case Material:	Brass -Necked
Total Grains:	768

OVERALL: 2.183"

It is most difficult to identify the manufacturer of this old round. The head does bear a strong resemblance to early U.M.C. .58 Berdan Musket ammunition. The head has a large wide raised rib found on other early folded head U.M.C. specimens.

The rim diameter of .755" is about the maximum measurement usually found on Snider ammunition.

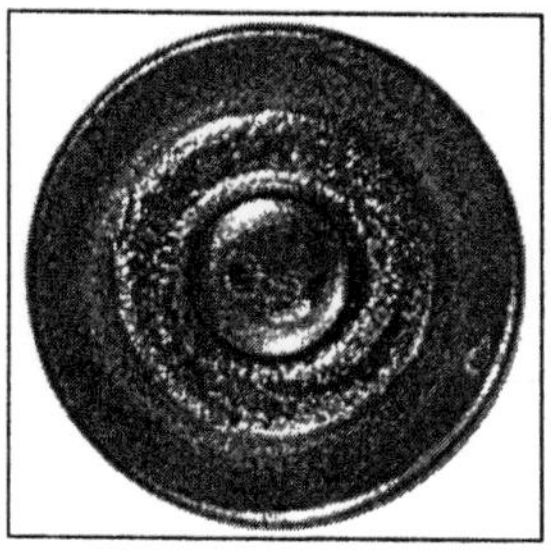

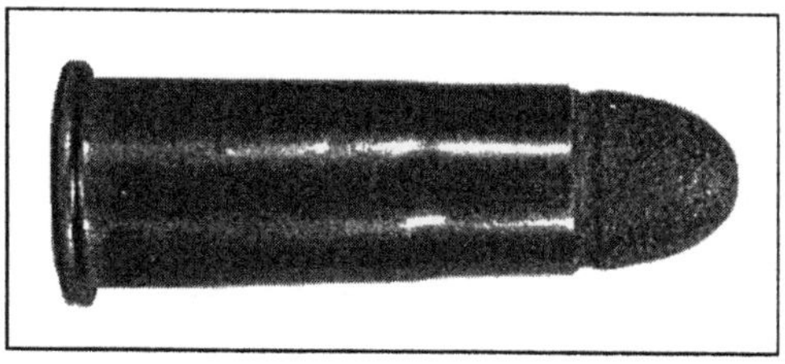

TITLE: .57 SNIDER-SPORTING

HEADSTAMP: Plain

DIMENSIONS:

Rim:	.743"
Head:	.654"
Neck:	.626"
Bullet:	.572"
Case:	1.615"

OTHER DETAILS:

Bullet Grains:	480
Powder Grains:	80
Bullet:	Tapered Point
Case Material:	Brass
Total Grains:	740

OVERALL: 2.063"

This is the sporting round version of the British Snider Military round manufactured in Canada. No satisfactory answer has ever been given why the sporting cartridge had a shorter case length.

The Snider rifles were extremely popular and for years, Kynoch and Dominion continued to produce the ammunition for these guns. Demand for the cartridges came from Africa and other areas throughout the world where Great Britain once controlled her colonies.

This round, although not head-stamped, is very likely an early product of the Dominion Cartridge Company.

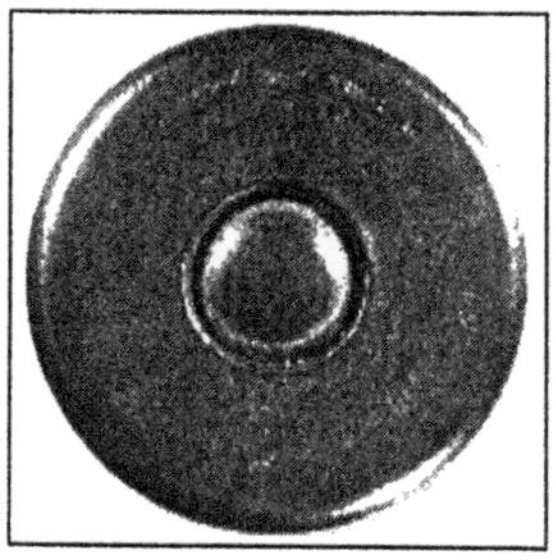

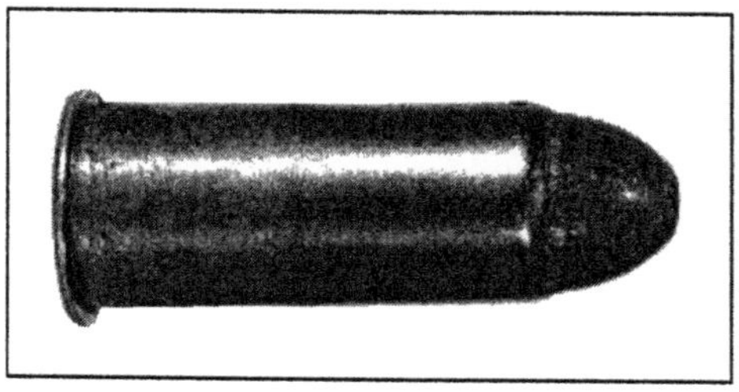

TITLE: .57 SNIDER-SPORTING

HEADSTAMP: D. C. Co. 57 SNIDER

DIMENSIONS:

Rim: .742"
Head: .659"
Neck: .601"
Bullet: .570"
Case: 1.617"

OTHER DETAILS:

Bullet Grains: 480
Powder Grains: 70
Bullet: Tapered, One Groove
Case Material: Brass
Total Grains: 674

OVERALL: 2.140"

The British Snider was a very popular rifle and was used in Canada by both target shooters as well as hunters for many years. Perhaps this case length was a compromise size for the great number of military arms left over after the war.

This round was manufactured by the Dominion Cartridge Co. of Canada and utilized a very small nickeled primer deeply set in the head of the case.

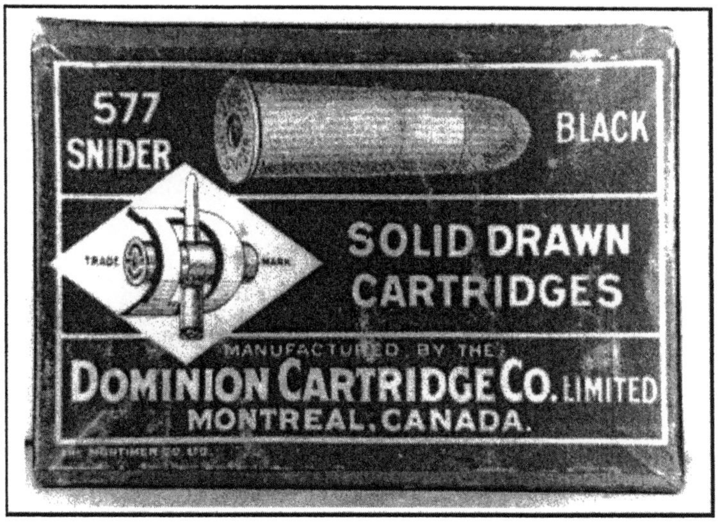

TITLE: .57 SNIDER-SPORTING-SHOT

HEADSTAMP: D.C. Co. 57 SNIDER

DIMENSIONS:		OTHER DETAILS:	
Rim:	.742"	Bullet Grains:	--
Head:	.660"	Powder Grains:	80 to 85
Neck:	.597"	Bullet:	Paper Sabot
Bullet:	.575"	Case Material:	Brass
Case:	1.620" to 1.625"	Total Grains:	580 to 610

OVERALL: 2.170" to 2.240"

It has been the practice for years to supply shot cartridges to go along with ball loads. Many hunters depended on one weapon to fulfill their needs and there would be a tremendous advantage in having the ability to adapt one's gun to the shotgun principal. This idea is still practiced.

The Dominion Cartridge Co. manufactured shot cartridges in different colored paper sabots-white, red and yellow. They were made to be used in the popular Canadian .57 Snider Rifle converted to sporting use.

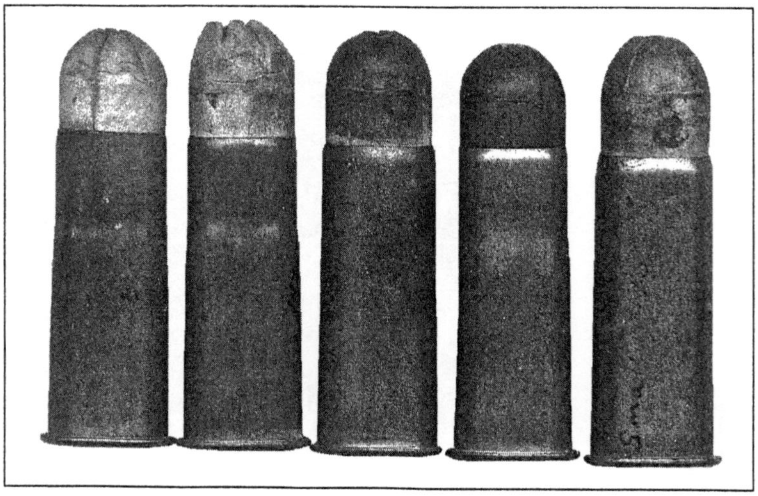

The author does not believe the colors are significant except that the total grains of the white specimen exceed the others by approximately thirty grains.

The average dimension of the rounds are listed above. The lettering of the head stamps found are small (rare), medium and large.

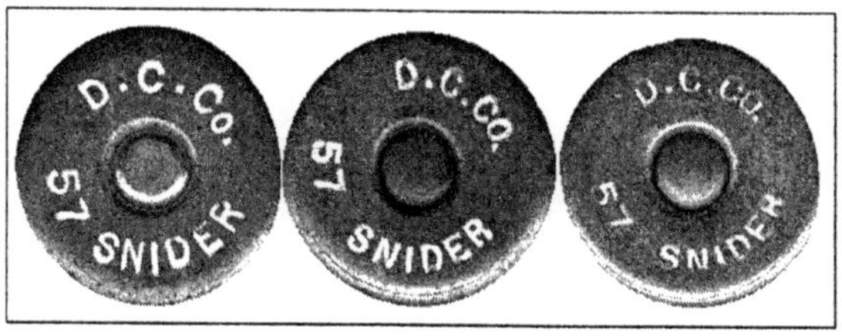

TITLE: SNIDER SHOT-BOARD DUMMY

HEADSTAMP: D.C.Co. 57 SNIDER

DIMENSIONS:

Rim: .753"
Head: .670"
Neck: .600"
Bullet: .574"
Case: 1.640"

OTHER DETAILS:

Bullet Grains: ---
Powder Grains: ---
Bullet: Paper Sabot
Case Material: Brass
Total Grains: 504

OVERALL: 2.085"

It was common practice for ammunition manufacturers to fabricate cartridge boards for displaying a representative specimen of each of their rounds and the completed boards were then distributed to retail stores.

This is a two holed board dummy of D.C. Co.'s Snider shot cartridge.

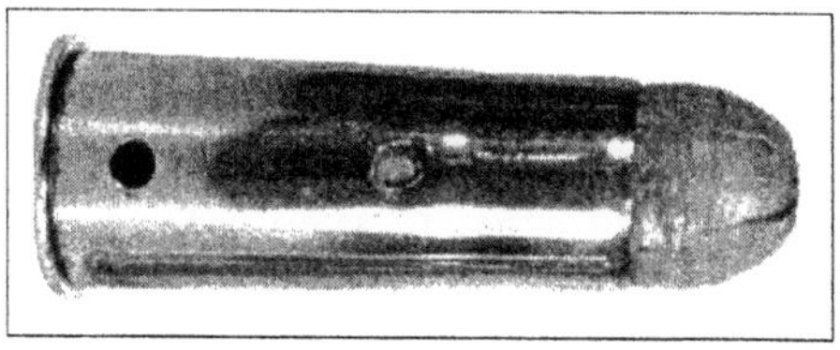

TITLE: .57 SNIDER SPORTING-PAPER PATCHED

HEADSTAMP: Plain

DIMENSIONS:

Rim:	.737"
Head:	.650"
Neck:	.626"
Bullet:	.590"
Case:	1.654"

OTHER DETAILS:

Bullet Grains:	480
Powder Grains:	70
Bullet:	Flat Nosed-Patched
Case Material:	Brass
Total Grains:	768

OVERALL: 2.332"

Generally, any of the Snider rounds found with paper patched bullets are to be considered to be of commercial make and were for public consumption.

Fouling of barrels from the use of black-powder was a common occurrence. In order to assist in keeping the barrel free of the black-powder residue build up, a patched ball was used, and it proved successful.

The very flat copper primer is unusual and seldom observed in the Snider cartridges.

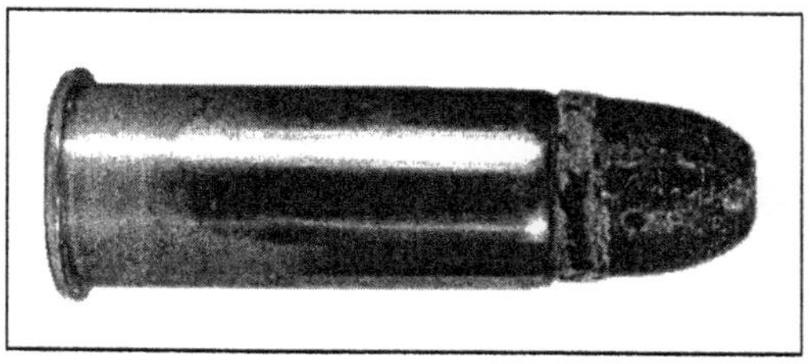

TITLE: 57 SNIDER-DUMMY

HEADSTAMP: Plain

DIMENSIONS:

Rim:	.743"
Head:	.655"
Neck:	.597"
Bullet:	.544"
Case:	1.662"

OTHER DETAILS:

Bullet Grains:	450
Powder Grains:	None
Bullet:	Round Point
Case Material:	Brass
Total Grains:	668

OVERALL: 2.111

Of unknown manufacturer. The case has an exaggerated tapered case mouth, and it contains a very small copper primer.

This round was originally part of the White & Munhall collection as evidenced by their code numbers scratched onto the side of the case-#1459 and #13504).

Neither of these numbers gives any clue as to its identity except "SNIDER".

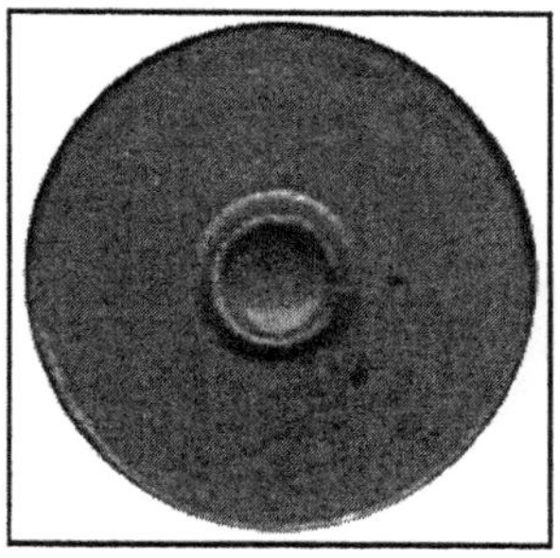

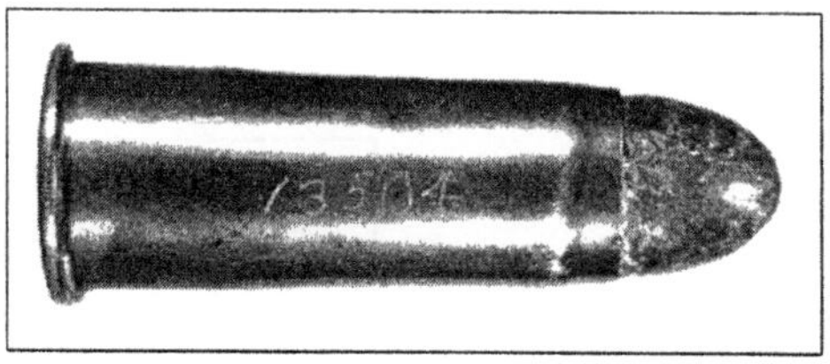

TITLE: .577 SNIDER

HEADSTAMP: Plain

DIMENSIONS:

Rim:	.739"
Head:	.656"
Neck:	.578"
Bullet:	.535"
Case:	1.665"

OTHER DETAILS:

Bullet Grains:	Unknown
Powder Grains:	Unknown
Bullet:	Lead, Tapered
Case Material:	Copper
Total Grains:	1140

OVERALL: 2.286

As was the case with many rounds in the White & Munhall collection, records only the briefest of description was used. In this instance only the words, .577 Snider model appear. Observe the unusual neck crimp. The author questions if the round was intended to be a dummy round or board display item but in fact was drilled out to empty the case in order to determine the amount and type of powder being used As was the case with many of the specimens in their collection, a code number 13506 was scratched near the drill hole. This specimen also has a second code number Y 4582. The primer is small, rounded and copperish in color.

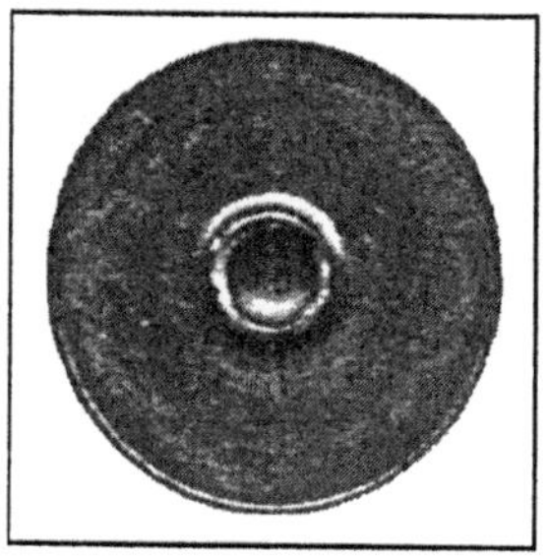

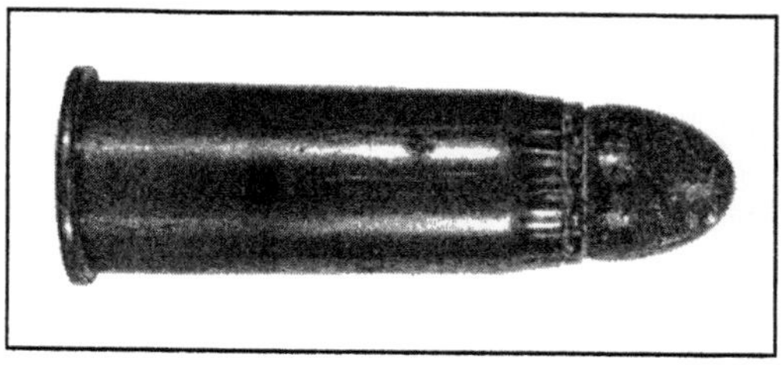

TITLE: SNIDER SPORTING

HEADSTAMP: Plain

DIMENSIONS:

Rim:	.740"
Head:	.656"
Neck:	.629"
Bullet:	.560"
Case:	1.670"

OTHER DETAILS:

Bullet Grains:	480
Powder Grains:	80
Bullet:	Tapered Lead
Case Material:	Brass
Total Grains:	745

OVERALL: 2.120"

A product of the Dominion Cartridge Company with a slightly longer case-length and over-all dimension.

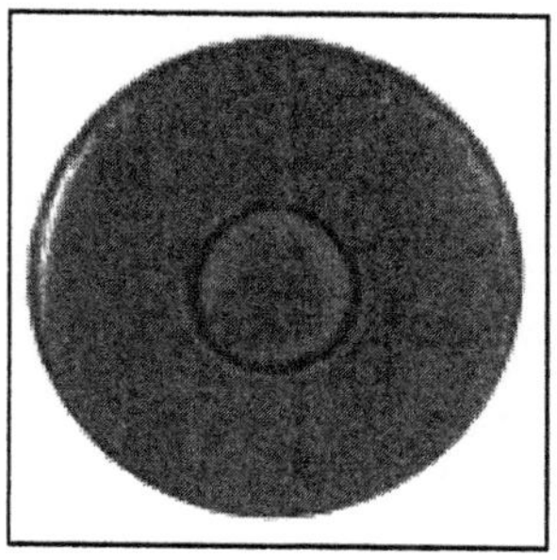

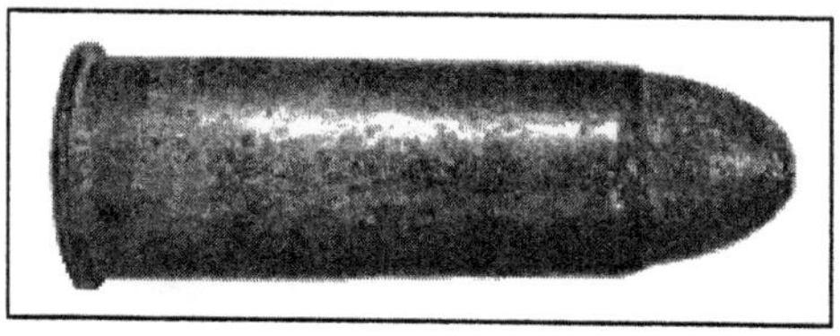

TITLE: .58 TURKISH SNIDER RIFLE

HEADSTAMP: Plain

DIMENSIONS:

Rim:	.735"
Head:	.656"
Neck:	.585"
Bullet:	.545"
Case:	1.675"

OTHER DETAILS:

Bullet Grains:	N/A
Powder Grains:	N/A
Bullet:	Lead
Case Material:	Brass
Total Grains:	650

OVERALL: 2.203"

The title of this round "Turkish Snider Rifle" is questionable. Over the years, references have been made to a rifle size as well as a carbine size.

The configuration of this round is similar to the common 1.565" case length. It could well be that the smaller length is a carbine and this cartridge with the 1.675" (42mm) case length is the rifle round.

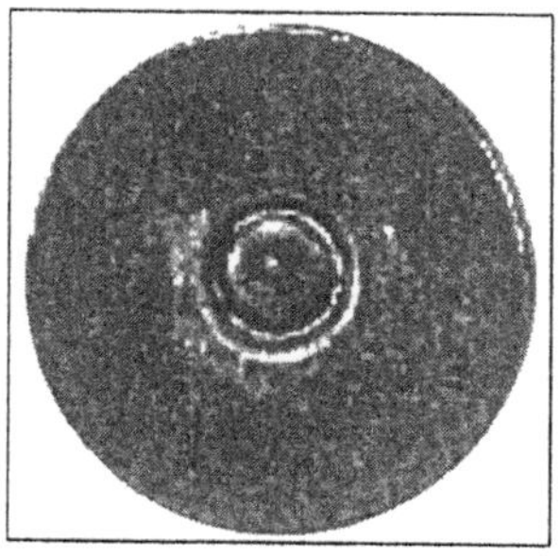

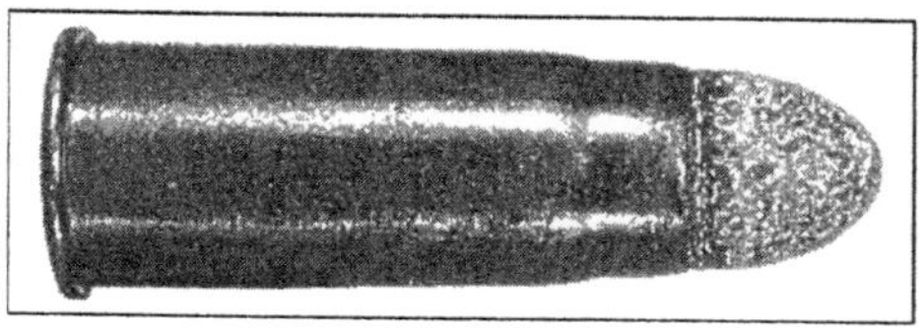

TITLE: .57 SNIDER SPORTING-SHOT

HEADSTAMP: KYNOCH .577

DIMENSIONS:

Rim: .739"
Head: .657"
Neck: .542"
Bullet: None
Case: 1.682"

OTHER DETAILS:

Bullet Grains: -0-
Powder Grains: ?
Bullet: Shot Load
Case Material: Brass
Total Grains: 350

OVERALL: 1.682"

This is an interesting variation in case construction for a shot load. The six-petal case mouth crimp and heavy crimp line sets it apart. The round is likely interchangeable with the original .577 and later .577/.450 Martini Henry Rifles. The head stamp uses uncommonly small letters.

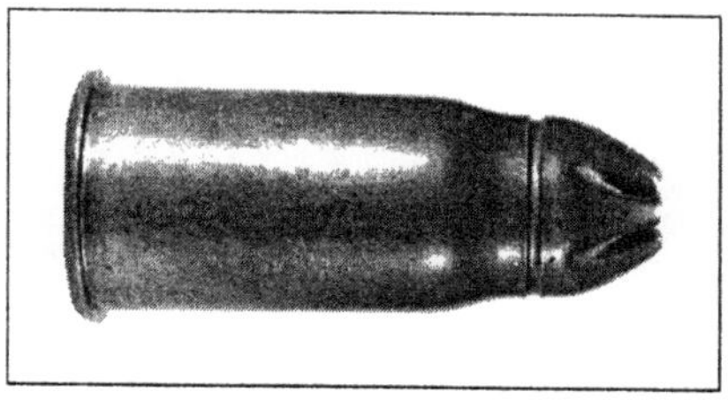

TITLE: .577 SNIDER

HEADSTAMP: Plain

DIMENSIONS:

Rim: .748"
Head: .676"
Neck: .588
Bullet: .563"
Case: 1.730

OTHER DETAILS:

Bullet Grains:	480
Powder Grains:	80
Bullet:	Flat Nose
Case Material:	Brass
Total Grains:	701

OVERALL: 2.219"

Another one of the oddities from the White & Munhall Collection.

The code numbers scratched on the side of this case is Y495B and #8464. It is simply recorded in their record book as ".577 Snider". The round has a bit smaller bullet diameter, and a flat plain head with a brass primer.

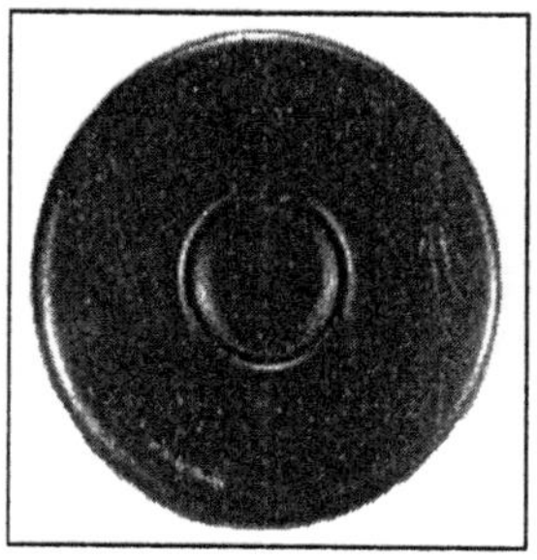

TITLE: .58 UNIDENTIFIED

HEADSTAMP: Plain

DIMENSIONS:

Rim:	.740"
Head:	.633"
Neck:	.614"
Bullet:	.575"
Case:	1.838"

OTHER DETAILS:

Bullet Grains:	
Powder Grains:	
Bullet:	Lead, Tapered Point
Case Material:	Brass
Total Grains:	666

OVERALL: 2.348"

This could be a variant Snider round but presently is unidentified.

A line runs from the rim to the case mouth indicating the possibility of this being a two piece case.

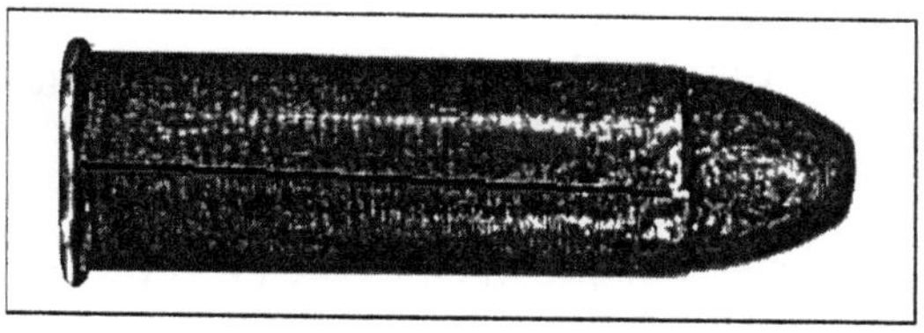

TITLE: .577 SNIDER WHITNEY MARK D

HEADSTAMP: Plain

DIMENSIONS:

Rim: .744"
Head: .655"
Neck: .618"
Bullet: .568"
Case: 1.850"

OTHER DETAILS:

Bullet Grains:	450
Powder Grains:	70
Bullet:	Smooth, tapered
Case Material:	Brass
Total Grains:	680

OVERALL: 2.404'

A product of the Colonial Ammunition Company of New Zealand. It was manufactured under the 1886 New Zealand patent issued to Ala Whitney, one of the founders of Colonial Ammunition Co.

This early round employs a brass rim instead of the more common steel type. The case is a brass seamless tube rather than being constructed of a brass coiled wrapped case. A roll crimp holds the smooth and lighter (450 grains) lead bullet. The brass rim is attached to the case body by the brass rivet of the copper battery cup primer.

The cartridge is described on the 1890's packet as a Mark D. This particular round has a White and Munhall designation 13511.

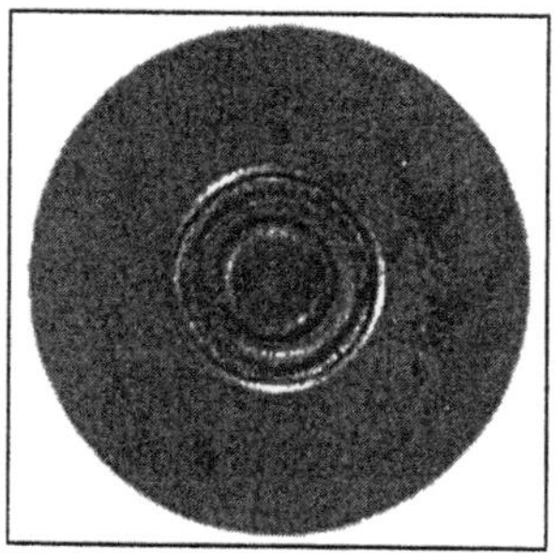

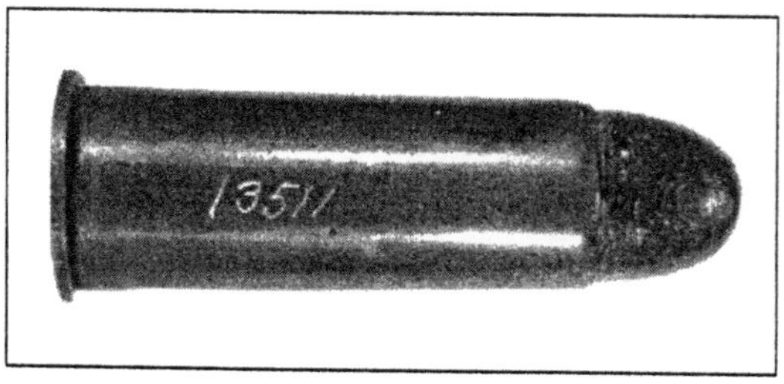

TITLE: SNIDER – WHITNEY MK III

HEADSTAMP: Plain

DIMENSIONS:

Rim:	.740”
Head:	.669”
Neck:	.610”
Bullet:	.571”
Case:	1.875”

OTHER DETAILS:

Bullet Grains:	480
Powder Grains:	70
Bullet:	Round Nose
Case Material:	Drawn Brass
Total Grains:	1220

OVERALL: 2.330”

An early Snider round manufactured by the Colonial Ammunition Company, New Zealand.

A drawn brass case turned over at the base and attached to an iron base disk by the primer pocket clinched over as internal wad pressed down around the pocket and over the exterior of the base disk. An 1898 dated packet identifies the round as the M K A III loaded by the Colonial Ammunition Company, Auckland, New Zealand.

An earlier packet dated 1896 calls the brass disk specimen a Mark D. Note the four punched dots around the case mouth to secure the bullet in place.

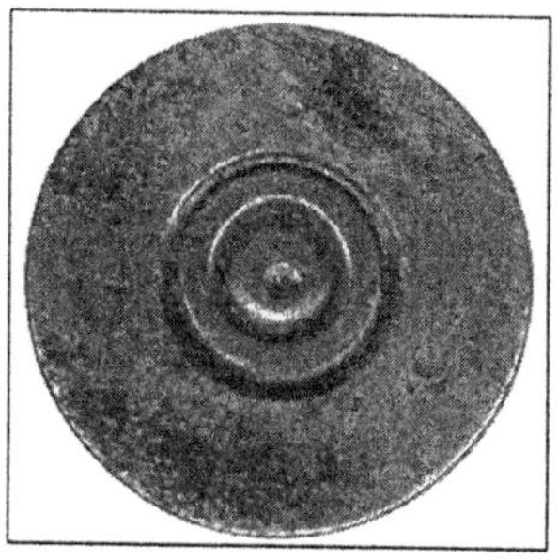

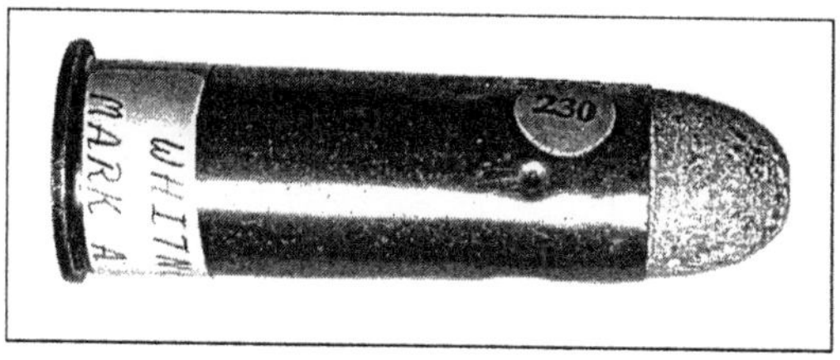

TITLE: 14.7mm x 49R (.577)
TURKISH SNIDER

HEADSTAMP: ARABIC CHARACTERS

DIMENSIONS:

Rim:	.738"
Head:	.646"
Neck:	.600"
Bullet:	.552"
Case:	1.892"

OTHER DETAILS:

Bullet Grains:	
Powder Grains:	
Bullet:	Lead, Round Nose
Case Material:	Brass
Total Grains:	925

OVERALL: 2.275"

Turkey was quick to switch from the muzzle-loading system to breech-loading weapons. While she did not manufacture her own arms during this period of time, extensive purchases were made from European and American makers.

Although the head stamp is of Turkish characters, the ammunition was likely to have been made by some European cartridge manufacturer.

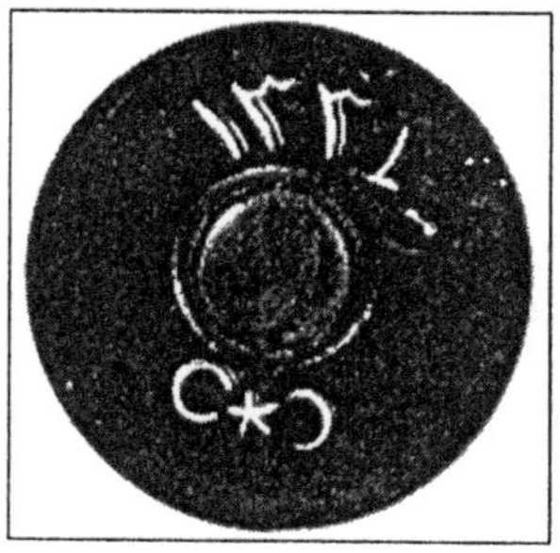

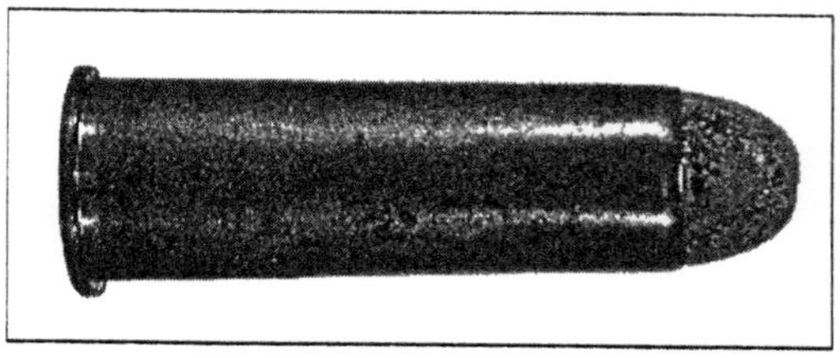

TITLE: SNIDER-COPPER CASE

HEADSTAMP: Plain

DIMENSIONS:

Rim:	.750"
Head:	.660"
Neck:	.606"
Bullet:	.568"
Base:	1.900"

OTHER DETAILS:

Bullet Grains:	45-
Powder Grains:	70
Bullet:	Grvd. & Ribbed
Case Material:	Copper
Total Grains:	1170

OVERALL: 2.415"

A drawn copper case with a brass, attached head. The head is attached to the main body of the case by four stab crimps. The numbers T155 and 10334 indicate the round was once in the White & Munhall collection. The unusual ribbed and grooved bullet is supported in the case by a neck crimp.

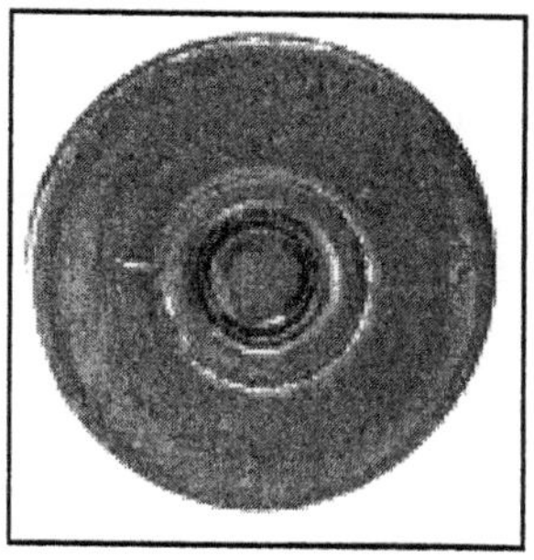

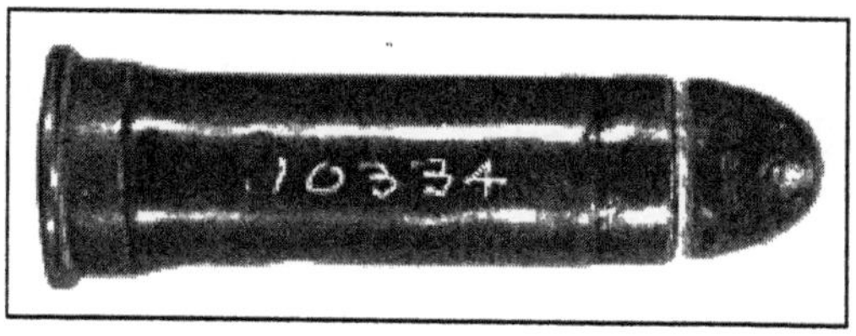

TITLE: SNIDER

HEADSTAMP: * SCB *
95

DIMENSIONS:

Rim: .738"
Head: .661"
Neck: .598"
Bullet: .580"
Case: 1.900"

OTHER DETAILS:

Bullet Grains: 480
Powder Grains: 70
Bullet: Lead
Case Material: Brass
Total Grains: 712

OVERALL: 2.515"

This cartridge was manufactured by Societe Cartoucherie Belge. Very little is known about this manufacturer, and rounds with this head stamp are seldom seen.

The bullet appears to be of more modern configuration. It has an unusual knurled groove. Perhaps the * 95 * on the head-stamp indicates the year of manufacture (1895).

The numbers (14058 and Y457) inscribed on this drilled case indicate it was part of the White & Munhall collection.

TITLE: SNIDER-MARK IX

HEADSTAMP: Plain

DIMENSIONS:

Rim: .748"
Head: .660"
Neck: .621"
Bullet: .574"
Case: 1.903"

OTHER DETAILS:

Bullet Grains:	480
Powder Grains:	70
Bullet:	Lead, Spun Over
Case Material:	Paper Covered
Total Grains:	716

OVERALL: 2.465"

The last official British issue of the Snider cartridge was approved August 16, 1871. The round is easily identified by the single red band painted around the paper-covered case and has a brass primer pocket and higher base cup. The bullet has three grooves instead of four.

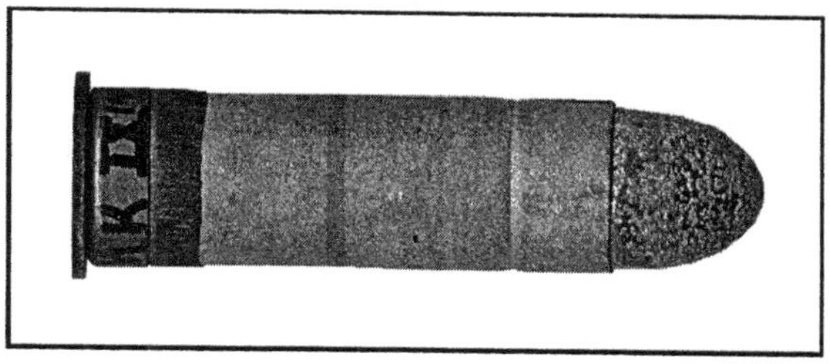

TITLE: SNIDER-MARK VI

HEADSTAMP: Plain

DIMENSIONS:

Rim:	.742"
Head:	.663"
Neck:	.609"
Bullet:	.572"
Case:	1.904"

OTHER DETAILS:

Bullet Grains:	480
Powder Grains:	70
Bullet:	Lead, Plug in Nose
Case Material:	Brass & Paper
Total Grains:	720

OVERALL: 2.474"

Identified by the black ring on the case, the round was introduced on June 12, 1868. The cartridge had a much stronger case. The sheet brass thickness was increased to .005" instead of .003", formed one turn, plus an overlap of .04", no application of spirits and wax between cup and the cup base.

All this resulted in better extraction and without the wax, the firing pin did not get clogged and hung up.

The bullet has the exposed wood plug.

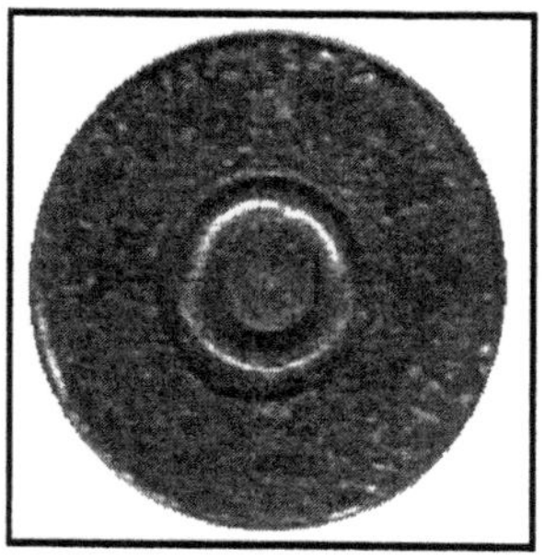

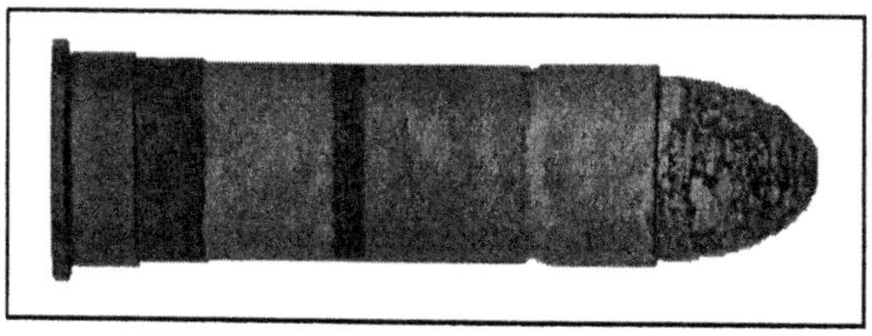

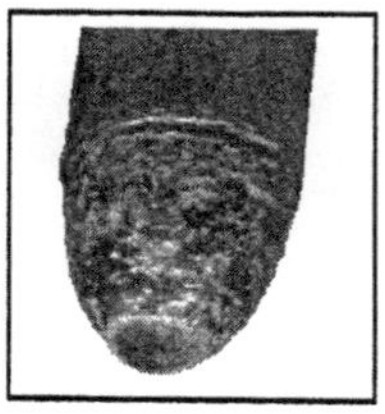

TITLE: .577 SNIDER-TRANSITION

HEADSTAMP: KYNOCH'S PATENT

DIMENSIONS:

Rim:	.745"
Head:	.661"
Neck:	.620"
Bullet:	.565"
Case:	1.909"

OTHER DETAILS:

Bullet Grains:	480
Powder Grains:	80
Bullet:	Lead
Case Material:	Brass
Total Grains:	

OVERALL: 2.413"

This round was made for coiled cased chambers. It is obviously one of the very early cartridges because the head or cup is attached to the brass tube and not simply "formed" by drawing the case to form a solid head.

The chambers of the rifles had space provided to accept the base cups of coiled wrapped cases. When a drawn case was made, they copied the Boxer style of case construction. This cartridge was initially designed utilizing an applied head to the cartridge base. This procedure (or technique) was shortly followed by a bulge around the base as on this specimen.

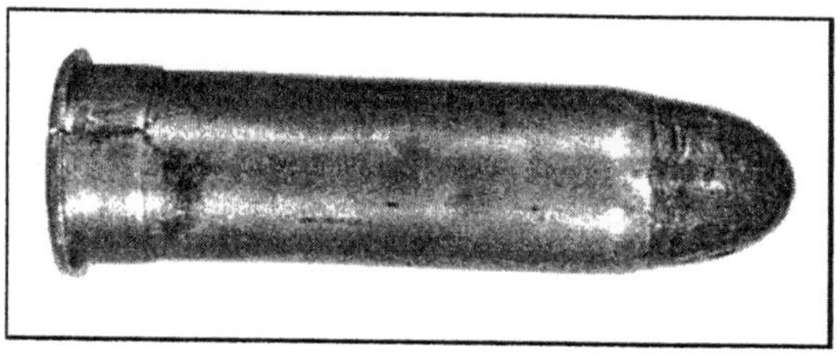

TITLE: .577 SNIDER COPPER CASE

HEADSTAMP: Plain

DIMENSIONS:

Rim: .748”
Head: .653” .610”
Neck: .606”
Bullet: .545”
Case: 1.912”

OTHER DETAILS:

Bullet Grains: 80
Powder Grains: N.A.
Bullet: Lead
Case Material: Copper
Total Grains: 1400

OVERALL: 2.123”

Another White & Munhall specimen #10333 – T156 called Snider. To make more space for the powder beneath the rim is a raised ring of metal surrounding the case head. Although this is a drilled hole the author believes the hole was for observation purposes only and not to create a dummy round.

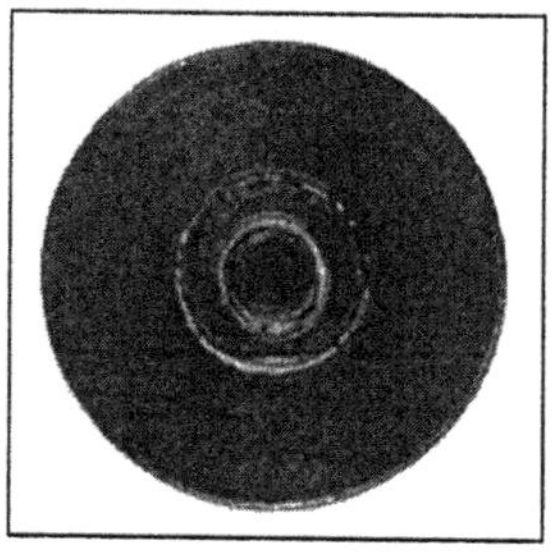

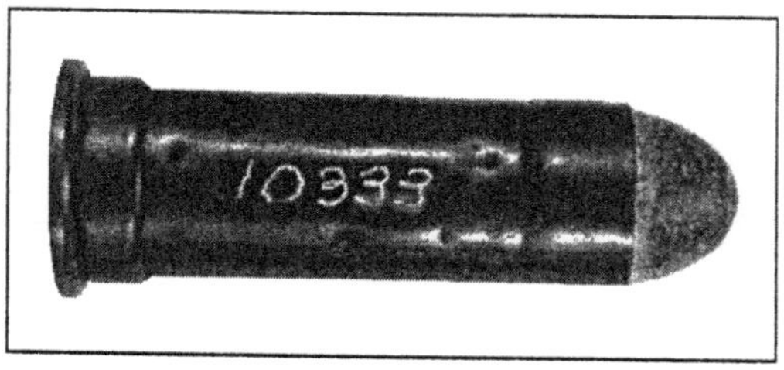

TITLE: .577 SNIDER

HEADSTAMP: KYNOCH & CO. BIRMINGHAM (Raised)

DIMENSIONS:

Rim: .741"
Head: .655"
Neck: .600"
Bullet: .573"
Case: 1.912"

OTHER DETAILS:

Bullet Grains: 480
Powder Grains: 80
Bullet: Smooth
Case Material: Brass
Total Grains: 749

OVERALL: 2.435"

This is another round made for use in the early coiled cased chambers.

The most outstanding feature of this cartridge is the raised head-stamp of. KYNOCH & CO.BIRMINGHAM.

This specimen was located in Australia.

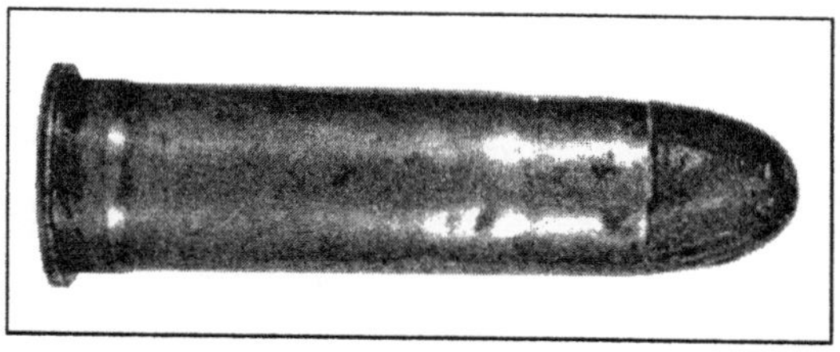

TITLE: SNIDER-MARK V

HEADSTAMP: Plain

DIMENSIONS:

Rim:	.745"
Head:	.662"
Neck:	.626"
Bullet:	.566"
Case:	1.928"

OTHER DETAILS:

Bullet Grains:	480
Powder Grains:	70
Bullet:	Lead, Spun Over
Case Material:	Brass, Brown Paper
Total Grains:	714

OVERALL: 2.430"

Identified by the brownish colored paper wrapper and the lighter weight base cup. The bullet with the spun over nose was changed to four saw-tooth grooves but still maintaining the same 480 grain weight.

Attempts to use lighter material were not satisfactory and the round was only in service from the late 1867 to August 1868 when it was replaced by the Mark VI.

All in all an attempt to economize that did not succeed.

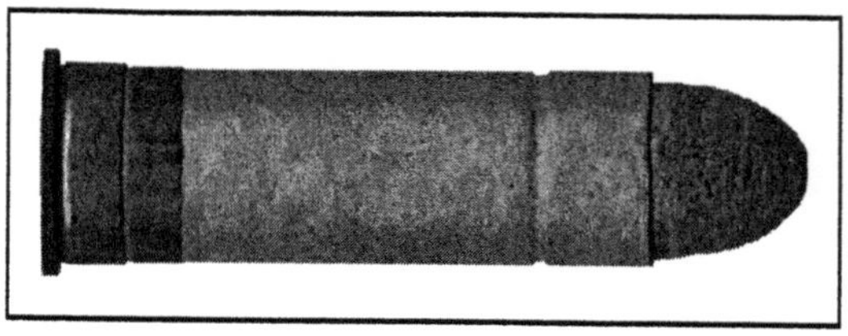

TITLE: SNIDER 15 X 49 Rmm

HEADSTAMP: SK***

DIMENSIONS:

Rim: .734"
Head: .656"
Neck: .610"
Bullet: .580"
Case: 1.928"

OTHER DETAILS:

Bullet Grains:	480
Powder Grains:	80
Bullet:	Plugged
Case Material:	Brass
Total Grains:	875

OVERALL: 2.517"

This is a .577 Snider cartridge manufactured by Keller & Co. of Austria. Perhaps it should be properly titled 15 x 49 Rmm, the metric equivalent. The bullet is unusual because the nose hollow has been filled with wax.

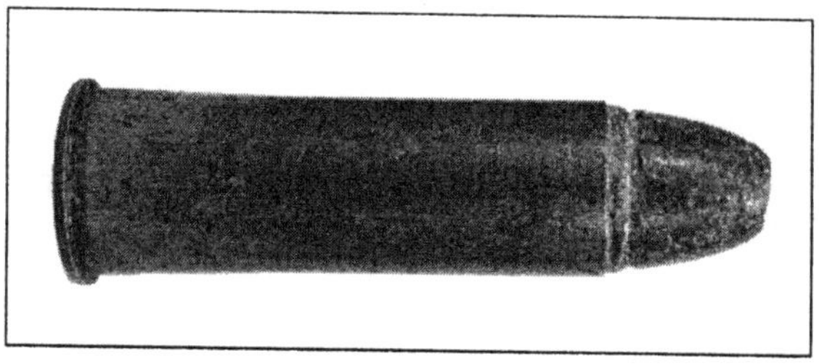

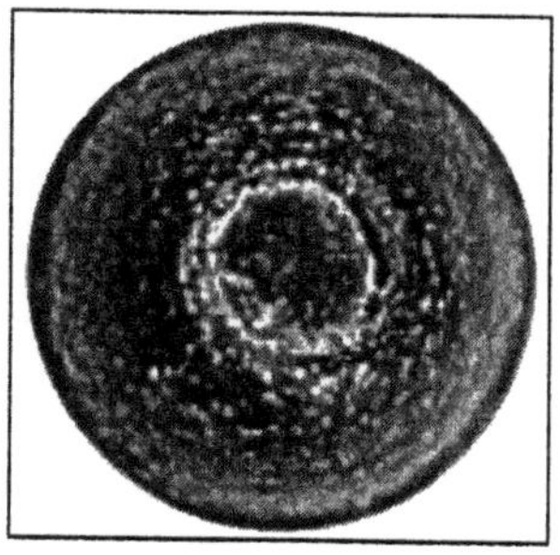

TITLE: DAW'S PATENT, 1867

HEADSTAMP: G H DAW'S PATENT NO. 26

DIMENSIONS:	
Rim:	.752"
Head:	.674"
Neck:	.588"
Bullet:	.573"
Case:	1.930"

OTHER DETAILS:	
Bullet Grains:	480
Powder Grains:	70
Bullet:	Lead, Patched
Case Material:	Foil
Total Grains:	772

OVERALL: 2.433"

Originally used in a shotgun, G. H. Daw eventually adapted the round for use in the converted British muskets.

English patent #600 was awarded Daw on March 8, 1867. The cartridge was constructed by winding a strip of brass foil in order to form a watertight tube. The base was formed by a metal cup pressed over the base end of the tube, inserted the metal primer pocket and flanged over the outside surface of the base cup. A papier-mache wad was pressed around the primer pocket. The anvil was a brass wire with four grooves running its length. The primer pocket was pierced at its closed end to admit the flame to the powder charge. The bullet has a paper patch extending to the end of the case mouth.

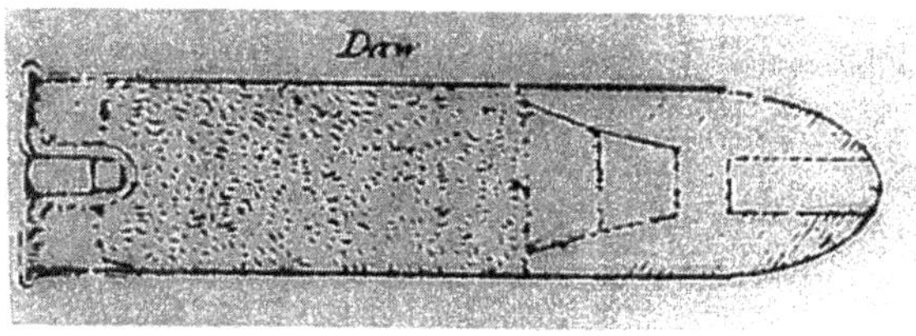

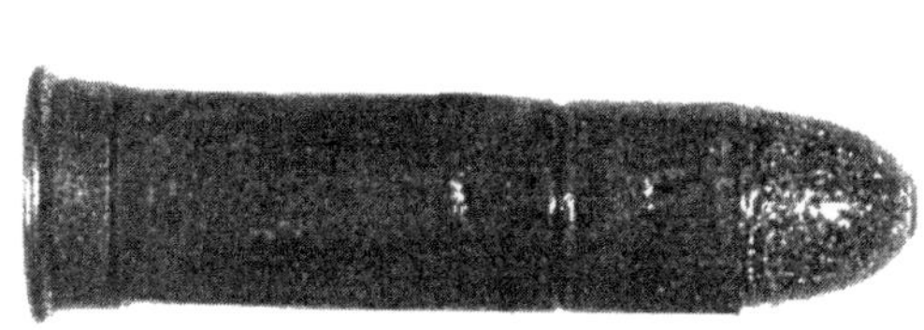

TITLE: .577 SNIDER MARK IV

HEADSTAMP: Plain

DIMENSIONS:

Rim: .748"
Head: .661"
Neck: .606"

Bullet: .562"
Case: 1.931"

OTHER DETAILS:

Bullet Grains: 480
Powder Grains: 70
Bullet:Nose - Plug outside - three saw tooth grooves on bullet
Case Material: Boxer
Total Grains: 708

OVERALL: 2.469"

The first Boxer cased cartridges used a base disk of brass. This technique was used on all cartridges until the Mark IV round.

In an effort to reduce manufacturing costs, an iron disk was substituted for the brass disk beginning with the Mark IV ammunition.

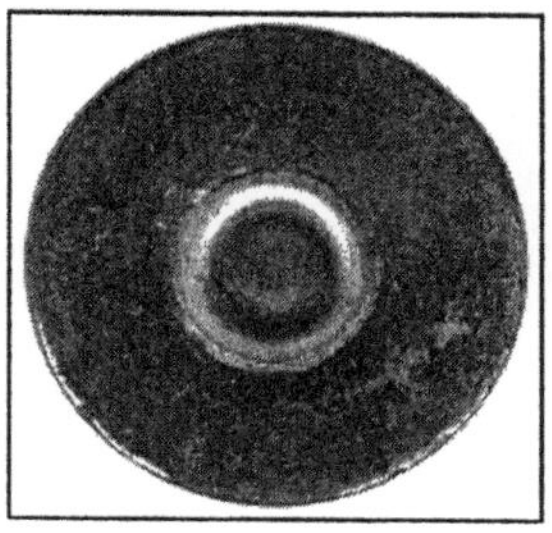

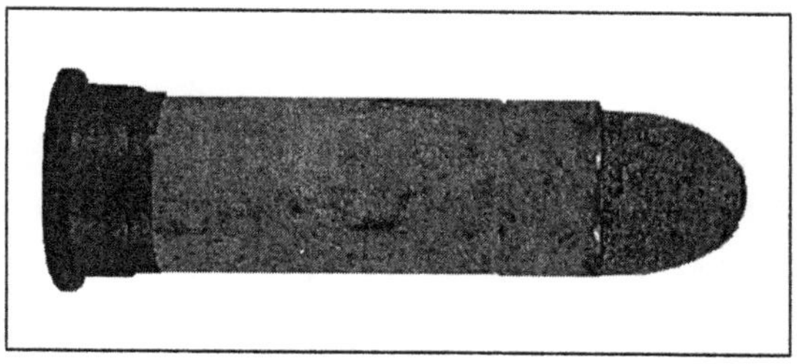

TITLE: .577 (14.5 x 50mm) CHINESE SNIDER

HEADSTAMP: S F M * GG *

DIMENSIONS:		OTHER DETAILS:	
Rim:	.746"	Bullet Grains:	480
Head:	.659"	Powder Grains:	80
Neck:	.601"	Bullet:	Lead
Bullet:	.568"	Case Material:	Brass
Case:	1.932"	Total Grains:	655

OVERALL: 2.447"

The head-stamp indicates the round was manufactured by Societe Francais des Munitions. It was probably made under a contract to supply China with ammunition for the Snider rifles, adopted by the Chinese Army.

The hole in the case indicates it was either an assembler's dummy round or used for display purposes.

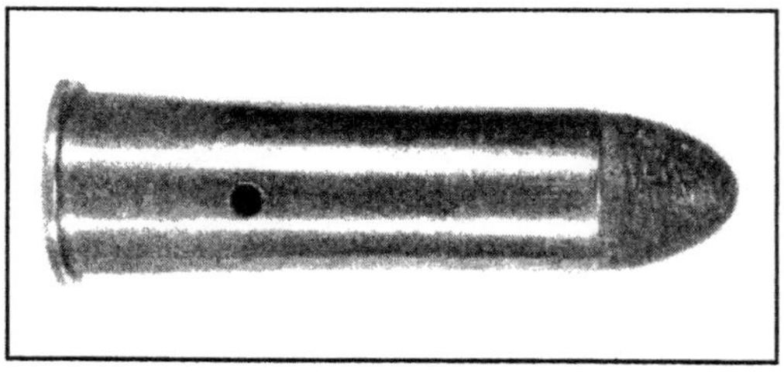

TITLE: .57 SNIDER-REMINGTON MANUFACTURER

HEADSTAMP: REM UMC 57 S

DIMENSIONS:		OTHER DETAILS:	
Rim:	.747"	Bullet Grains:	480
Head:	.658"	Powder Grains:	80
Neck:	.608"	Bullet:	Tapered Point
Bullet:	.564"	Case Material:	Brass
Case:	1.933"	Total Grains:	787

OVERALL: 2.459"

Due to the great number of Snider rifles in the hands of civilians; there was a large demand for ammunition to shoot in these guns.

The Remington Company apparently felt there was an American market for such ammunition and it would be financially worth their while to produce a limited number of rounds in this caliber.

The only American ammunition manufacturer to manufacture Snider cartridges was REM-UMC.

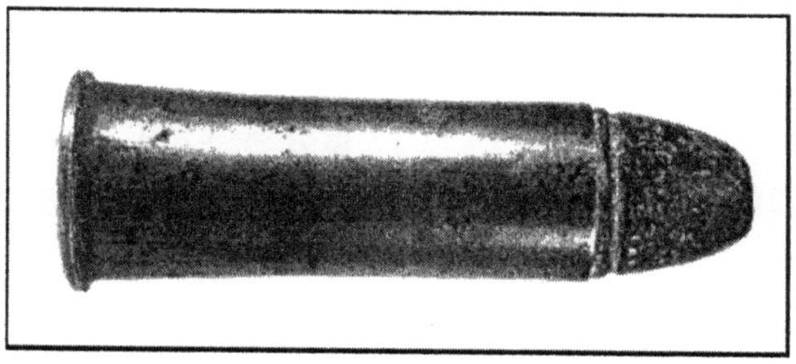

TITLE: .577 SNIDER BOXER-SPORTING

HEADSTAMP: Plain

DIMENSIONS:

Rim:	.744:
Head:	.664"
Neck:	.615"
Bullet:	.581"
Case:	1.933"

OTHER DETAILS:

Bullet Grains:	500 (?)
Powder Grains:	80
Bullet:	Patched, Tubed
Case Material:	Brass Wrapped
Total Grains:	813

OVERALL: 2.612"

This specimen was used for hunting or commercial purposes as evidenced by the paper patched, heavy copper tubed bullet. This hunting round was referred to as an "Express" cartridge; a catchy phrase meaning a lead projectile with a copper tube in the nose cavity. The copper tube held no explosive, but some rounds have been loaded with an explosive material.

The cartridge has some unusual features. It has a high brass (16mm) reinforcing cup, two stab crimps holding the bullet in place and a "fast" turn or lap to the brass sheet coil instead of the very gradual lap we see on these rolled cases.

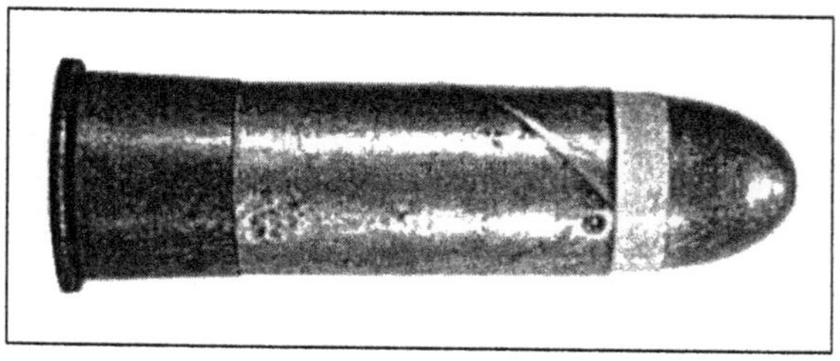

TITLE: SNIDER-MARK VIII

HEADSTAMP: Plain

DIMENSIONS:

Rim:	.745"
Head:	.660"
Neck:	.624"
Bullet:	.578"
Case:	1.940"

OTHER DETAILS:

Bullet Grains:	480
Powder Grains:	70
Bullet:	Spun Over
Case Material:	Paper Boxer
Total Grains:	722

OVERALL: 2.445"

The Mark VIII is easily identified by the two black lines painted on the paper-covered case.

The inside of the case is shellacked and covered with thin paper. The bullet is spun over and shellacked. The case has a copper primer pocket instead of brass.

The purpose of the thin paper and shellac was for water and corrosion proofing.

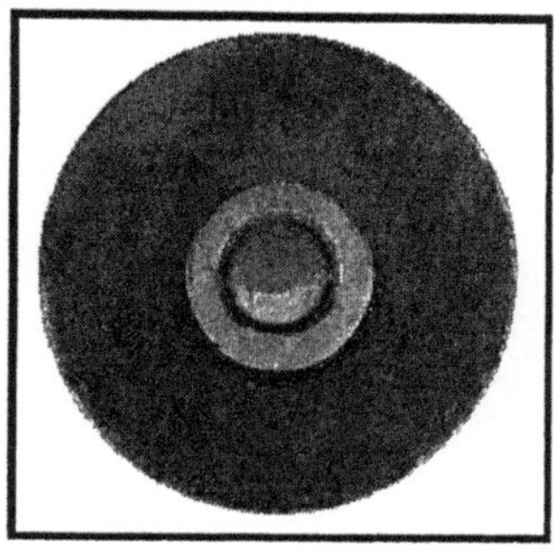

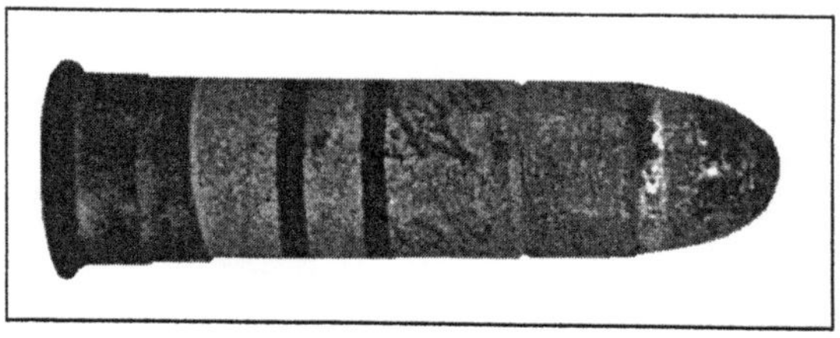

TITLE: .577 SNIDER KYNOCH PATENT BASE

HEADSTAMP: Plain

DIMENSIONS:

Rim:	.745"
Head:	.663"
Neck:	.601"
Bullet:	.571"
Case:	1.944"

OTHER DETAILS:

Bullet Grains:	480
Powder Grains:	80
Bullet:	Lead
Case Material:	Copper
Total Grains:	715

OVERALL: 2.447"

An early, solid drawn, copper case with a stepped base (to accommodate the chamber of those rifles originally firing the coiled cases with attached cups and riveted heads).

White & Munhalls records only indicate "SNIDER" for this round with the number (13512) scratched on the side of the case body.

Close examination of the case mouth reveals four very faint stab crimps to secure the bullet.

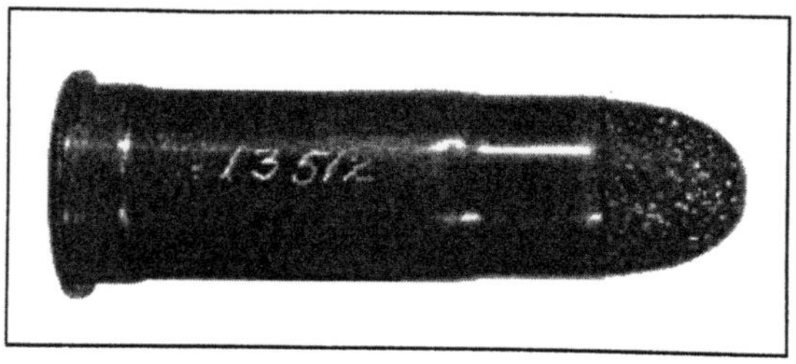

TITLE: SNIDER-BELGIUM COILED CASE

HEADSTAMP: Plain

DIMENSIONS:

Rim:	.743"
Head:	.664"
Neck:	.604"
Bullet:	.536"
Case:	1.952"

OTHER DETAILS:

Bullet Grains:	480
Powder Grains:	80
Bullet:	Tapered Point
Case Material:	Coiled Brass
Total Grains:	748

OVERALL: 2.360"

This round is thought to be of Belgian origin.

There are a number of distinguishing features of this cartridge which set it apart from the common brass wrapped Boxer case. The case has been rolled in the opposite direction, it has only one reinforcing cup and the other cup has been formed by another roll of brass about 15mm long. The iron disk contains a very large (.389") diameter brass reinforcing ring which contains an equally large (.236") diameter brass primer.

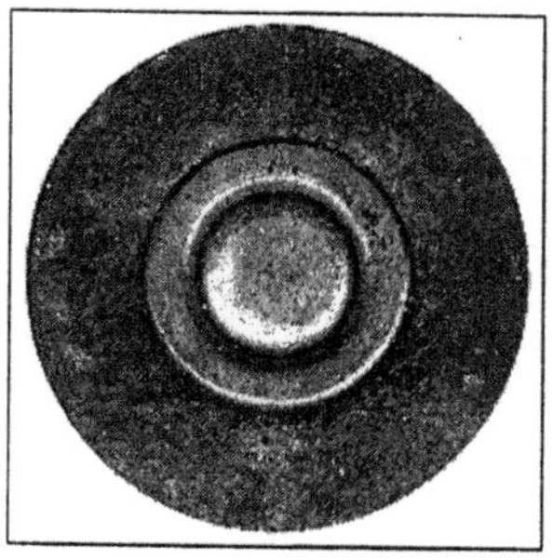

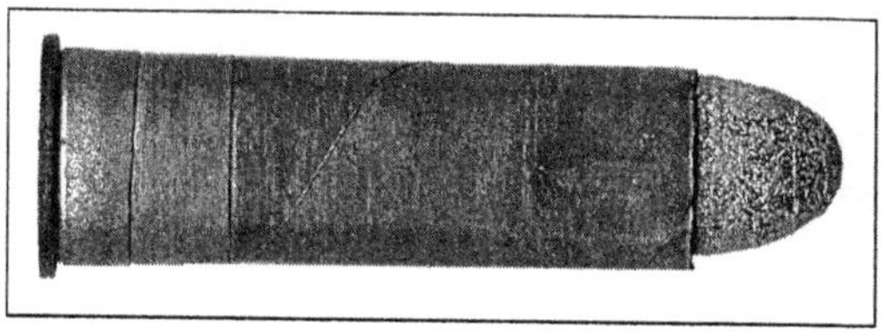

TITLE: SNIDER-CANADIAN-MARK IX

HEADSTAMP: Plain

DIMENSIONS:

Rim:	.742"
Head:	.660"
Neck:	.604"
Bullet:	.565"
Case:	1.953"

OTHER DETAILS:

Bullet Grains:	480
Powder Grains:	70
Bullet:	Lead
Case Material:	Brass Wrapped
Total Grains:	723

OVERALL: 2.445"

This round was manufactured (circa 1886) at the Canadian Government Cartridge factory under the Mark IX configuration and described as the Second Pattern.

The First Pattern, which was paper wrapped, functioned poorly when exposed to moisture, so to correct the problem, the paper covering was eliminated on this Second Pattern.

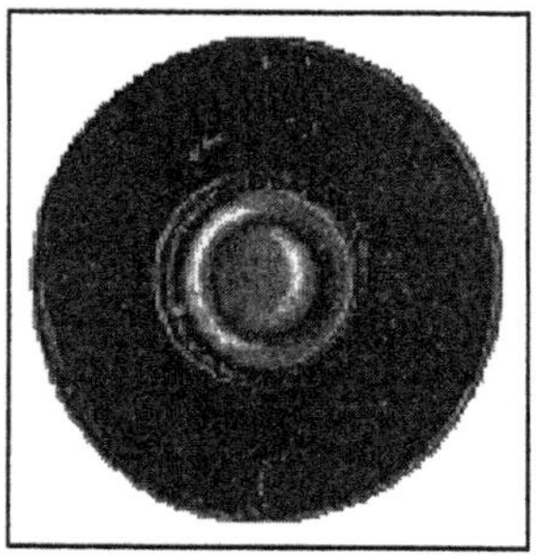

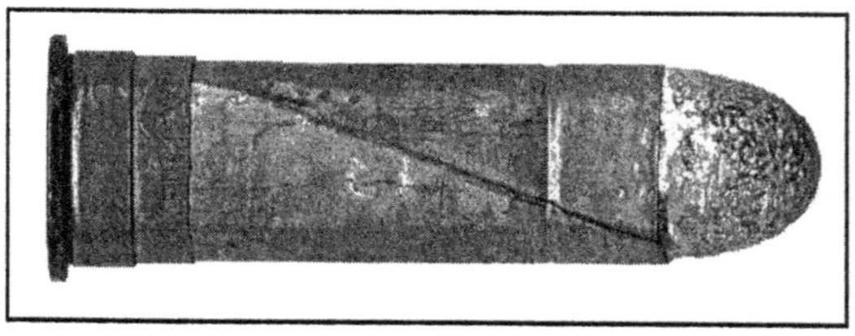

TITLE: .577 SNIDER

HEADSTAMP: Plain

DIMENSIONS:

Rim: .748"
Head: .664"
Neck: .626"
Bullet: .585"
Case: 1.958"

OTHER DETAILS:

Bullet Grains: 500
Powder Grains: 90
Bullet: Lead, Tapered, Paper Patched
Case Material: Boxer, Brass Coil
Total Grains: 790

OVERALL: 2.500"

This round has two distinguishing features:

A heavy tapered paper patched bullet.

Observe the reverse wrapping of the brass case.

There is speculation that this round could have been created for a commercial market rather than for the military.

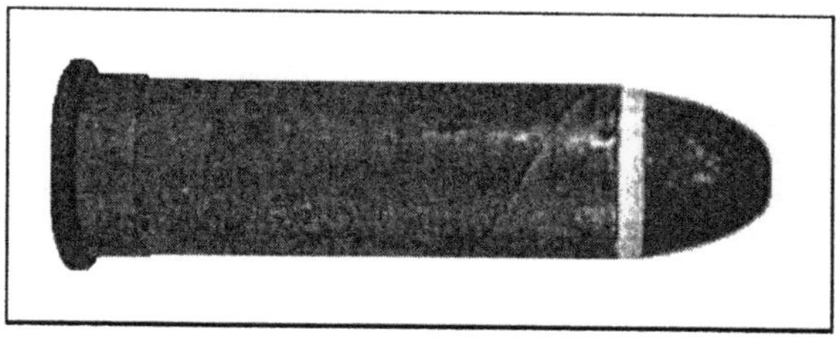

TITLE: SNIDER-MARK II

HEADSTAMP: Plain

DIMENSIONS:

Rim:	.750"
Head:	.665"
Neck:	.602"
Bullet:	.566"
Case:	1.963"

OTHER DETAILS:

Bullet Grains:	525
Powder Grains:	70
Bullet:	Lead Tapered
Case Material:	Brass & White Paper
Total Grains:	765

OVERALL: 2.425"

The Mark II and Mark III were approved during August 1866 and April 1867. It is different from the Mark I by the use of a separate brass base disk, base cup and reinforcing strip, all with a central hole into which the cup chamber was inserted. The cap chamber is crimped over a papier-mâché wad and the outer surface of this brass disk thereby forming a rivet which assisted in securing the case and head assembly together.

The four groove bullet weighed 525 grains with a hollow base containing pressed clay. The nose was hollow and filled with a sycamore wood plug.

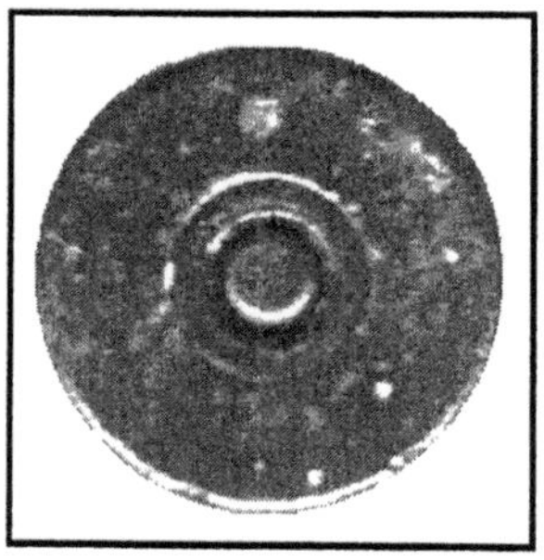

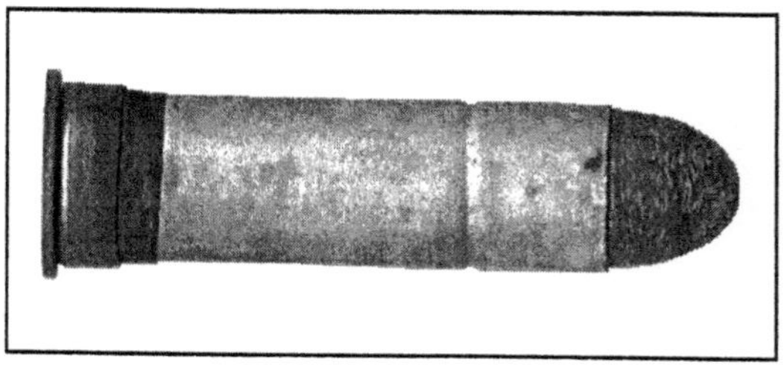

TITLE: .577 SNIDER KYNOCH PATENT BASE

HEADSTAMP: Plain

DIMENSIONS:

Rim: .754"
Head: .662"
Neck: .602
Bullet: .575"
Case: 1.967"

OTHER DETAILS:

Bullet Grains: 480
Powder Grains: 80
Bullet: Blunt-Tapered
Case Material: Brass
Total Grains: 777

OVERALL: 2.365"

The case is labeled "for coiled case chambers". The transition from the coiled case to a solid drawn cartridge case obviously presented problems to ammunition manufacturers because of variations in chamber measurements. This case has a large thin rim (.754" Dia.) with a raised ridge under the rim to accommodate the earlier coiled case chamber specifications. The average rim diameter is from .735" to.740". The raised ridge can be found on a number of other early or transitional specimens pictured in this book.

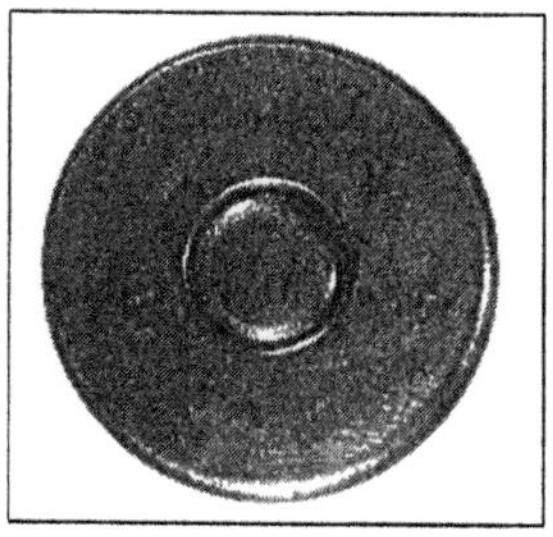

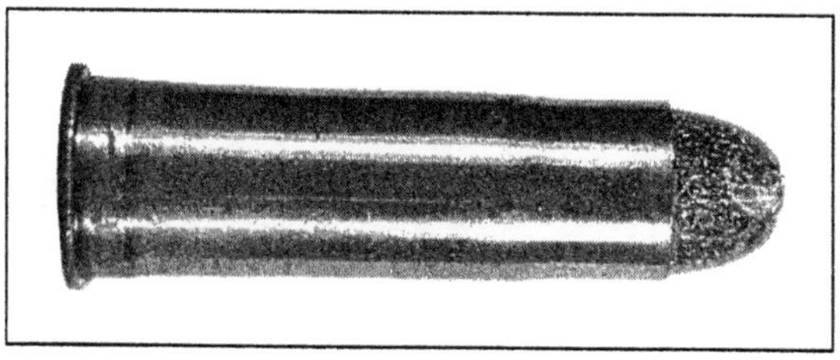

TITLE: .577 SNIDER

HEADSTAMP: KYNOCH .577.S

DIMENSIONS:

Rim:	.749"
Head:	.660"
Neck:	.601
Bullet:	.558
Case:	1.983"

OTHER DETAILS:

Bullet Grains:	480
Powder Grains:	80
Bullet:	Round Nose
Case Material:	Brass
Total Grains:	890

OVERALL: 2.465"

A late manufacture by Kynoch of the 2" cased Snider Cartridge.

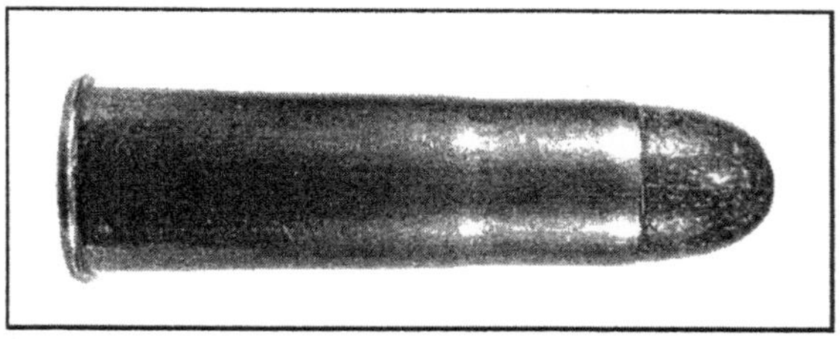

TITLE: .577 SNIDER - MODERN EMPTY CASE

HEADSTAMP: 577 SNIDER (with a rifle pictured)

DIMENSIONS:

Rim:	.744"
Head:	.660"
Neck:	.634"
Bullet:	none
Case:	2.000"

OTHER DETAILS:

Bullet Grains:	N.A.
Powder Grains:	N.A.
Bullet:	N.A.
Case Material:	Brass
Total Grains:	N.A.

OVERALL: 2.000"

A modern brass case sold for reloading purposes by modern day shooters.

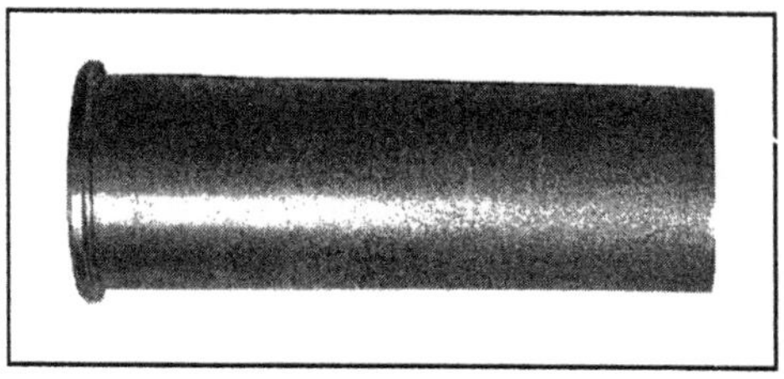

TITLE: .577 SNIDER SPORTING

HEADSTAMP: ELEY .577.LONDON

DIMENSIONS:		OTHER DETAILS:	
Rim:	.742"	Bullet Grains:	?
Head:	.662"	Powder Grains:	80
Neck:	.620"	Bullet:	Patched, tapered
Bullet:	.580"	Case Material:	Brass
Case:	2.000"	Total Grains:	968

OVERALL: 2.628"

The most distinguishing features of this round are the heavy patched bullet and the total grain weight of 968 grains.

This sporting version of the Snider series was made by the well-known firm of Eley Bros. Of London.

The number (14935) inscribed on the side of the case indicate it originally was in the White Munhall collection.

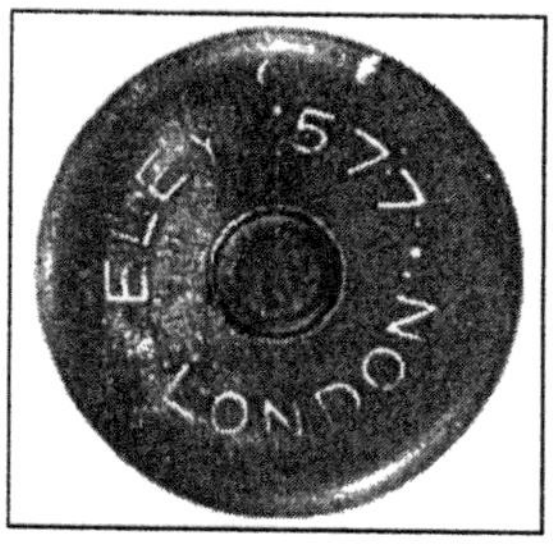

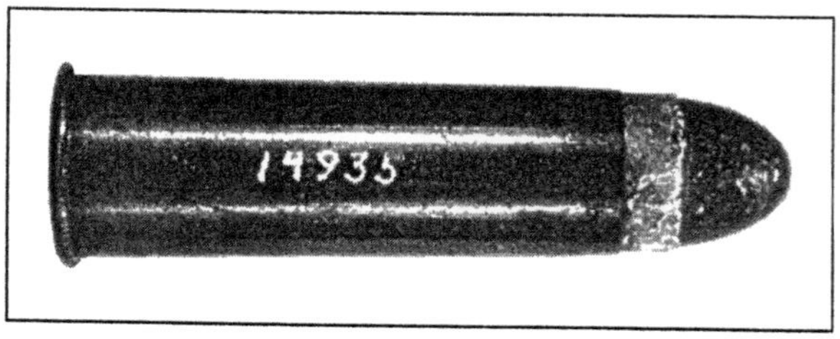

TITLE: .577-.577/450 BLANK-PETAL CRIMP

HEADSTAMP: KYNOCH 577/450

DIMENSIONS:

Rim:	.749"
Head:	.665"
Neck:	.664"
Bullet:	None
Case:	2.065"

OTHER DETAILS:

Bullet Grains:	-0-
Powder Grains:	?
Bullet:	-0-
Case Material:	Brass
Total Grains:	305

OVERALL: 2.065"

While the round is head-stamped .577/.450, undoubtedly the round was interchangeable with the older .577" chambered weapons.

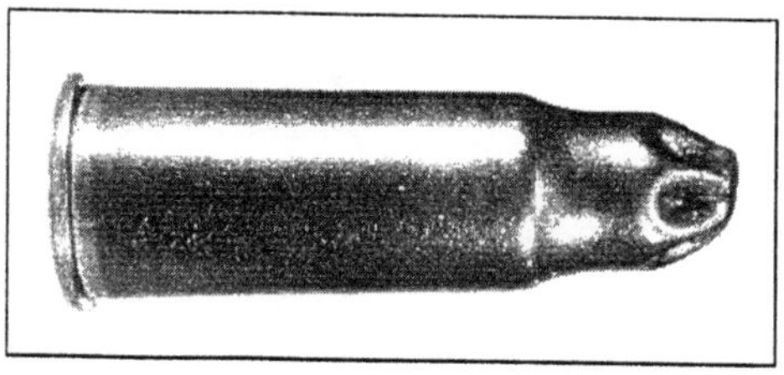

TITLE: POTTET (14.7 x 58R)

HEADSTAMP: Plain

DIMENSIONS:		OTHER DETAILS:	
Rim:	.722"	Bullet Grains:	not known
Head:	.675"	Powder Grains:	not known
Neck:	.645"	Bullet:	Lead
Bullet:	.450"	Case Material:	Cardboard
Case:	2.175"	Total Grains:	1,216

OVERALL: 2.575"

The French Schneider cartridge patent No. 26901 applied the idea developed by Pottet at an earlier date for a metal cup on a paper case body which is similar to that used in the manufacture of shotgun shells.

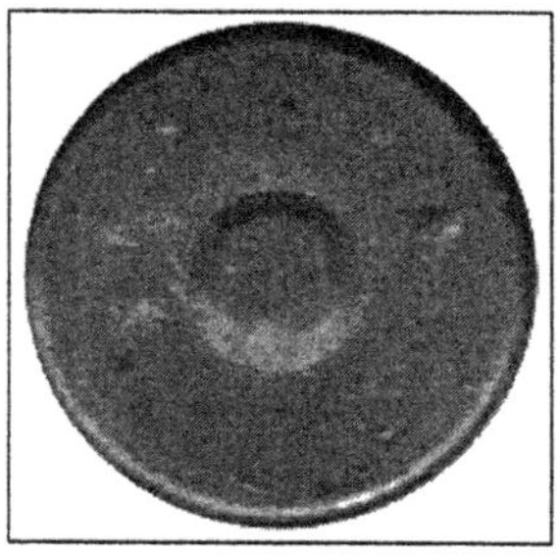

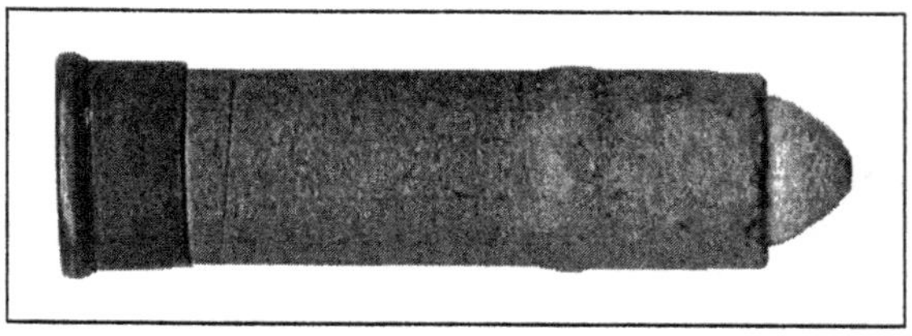

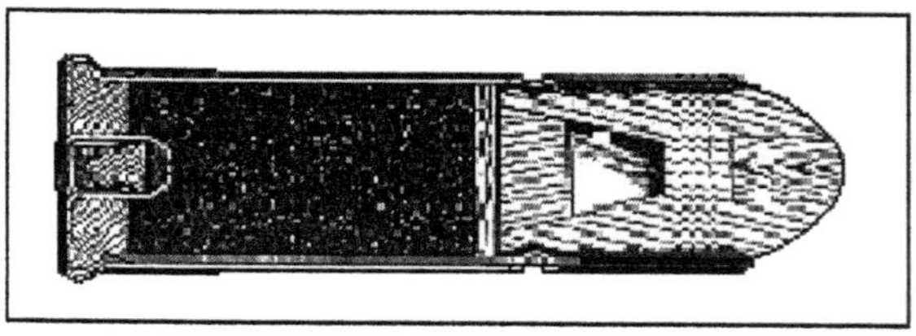

George Henry Daw, a London gun maker, was credited with the introduction of the first center fire cartridge in England during the 1860s.

The round was designed as an inside primed system. This composite cartridge consisted of five parts plus the lead projectile. The round consisted of a piece of sheet brass or copper formed into a flanged cup, an anvil, a primer pocket, a cupped primer and a pulp board case.

While successful for sporting purposes, the military experienced problems with the weak copper head and swelling of the paper case which made chambering the round difficult and extraction troublesome.

A quote from the September 1866 issue of the Cornhill Magazine provides some interesting remarks regarding the Schneider cartridge:

> "The first pattern of cartridge, with the Potet base, was found to be unreliable and one or two breech blocks were blown open in Canada, to which station the bulk of the ammunition of this pattern had been sent. But the defect had been noticed and corrected before these mishaps occurred, although too late to admit of strengthened cartridges being supplied Canada, on account of the communications having become closed by the ice. In the spring of this year the unsafe ammunition was recalled; and it is improbable that a single round of it exists, except in the form of blank cartridges, into which it was converted on its return to this country."

TITLE: .577 BLACK POWDER-3 1/4"

HEADSTAMP: KYNOCH.577

DIMENSIONS:		OTHER DETAILS:	
Rim:	.750"	Bullet Grains:	560
Head:	.668"	Powder Grains:	Black-160
Neck:	.609"	Bullet:	Lead
Bullet:	.579"	Case Material:	Brass
Case:	2.250"	Total Grains:	----

OVERALL: 3.858"

This is the largest of the .577 series. They started at two and one-half inches, and were made in various lengths to three and one quarter inch case. With the advent of cordite powder, the necessity for large capacity cases soon made the round obsolete.

After 1903, Kynoch no longer listed this round in their catalog while Westley Richards & Co. still listed them for sale as late as 1925.

Interestingly, John Taylor, author of "AFRICAN RIFLES AND CARTRIDGES" never mentioned this 3 1/4" round.

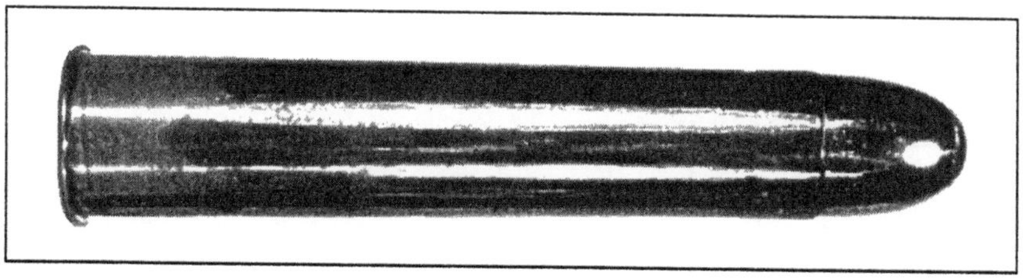

TITLE: .577 SNIDER SHOT-COLONIAL CONTRACT

HEADSTAMP: Plain

DIMENSIONS:

Rim: .740"
Head: .666"
Neck: .617"
Bullet: None
Case: 2.324

OTHER DETAILS:

Bullet Grains: None
Powder Grains: 70-80
Bullet: Shot
Case Material: Brass
Total Grains: 738

OVERALL: 2.324"

This is a non-military round probably issued to police, prison guards or the militia. Its use was most likely intended for mob or riot control. This case is the early Whitney Patent construction (iron disc)

As with other ammunition produced in the British colonies (and pictured elsewhere in the book), this case is a brass tube with an iron disk attached at the base. The copper primer is large, and it is inserted in a brass reinforcing ring.

The neck has twelve petal crimps holding the wad that is imprinted:

THE COLONIAL AMMUNITION CO. No. 4

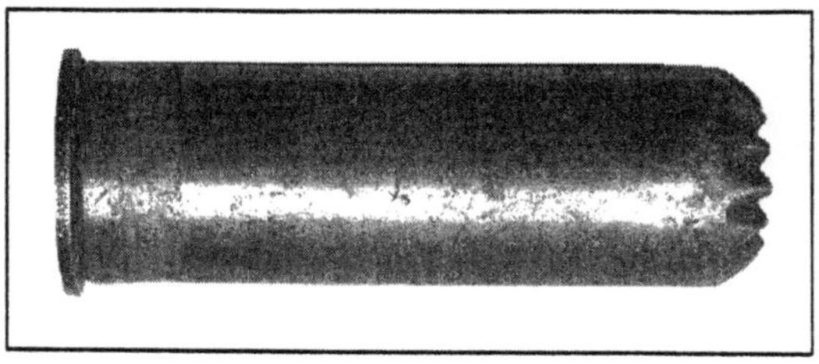

TITLE: 20/577 ALEXANDER HENRY

HEADSTAMP: ELEY BROs LONDON

DIMENSIONS:

Rim: .750"
Head: .708"
Neck: .621"
Bullet: .576"
Case: 2.738"

OTHER DETAILS:

Bullet Grains: 560/570
Powder Grains: 167 Black. Powder
Bullet: Lead, Patched
Case Material: Bottle Neck, Brass
Total Grains: 985

OVERALL: 3.300"

Alexander Henry, an Edinburg, Scotland gunsmith, originated this cartridge for use in a double barrel gun. It is based on a twenty gauge shotgun case necked down to .577 and drawn to two and three quarters inches in length. The case was loaded with a 560 grain; paper patched lead bullet with what appears to be a .22 caliber rimfire case inserted into the nose . Two tiny stab crimps secure the large copper tube lead bullet. A small copper primer provides ignition.

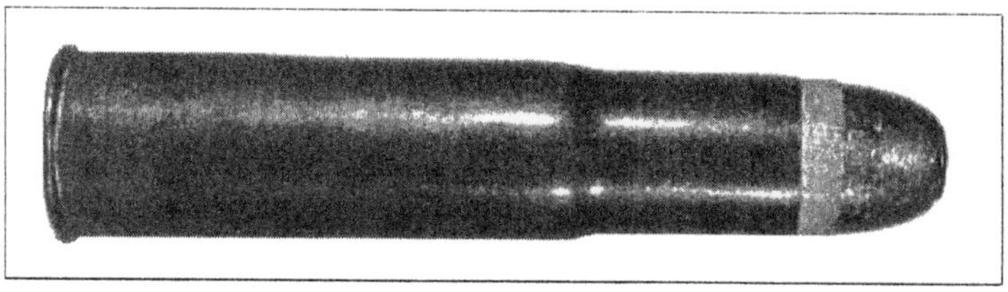

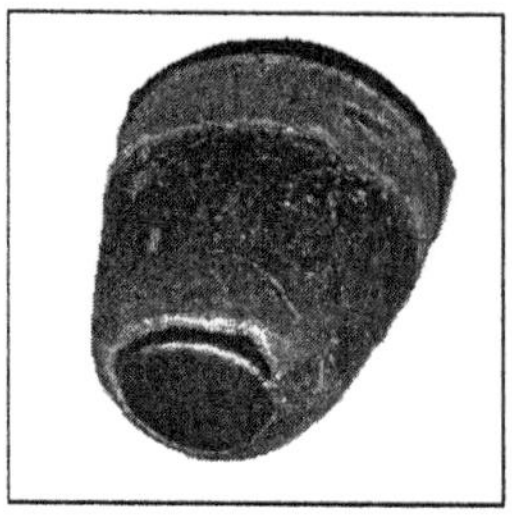

TITLE: 20/577 ALEXANDER HENRY

HEADSTAMP: ELEY BROs LONDON

DIMENSIONS:		OTHER DETAILS:	
Rim:	.753"	Bullet Grains:	?
Head:	.706"	Powder Grains:	?
Neck:	.621"	Bullet:	Lead
Bullet:	.578"	Case Material:	Brass, Necked
Case:	2.741"	Total Grains:	Over 1,000

OVERALL: 3.892"

Alexander Henry, an Edinburg, Scotland gunsmith, originated this cartridge for use in his dropping block rifle. The projectile has four grooves and certainly provides the observer with a lasting impression of its potency.

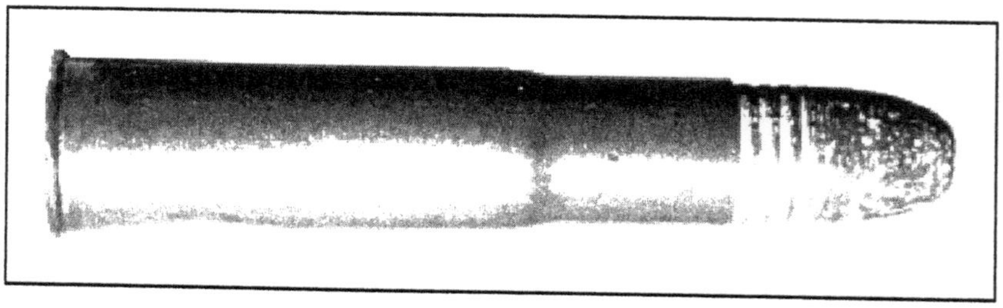

TITLE: .577 EXPRESS 2 3/4" CASE

HEADSTAMP: ELEY '577 S.

DIMENSIONS:

Rim: .742"
Head: .665"
Neck: .611"
Bullet: .587"
Case: 2.750"

OTHER DETAILS:

Bullet Grains: 550
Powder Grains: 90
Bullet: Blunt-Solid
Case Material: Brass
Total Grains: 793

OVERALL: 3.206"

Another of the variety loads for the scarce 2.75" Express Hunting Rifle.

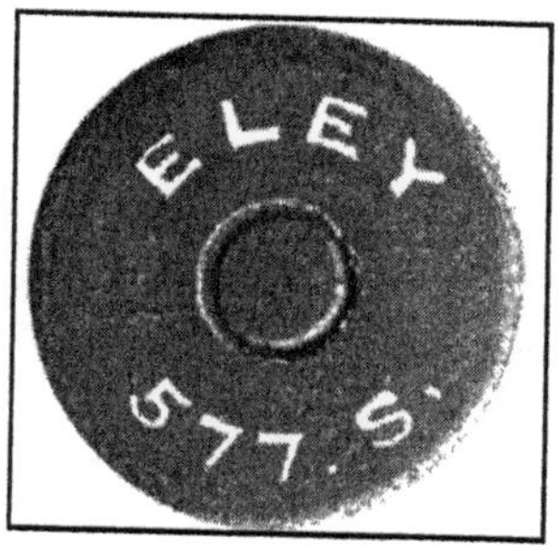

TITLE: .577-2 3/4"BOXER COILED CASE

HEADSTAMP: Plain

DIMENSIONS:

Rim: .744"
Head: .669"
Neck: .605"
Bullet: .572"
Case: 2.750"

OTHER DETAILS:

Bullet Grains: 520
Powder Grains: 106
Bullet: Patched, Copper Tubed
Case Material: Coiled, Brass
Total Grains: 895

OVERALL: 3.253"

A variation of the standard loading. The cartridge has the copper tubed (Express Bullet) long patched bullet.

The two and three quarter inch case round was popular for only a short period of time. It was soon replaced by the more powerful three inch case.

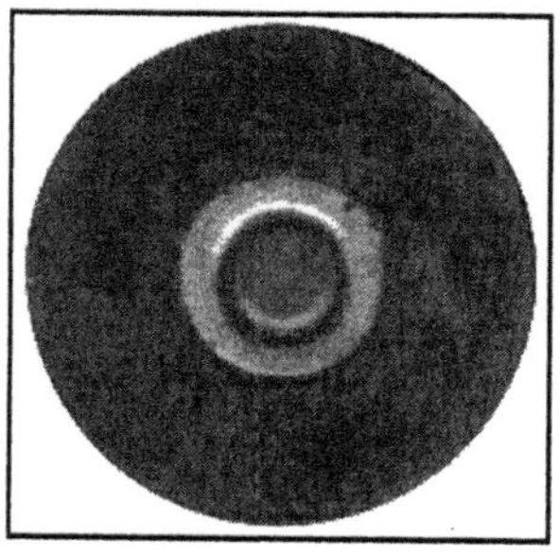

TITLE: .577-2 3/4 BLACK POWDER
EXPRESS-COILED CASE

HEADSTAMP: Plain

DIMENSIONS:

Rim:	.748"
Head:	.670"
Neck:	.618"
Bullet:	.570"
Case:	2.750"

OTHER DETAILS:

Bullet Grains:	520
Powder Grains:	160
Bullet:	Patched, Copper Tubed
Case Material:	Coiled Brass
Total Grains:	890

OVERALL: 3.275"

The predecessor of the British solid drawn brass case was this early coiled brass wrapped specimen with iron head.

These rounds were sold commercially for hunting as evidenced by the so-called Express Bullet with a copper tube inserted in the nose.

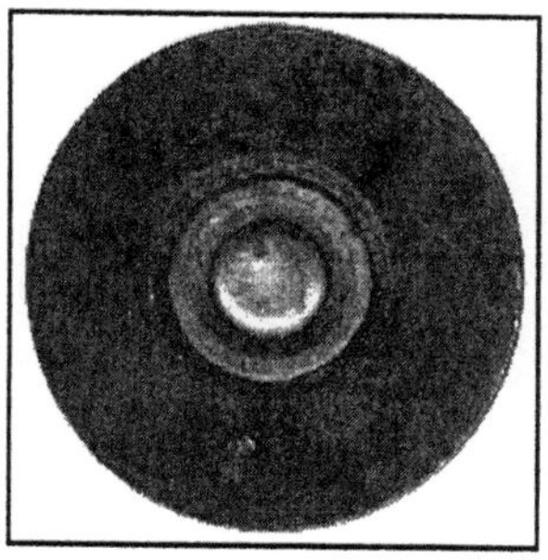

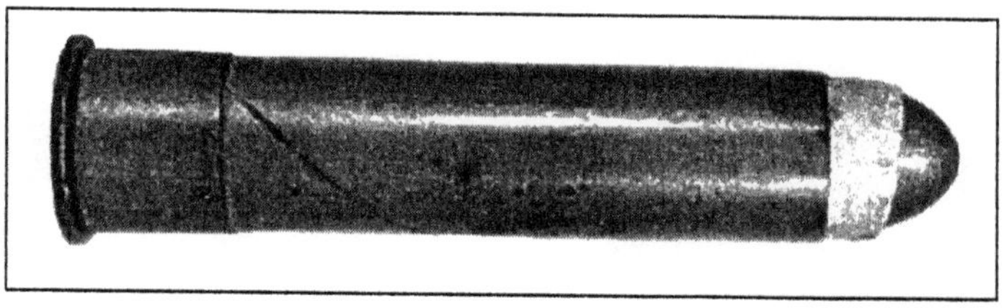

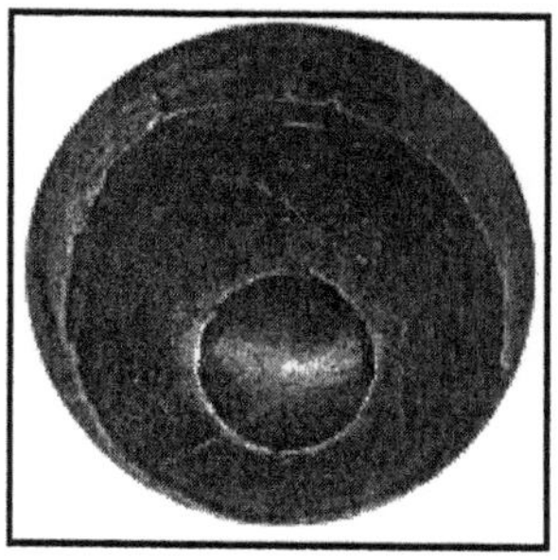

TITLE: .577 COILED 2 3/4" CASE

HEADSTAMP: Plain

DIMENSIONS:

Rim:	.748"
Head:	.663"
Neck:	.610"
Bullet:	.579"
Case:	2.750"

OTHER DETAILS:

Bullet Grains:	520
Powder Grains:	106
Bullet:	Patched, Hollow Point
Case Material:	Boxer-Coiled
Total Grains:	876

OVERALL: 3.366"

While this cartridge is not of the Snider series, it represents the logical trend for sporting purposes. The Snider 2" case was increased to the 2 1/4". 2 1/2" and 2 3/4" case lengths. Sportsmen demanded more stopping power, so the solution was to extend chambers and provide rounds with a greater charge capacity to fit the hunting rifles.

The charge was increased from the Snider load of 70-80 grains of black powder to 160 grains. Bullet weight was increased to 520 grains from 450 grains.

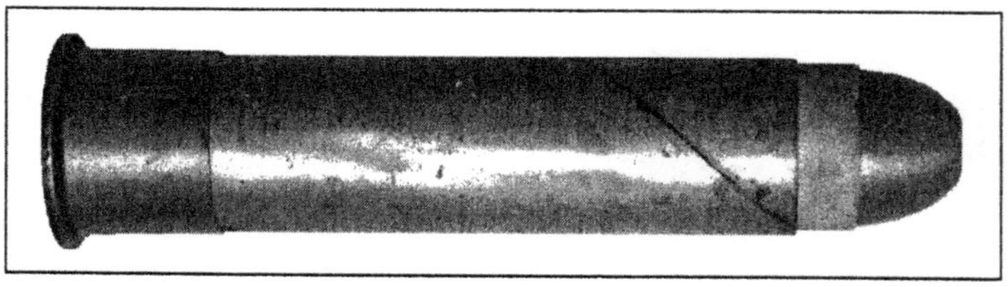

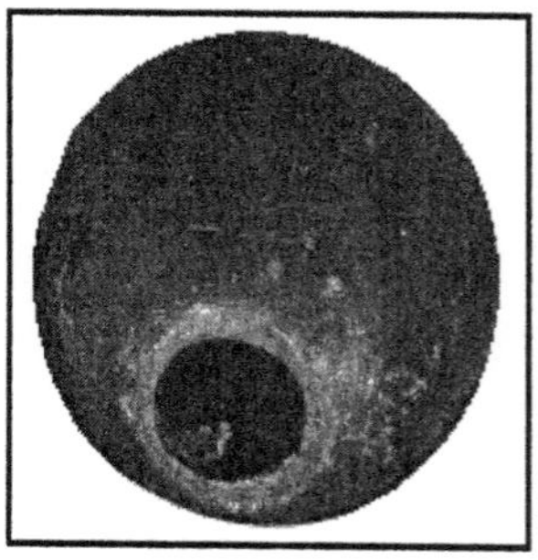

TITLE: .577 NITRO-2 3/4"

HEADSTAMP: KYNOCH '577 NITRO

DIMENSIONS:

Rim: .737"
Head: .663"
Neck: .608"
Bullet: .580"
Case: 2.750"

OTHER DETAILS:

Bullet Grains: 650
Powder Grains: 90
Bullet: Jacketed, Soft Point
Case Material: Brass
Total Grains: 1,050+

OVERALL: 3.337"

This solid brass cased cartridge followed the development of the Boxer cased round. Originally the Kynoch rounds were loaded with 160 grains of black powder and a lead, copper-tubed or solid lead 520-560 grain bullets. The impressive title given to these early weapons was "Black Powder Express".

Apparently few weapons were chambered for this 2 3/4" case. John Taylor, author of African Rifles and Cartridges, mentions that in all his years as a professional hunter, he only saw one rifle chambered and regulated for the 2 3/4" cartridge. He personally owned a number of double guns chambered for the longer 3" case, but not one of the shorter chambered weapons.

With the advent of the much more popular 3" case loadings, the 2 3/4" was quickly obsolete due to the greater power available in the bigger case.

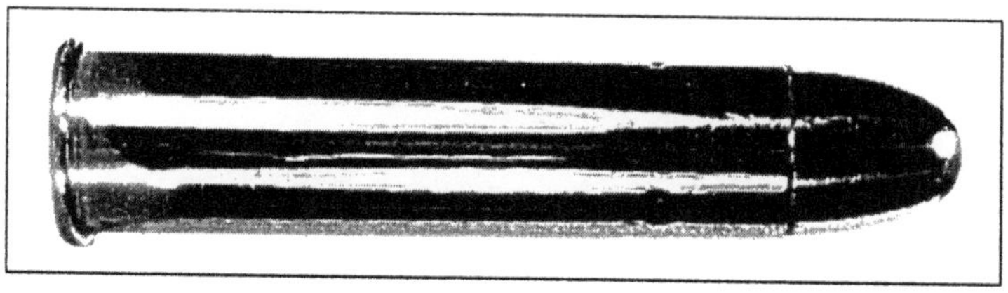

TITLE: .577 2 3/4" BLACK POWDER EXPRESS

HEADSTAMP: ELEY .577. LONDON

DIMENSIONS:

Rim: .744"
Head: .661"
Neck: .610"
Bullet: .584"
Case: 2.750"

OTHER DETAILS:

Bullet Grains: 550
Powder Grains: 90
Bullet: Patched, Copper Tube
Case Material: Brass
Total Grains: 1010

OVERALL: 3.455"

When the manufacture of solid brass cases was perfected, it followed that the old style coiled case Boxer cartridge would be replaced with this more durable design.

No explosive is contained in the copper tube inserted in the nose of the bullet, but it did create the effect of an expanding bullet, highly effective on soft skinned animals.

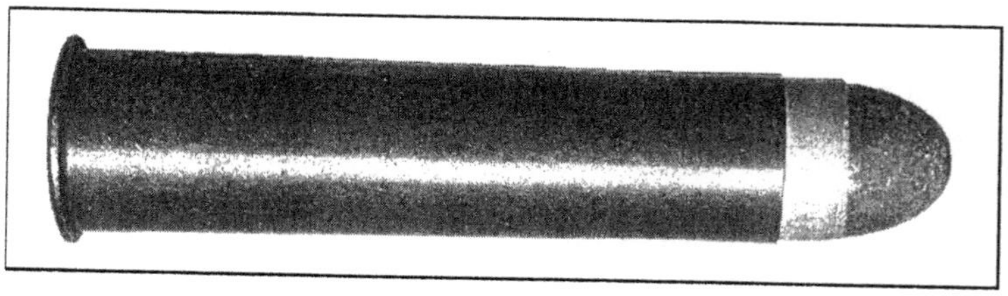

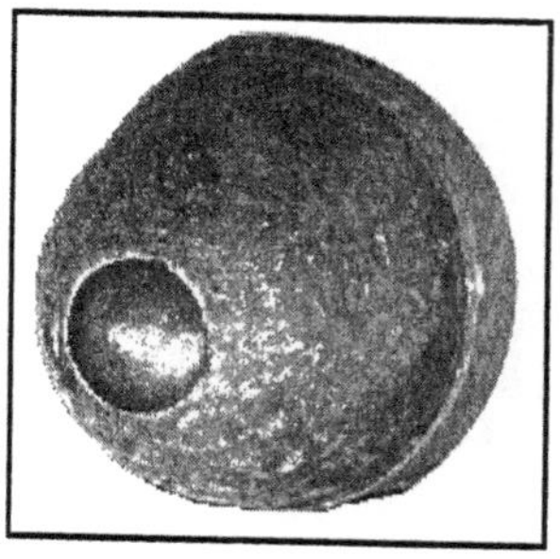

TITLE: .577 GREENER POLICE GUN

HEADSTAMP: GREENER E K POLICE GUN

DIMENSIONS:

Rim:	.895"	.842"
Head:	.795"	.814"
Neck:	.738"	.752"
Bullet:	Shot	Shot
Case:	2.860"	2.483"

OTHER DETAILS:

Bullet Grains:	?
Powder Grains:	80-90
Bullet:	#4 Shot (both)
Case Material:	Brass
Total Grains:	1000+/730

OVERALL: 2.860"/2.483"

This specimen is of recent manufacture. The cartridge was meant to be used in .577 chambers measuring one inch or longer. The case has been reduced so that the shell will fit a number of different weapons.

The cartridge is loaded with shot and the case mouth has a ten petal crimp to hold the load. It is obvious from the head-stamp that the shell was intended for guard duty or mob control.

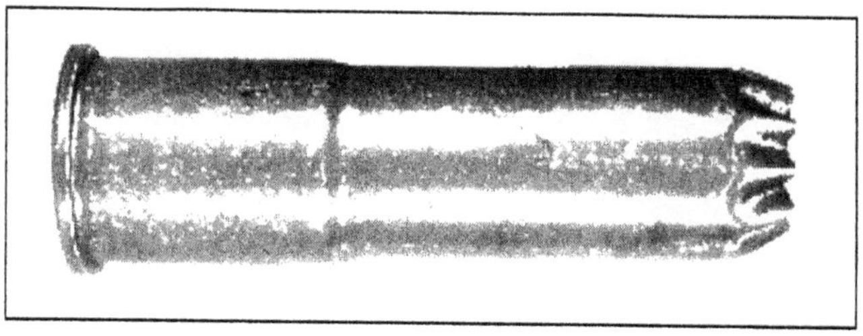

Modern version pictured below with additional measurements listed above.

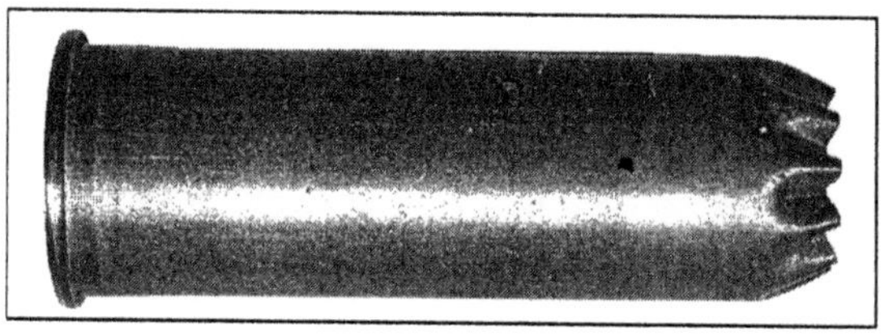

TITLE: .577 – 3" EXPRESS

HEADSTAMP: KYNOCH L 577 C

DIMENSIONS:

Rim:	.731"
Head:	.656"
Neck:	.606"
Bullet:	.566"
Case:	3.000"

OTHER DETAILS:

Bullet Grains:	610 grain-solid
Powder Grains:	167 grains-N C
Bullet:	Lead, tapered, 3 crimps
Case Material:	Brass
Total Grains:	Over 1,000 grains

OVERALL: 3.601"

A very popular cartridge with numerous loadings, Kynoch, Holland & Holland, Eley Nitro, and coiled cases were available. The more common head stamp had a double C C while the earlier rounds were stamped L C for light loading of reduced load of nitro cellulose powder.

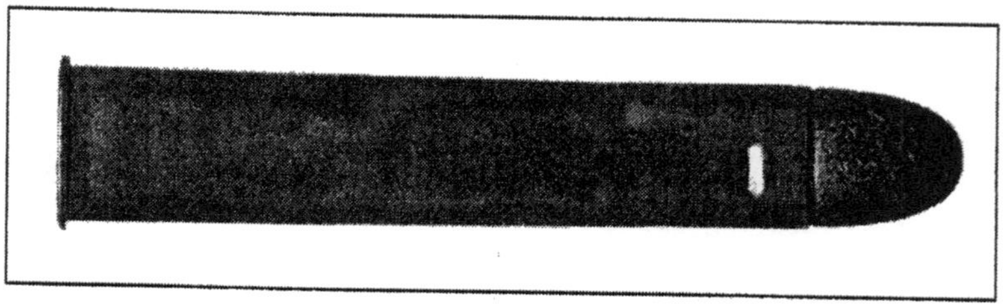

TITLE: .577-3"

HEADSTAMP: C KNYOCH C '577

DIMENSIONS:

Rim: .739"
Head: .660"
Neck: .608"
Bullet: .581"
Case: 3.000"

OTHER DETAILS:

Bullet Grains: 750
Powder Grains: 100
Bullet: Jacketed, Soft Point
Case Material: Brass
Total Grains: 1000+

OVERALL: 3.624"

European professional hunters demanded powerful weapons. This .577-3" was second in stopping power only to the .600" Nitro. The double C in the head stamp denotes a full power cordite load capable of exerting 5,000 ft. lbs. Of energy at 100 yards. The bullet had a velocity of 1,750 ft. per second at this distance because the bullet was tapered more than the .600 Nitro. It was considered by market hunters to be a better game killer of elephants and other large game than the acclaimed .600 Nitro that had less penetration.

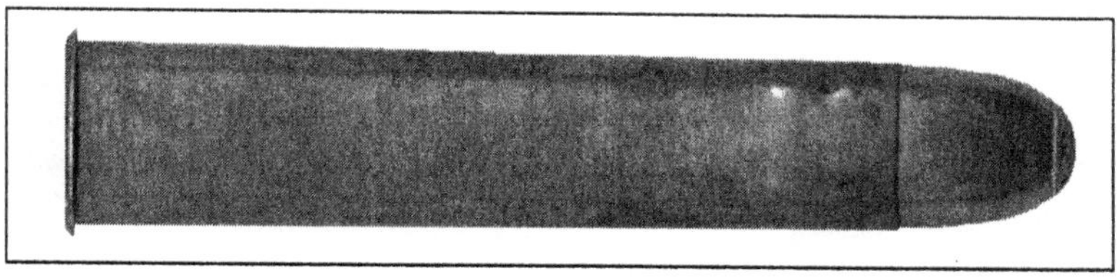

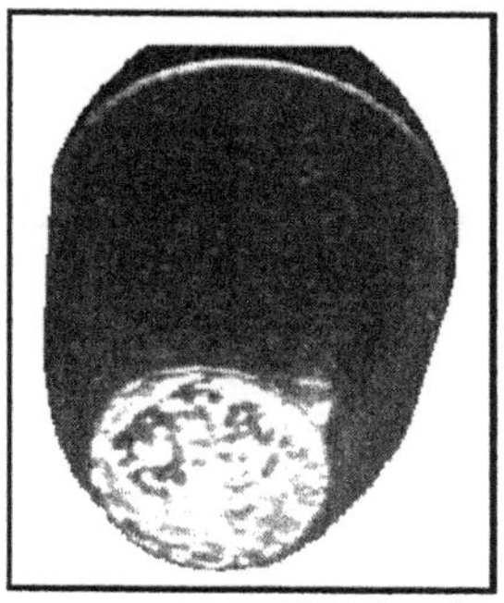

TITLE: .577 - 3" EXPRESS

HEADSTAMP: KYNOCH L 577 C

DIMENSIONS:

Rim:	.732"	.731"
Head:	.656"	.659"
Neck:	.618"	.615"
Bullet:	.569"	.573"
Case:	3.000"	2.980"

OTHER DETAILS:

Bullet Grains:	610
Powder Grains:	167 black powder
Bullet:	Lead, tapered, two grooves
Case Material:	Brass, stab crimps
Total Grains:	Over 1,000 grains

OVERALL: 3.625"/3.610"

These 3 inch cased rounds were loaded with a 610 grain lead bullet with a so called express projectile or pegged type. This indicates that a small square piece of wood was inserted into the nose of the bullet. This style predated the copper tubed variety. Three stab crimps were applied to secure the rounds in the cases.

The L C in the head stamp was to identify a lighter load of cordite loaded in the case.

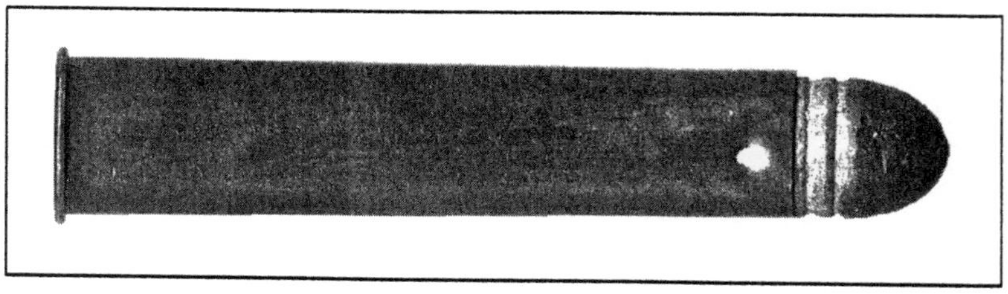

TITLE: .577 R E W A

HEADSTAMP: R E W A .577 K

DIMENSIONS:

Rim:	.806"
Head:	.696"
Neck:	.619"
Bullet:	.588"
Case:	3.000"

OTHER DETAILS:

Bullet Grains:	750
Powder Grains:	110-Cordite
Bullet:	Jacketed, Tapered
Case Material:	Brass
Total Grains:	Over 1000

OVERALL: 3.840"

This cartridge was made in 1929 for the Maharaja of Rewa by the Kynoch Works, Witton, Birmingham. It is estimated that 1000 rounds were made for use in a special Holland and Holland rifle. The cases were formed from the three inch .600 Nitro necked down to .577". The 750 grain bullet reached a muzzle velocity of 1960 feet per second. The box label reads:

10 SMOKELESS 10
KYNOCH
577
REWA
NITRO EXPRESS
CARTRIDGES
CASE 3"
FOR USE IN HOLLAND RIFLES
750 GRAIN NICKEL CAPPED BL'T
Manufactured At The Kynoch Factories
Imperial Chemical Industries Lmtd.
Birmingham, England

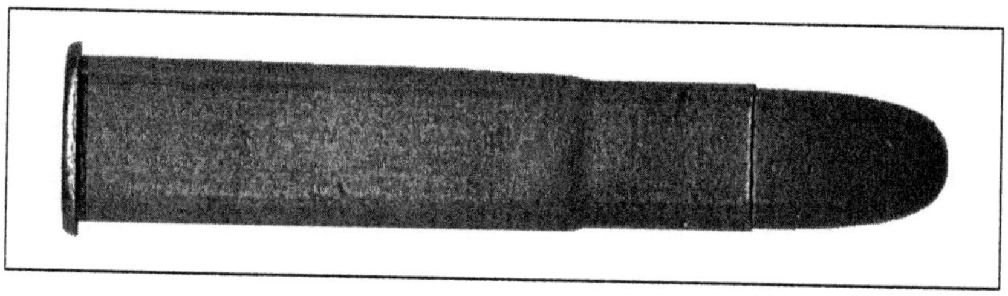

A HISTORY OF AMMUNITION MANUFACTURE IN NEW ZEALAND

by G. Sherriff and Dave Andrews

The history of firearms and their use in early New Zealand seems to be something of a neglected subject. This article will hopefully be the first of several covering an area that should be of interest to any reader concerned with shooting or arms collecting.

The initiative to commence ammunition manufacture in this country came from the Whitney family in particular father John and eldest son, Asa. They were also responsible at a later date for setting up the first ammunition manufacturing plant in Australia.

Father John Whitney, his wife and six children were a middle class English family from Oban Argyllshire. They later spent10 years living in County Kerry, Ireland before returning to England in 1869. John was a country gentleman with a great interest in shooting, horses, fishing. He received a 12-gauge muzzleloader, as a gift from his grandfather on his 9th birthday. This shotgun would be quite a handful for a lad of that age.

Asa left the family when he was 15 and entered the merchant marines as an officer trainee. He was a bright lad and gained his commission in 1882. In 1884 the family emigrated to Auckland, arriving in the ship "Waiora". On his arrival John rented a house called the Deanery in which he installed his family, later moving to a new residence "Woodlands" some 30 miles from Auckland.

It is said that John first became interested in ammunition manufacture while talking to Sir George Whitmore in the Auckland Country Club. Col. Whitmore commanded Government forces during the later half of the New Zealand wars, from 1866 until 1872 when Te Kooti fled into oblivion. Sir George commented "It is a pity no one will start making ammunition in New Zealand and John replied "I will".

So John joined up with Mr. Hazard, an Auckland Gun and cartridge importer. It was, however, a haphazard venture as neither had any experience in making cartridges. They purchased machinery from Mr. Holroyd in Auckland, put up buildings and started manufacture of the .577 Snider round.

We are unaware of the cartridge details here, however the first sample delivery was to government at the Wangannie rifle matches. Unfortunately many of the cartridges burst and the whole consignment was rejected. With this failure the partnership dissolved leaving John 600 pounds in debt to the bank.

At this moment Asa returned to Auckland to sit for his engineers examination. Seeing the difficulties facing his father he decided to help and was made plant manager. The new firm registered as Whitney and Sons with share holders consisting of John Whitney and sons Asa, Cecil and Robin.

At this time Snider cases were of Boxer construction having the walls made from sheet brass wound into a coil, the rim base cups, internal wads etc. all being held together by the primer pocket which acted as a rivet. Asa and John designed a new case which utilized a one piece brass drawn body to replace the foil coil of the Boxer round. The new case was still fabricated, however, fewer parts were required in its construction. A further advantage of the case was that it could be reloaded, something not practical with the Boxer constructed case.

In July 1886 John Whitney applied for a patent on "The non corrosive metal rifle cartridge case" and called it "The New Zealand cartridge case."

With a saleable cartridge ready for production, samples were sent to Messers. Greeenwood and Baily of Leeds, England who were at that time large manufacturers of both ammunition and ammunition making machinery. They were impressed and a company was set up in England to manufacture the cartridge in New Zealand. The company was called the "Colonial Cartridge Company" or C.A.C, a name that is still with us to this present day.

Sample production was soon underway and a quantity of loaded ammunition was made ready for trials that were held in January 1887. These trials conducted by the Defense Department under the supervision of a Col. Baily compared the performance of the Whitney round to that of current English production. The results were considered very satisfactory except for a number of rounds having faulty primers and also oversize cases making chambering difficult. These problems were not thought to be too serious being able to be corrected at the factory.

International politics also helped gain orders for the new factory. In 1887 we had the Russian scare when the public fully expected an armed Russian invasion at any moment. This fear was a mixture of rumours and fantasy, however it was not only responsible for a strong demand to have our own ammunition manufacturing plant but also initiated the construction of defense works around our main ports. The remnants of these are still visible today.

During this period though the Snider rifle and cartridge was used by our volunteer units, it was long obsolete and the Kynoch ammunition on issue was of very low quality. Because of this, local shooters were very dubious of the local product.

The problems with Kynoch ammo can best be described in this report from the 1887 New Zealand Rifle Assn. meeting. "Kynoch's ammo was faulty in the extreme, the best shots gained only 55 out of 100. Breach bolts were blown open, many head separations, misfires, cases would not chamber, and many were short on powder". In their frustration, the competitors ceremoniously burned an effigy of Kynoch at the conclusion of the meeting. During 1888 the Volunteers obtained a number of Remington Lee Bolt action rifles in 43 Spanish caliber. Although an immense improvement on the Snider these also gave much trouble with faulty ammunition. These rifles were withdrawn after only a few months' use and Sniders reissued. The shooters could however only be persuaded to use these old weapons with much difficulty and the decision was made to purchase the Martini Henry.

Until the issue of Martini rifles in 1890, Sniders continued to be used. The ammunition supplied in 1888 and 1889 was predominately of Colonial Ammunition Company (C.A.C) manufacture but the quality varied greatly. A report from the 1889 Dunedin meeting makes interesting reading, "Had to use Snider ammunition manufactured by Whitney Auckland, Shooting very poor, ammunition inferior to Kynoch, bad loading, rims too thick, difficulties in extraction, After the splendid ammunition of 1884 (Woolwich manufacture) came the Kynoch issue when one man was always employed in rear to see where the shot went: with Whitney issue the man in rear is employed to load and unload the rifle".

End of report which originally appeared in the pages of "New Zealand Wildlife. Grateful acknowledgement goes to the magazine, the author G. Sheriff, and Dave Andrews, Ammunition Research Consultant, Contributing Editor – "The American Rifleman".

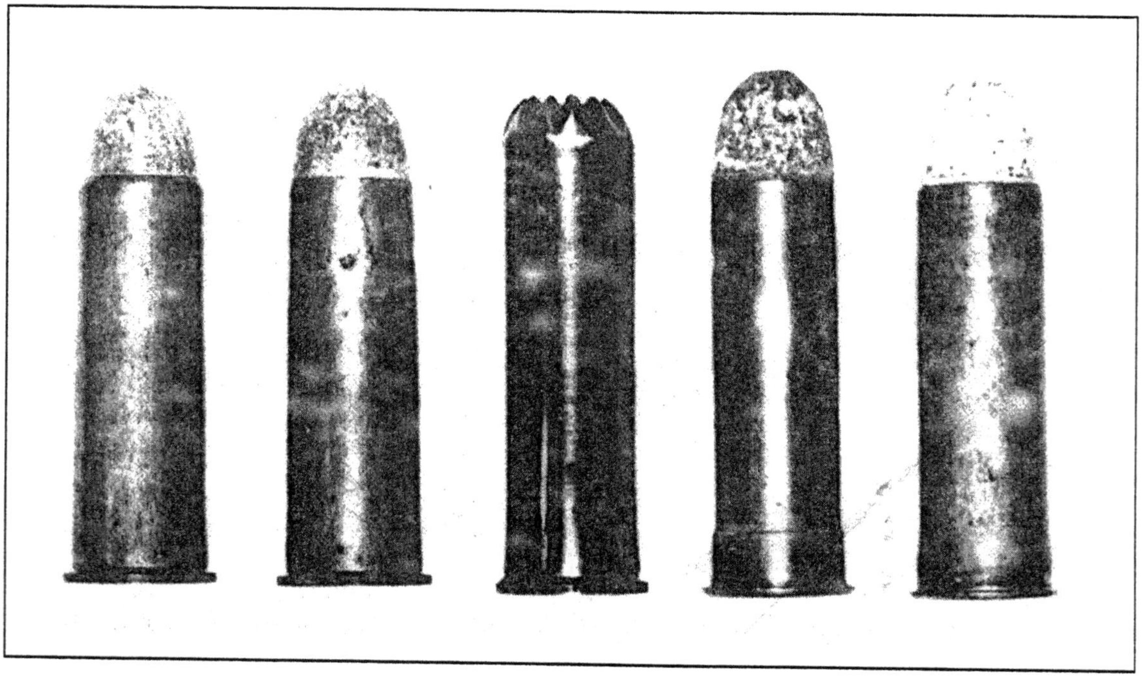

THE .577 REVOLVER AND HOWDAH PISTOL

During the mid 1860s, the Webley Gun Manufacturing Company of England manufactured a six shot revolver which chambered this large cartridge. The arm was known as the Webley-Pryse Model of 1866 revolver. Another weapon, which appeared about the same period of time, was a five shot revolver and was called the Bland-Pryse. The predecessor of the .577 revolver was an experimental pin-fire weapon that never went into production. It is quite possible the ammunition for this gun was the fifteen-millimeter pin-fire cartridge.

The first Webley revolvers used a Boxer type cartridge similar in construction to the military Snider rifle ammunition. Later on improvements were made in the revolver and during 1877 the guns were chambered for the solid brass cased ammunition which is illustrated here with the ELEY-LONDON 577 headstamp.

The British soldier learned to his chagrin that the Enfield .476 revolver issued to him was an ineffectual man-stopper when facing the Zulus, Afghan and Dervish native fanatics. A larger caliber and more powerful weapon was needed and the .577 seemed to be the answer. Unquestionably the cartridge had tremendous knock down power in spite of the fact that the case contained only 30 grains of black powder.

Another use for the larger caliber revolver and the various two barreled pistols available during the late 1860s was for a back up pistol while tiger hunting from the backs of elephants. The weapons became known as Howdah pistols because they were kept readily available in the Howdahs where the hunters perched on the back of the elephant. It was common for a wounded tiger to lie in wait for the elephant and its rider to pass close by in the dense brush and then attack by leaping upon the animal's back. Hence the need of a large powerful short range weapon to back up the hunter with an empty hunting rifle.

Ammunition was loaded in France and Germany as well as in England and was discontinued during the early 1920s. Case materials were the Boxer type, i.e. paper or brass wrapped with iron bases, or solid brass. Bullets were either round or flat nose and the primers were Berdan or Battery cup primed.

TITLE: .577 PISTOL-SHOT LOAD

HEADSTAMP: ELEY .577 LONDON

DIMENSIONS:

Rim: .736"
Head: .659"
Neck: .640"
Bullet: ----
Case: .596"

OTHER DETAILS:

Bullet Grains: None
Powder Grains: 28-30
Bullet: None
Case Material: Brass
Total Grains: 450

OVERALL: 1.21"

A very scarce loading in shot rather than the conventional lead ball bullet.

One can only assume that the cartridge was intended to be used at short distances and would be adequate for either fowl or small game.

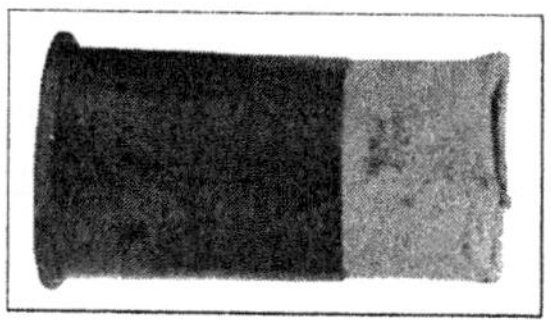

TITLE: .577 REVOLVER-DUMMY ROUND

HEADSTAMP: Plain

DIMENSIONS:

Rim:	.733"
Head:	.660"
Neck:	658"
Bullet:	.620"
Case:	.761"

OTHER DETAILS:

Bullet Grains:	450
Powder Grains:	----
Bullet:	Pointed
Case Material:	Brass
Total Grains:	577

OVERALL: 1.200"

This round is obviously a dummy or assembler's functioning dummy for use in the Webley Revolvers. Note the deeply recessed copper primer and wooden insert to prevent the projectile from being pushed back into its case when chambered.

In addition to the one large hole, two other much smaller holes appear in different locations on the case. The maker is unknown.

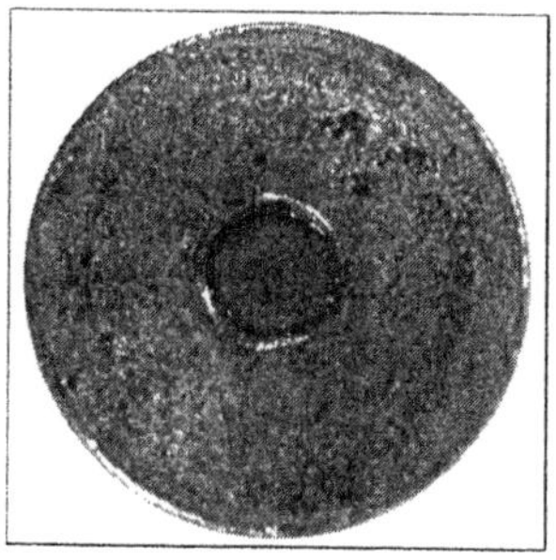

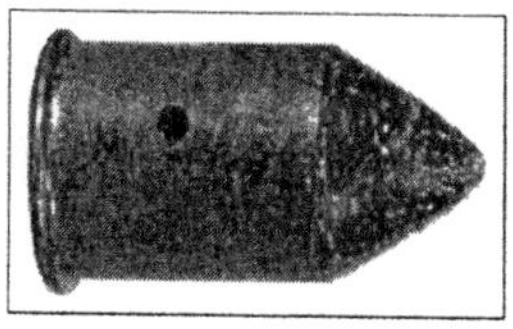

TITLE: .577 PISTOL

HEADSTAMP: Plain

DIMENSIONS:

Rim:	.746"
Head:	.660"
Neck:	.635"
Bullet:	.594"
Case:	1.000"

OTHER DETAILS:

Bullet Grains:	300
Powder Grains:	30
Bullet:	Round Nose
Case Material:	Boxer
Total Grains:	480

OVERALL: 1.308"

Another variation of the Boxer cased pistol round for use in a Howdah pistol.

Because of the heavy recoil of these guns, the manufacturer saw fit to put three stab crimps around the case mouth to further secure the bullet.

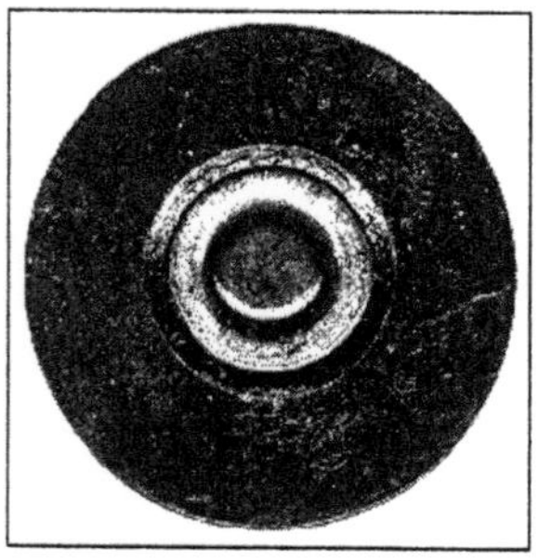

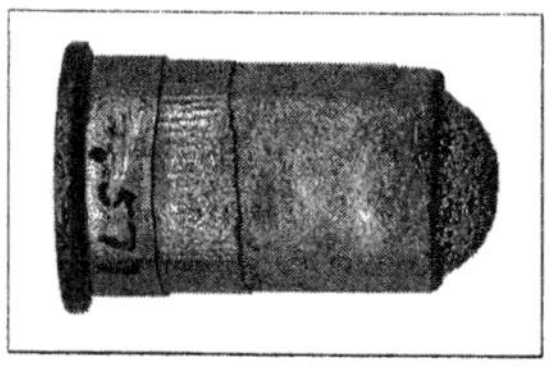

TITLE: .577 REVOLVER-BOXER TYPE

HEADSTAMP: Plain

DIMENSIONS:

Rim:	.745"
Head:	.663"
Neck:	.614"
Bullet:	.578"
Case:	1.000"

OTHER DETAILS:

Bullet Grains:	300
Powder Grains:	28
Bullet:	Round Ball
Case Material:	Paper and Brass
Total Grains:	450

OVERALL: 1.146"

This is the earliest type of this round constructed on the Boxer principle of paper and brass wrapped case with the Battery Cup Primer set in an iron head or washer.

The Webley Model 1866 revolver was chambered for this round.

Blank rounds were also manufactured but none could be located at this time.

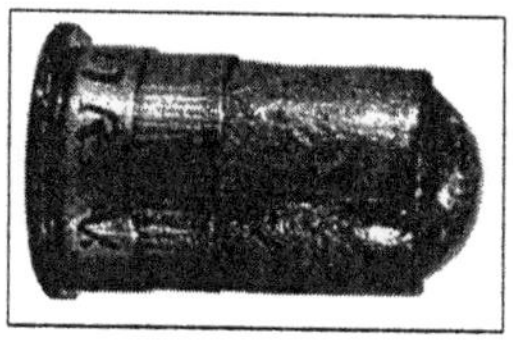

TITLE: .577 HOWDAH PISTOL

HEADSTAMP: 24 G J

DIMENSIONS:

Rim:	.722"
Head:	.650"
Neck:	.647"
Bullet:	577"
Case:	1.312"

OTHER DETAILS:

Bullet Grains:	300
Powder Grains:	35
Bullet:	Round Ball
Case Material:	Brown heavy paper Copper rim at base
Total Grains:	432

OVERALL: 1.453"

I have used this title with some reservations due to the longer than usual case, attributable to the .577 Pistol round, but the round ball loading seems to bear out the possibility that this round would have been chambered in the numerous single and double barreled Howdah pistols carried on the backs of elephants to be used in an emergency.

This specimen manufactured by Societe Francaise des Munitions, Paris France and resembles a shotgun case as it has a paper case with a low-based copper head. The number 24 could refer to a shotgun shell size.

A more complete history of .577 pistol and revolver ammunition is to be found elsewhere in the book.

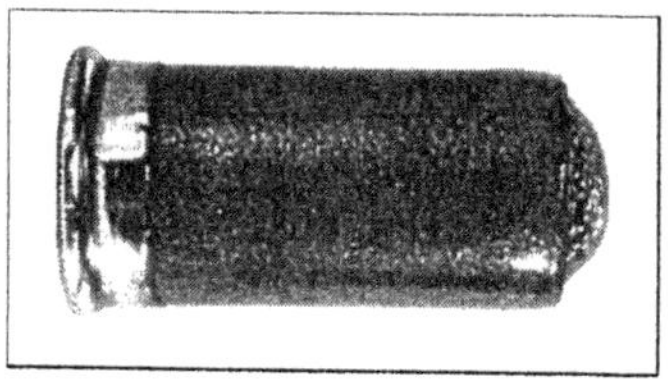

TITLE: .577 CARBINE

HEADSTAMP: Plain

DIMENSIONS:

Rim: .746"
Head: .666"
Neck: .628"
Bullet: .600"
Case: 1.315"

OTHER DETAILS:

Bullet Grains: 450 (?)
Powder Grains: ?
Bullet: Round Nose, Two Grooves
Case Material: Brown Paper
Total Grains: 770

OVERALL: 2.000"

A very questionable authentic round, but a close examination of the bullet and case would not indicate that the bullet was simply inserted into a cut-off case. Conceivably the round was intended for use in a carbine or even as a modification of the .577 revolver round.

The base has the usual iron disk and copper primer but is attached to a brass and brown paper wrapper instead of the more common white paper or brass wrapping.

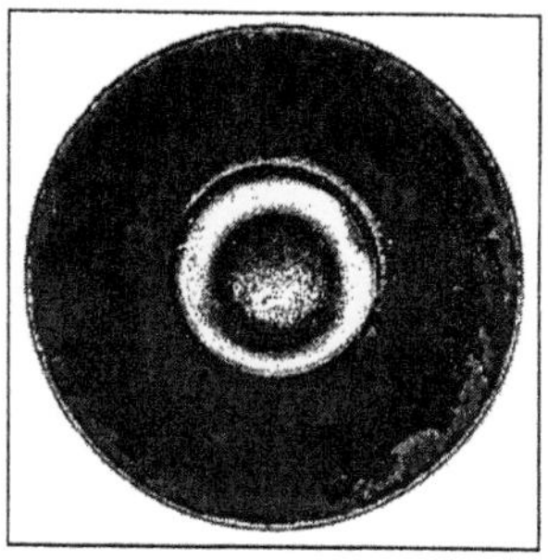

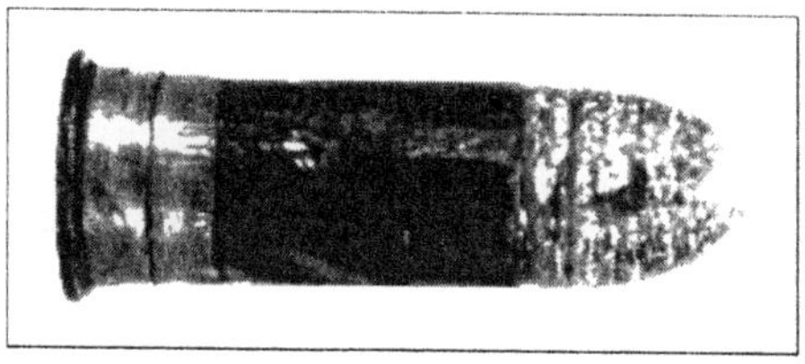

TITLE: .577 BOXER PISTOL

HEADSTAMP: Plain

DIMENSIONS:

Rim:	.745"
Head:	.662"
Neck:	.594"
Bullet:	.576"
Case:	1.530"

OTHER DETAILS:

Bullet Grains:	450
Powder Grains:	70
Bullet:	Lead, Tapered
Case Material:	Paper Wrapped
Total Grains:	656

OVERALL: 1.722"

Two barreled pistols chambered for a .577 caliber round were also carried in the Howdahs as a back-up weapon should the rifle misfire or a badly placed shot simply wound the tiger.

It would seem unlikely that this round, due to its longer over-all length could ever be used in a revolver. The cartridge could easily be chambered in one of the many two shot weapons used during the 1880 era.

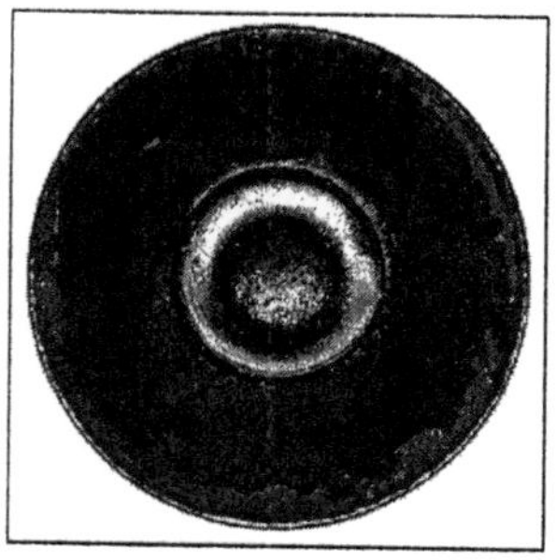

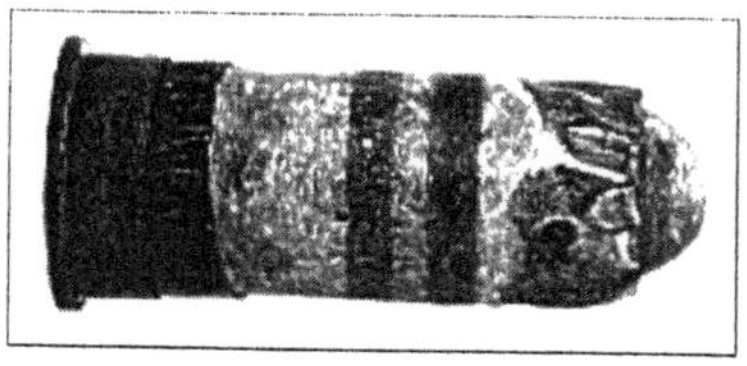

TITLE: .577 HOWDAH PISTOL

HEADSTAMP: .N. A. & A Co. SNIDER

DIMENSIONS:

Rim:	.745"
Head:	.660"
Neck:	.610"
Bullet:	.540"
Case:	1.650"

OTHER DETAILS:

Bullet Grains:	480
Powder Grains:	70
Bullet:	Lead
Case Material:	Brass
Total Grains:	732

OVERALL: 2.100"

It has been confirmed by an English cartridge collector that this round was used in a .577 Howdah Pistol. Although the weapon it was fired from is unknown, more than likely it was singled barreled.

A product of the National arms and Ammunition Company, it was produced circa 1883- about the time they went into bankruptcy.

The round has three distinguishing features; a rounded thin rim, the unusual "waffled" markings on the copper Boxer primer and the roll crimp at the case mouth. The cartridges were made without the inner cardboard liner usually found in drawn brass cases.

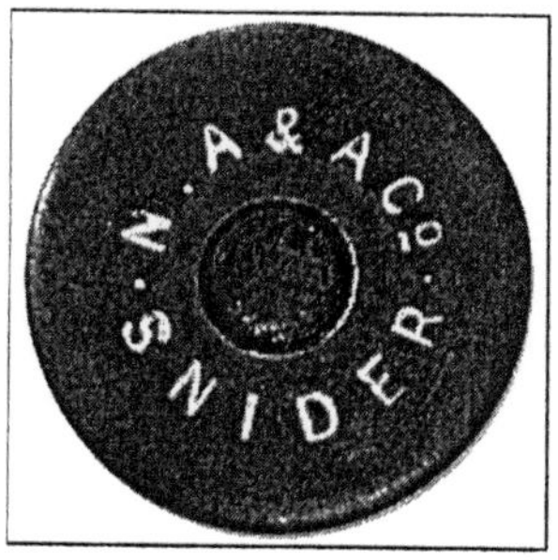

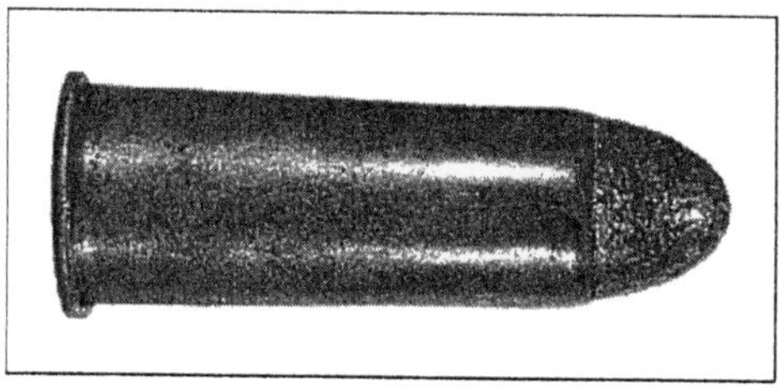

TITLE: .577 HOWDAH PISTOL

HEADSTAMP: Plain

DIMENSIONS:

Rim:	.743"
Head:	.655"
Neck:	.601"
Bullet:	.542"
Case:	1.944"

OTHER DETAILS:

Bullet Grains:	N/A
Powder Grains:	70
Bullet:	Lead, Round Ball
Case Material:	Boxer
Total Grains:	502

OVERALL: 2.170"

This is a round for the Wilkinson Double-Barrel Howdah pistol. Although a basically military style Boxer Case, no round ball load was manufactured for the British Snider Rifle.

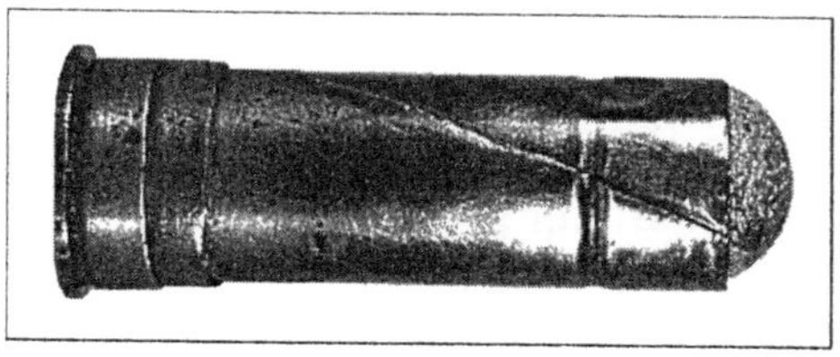

JOSIAH V. MEIGS
HIS RIFLE AND CARTRIDGES

"...In June of 1861 awaiting the coming of the 7th day of that month, I studied guns and ammunition as was very natural at the moment. I awaited alone the coming of the 8th of June to vote my first vote against the secession of Tennessee. As I became of age on the 7th of this month..."

This reminiscence is the final entry in a journal discovered in the effects of Captain Josiah V. Meigs some years ago, now in a private collection. The journal, and an engineering sketchbook entitled "On The Meigs Rifle." Give some fascinating insight into the construction and development of the Meigs cartridges illustrated on the following pages.

The Period 1865 – 1866

The Civil War interrupted Meigs' arms and ammunition investigations but he resumed this work in 1865, as most of the notations and drawings in these records are dated that year and 1866. The composit drawing shown on the next page he entitled "The Cartridge Study " is dated December 30, 1866. It is found in the notebook bearing the cover title "MEIGS 50 SHOOTING RIFLE."

Meigs' primary aim was to develop a centerfire cartridge which would be safe to use in an arm where the cartidges were contained in a magazine tube with the bullet of one in contact with the head of the next. He feared that recoil would tend to accidentally explode cartridges in the magazine unless the head was constructed to place the primer and anvil well within the body of the round. The drawings show several sizes of cartridges and a wire anvil bowed up and holding the primer away from the head. Details of other anvil ideas are illustrated: a double wire type, an iron one holding a glass tube containing the fulminate , others of perforated tin.

The cross slotted head cartridge was actually manufactured but was abandoned by Meigs in favor of the system shown at upper center and right wherein an inner priming repository, part of the case itself, was formed during the drawing and punching operation. Even this idea gave way to the external centerfire primer.

Still a mystery is the method contemplated for extracting the rimless - groove less cartridge case shown on the following pages. Its adoption was surely considered some still-unknown rifle. The rectangular indentation near the mouth of the fired case suggests that an extractor was built into the side of the chamber wall in a prototype action. In any case , the idea was not pursued beyond the experimental stage by the United States Cartridge Company who produced this ammunition.

"50 SHOOTING RIFLE"

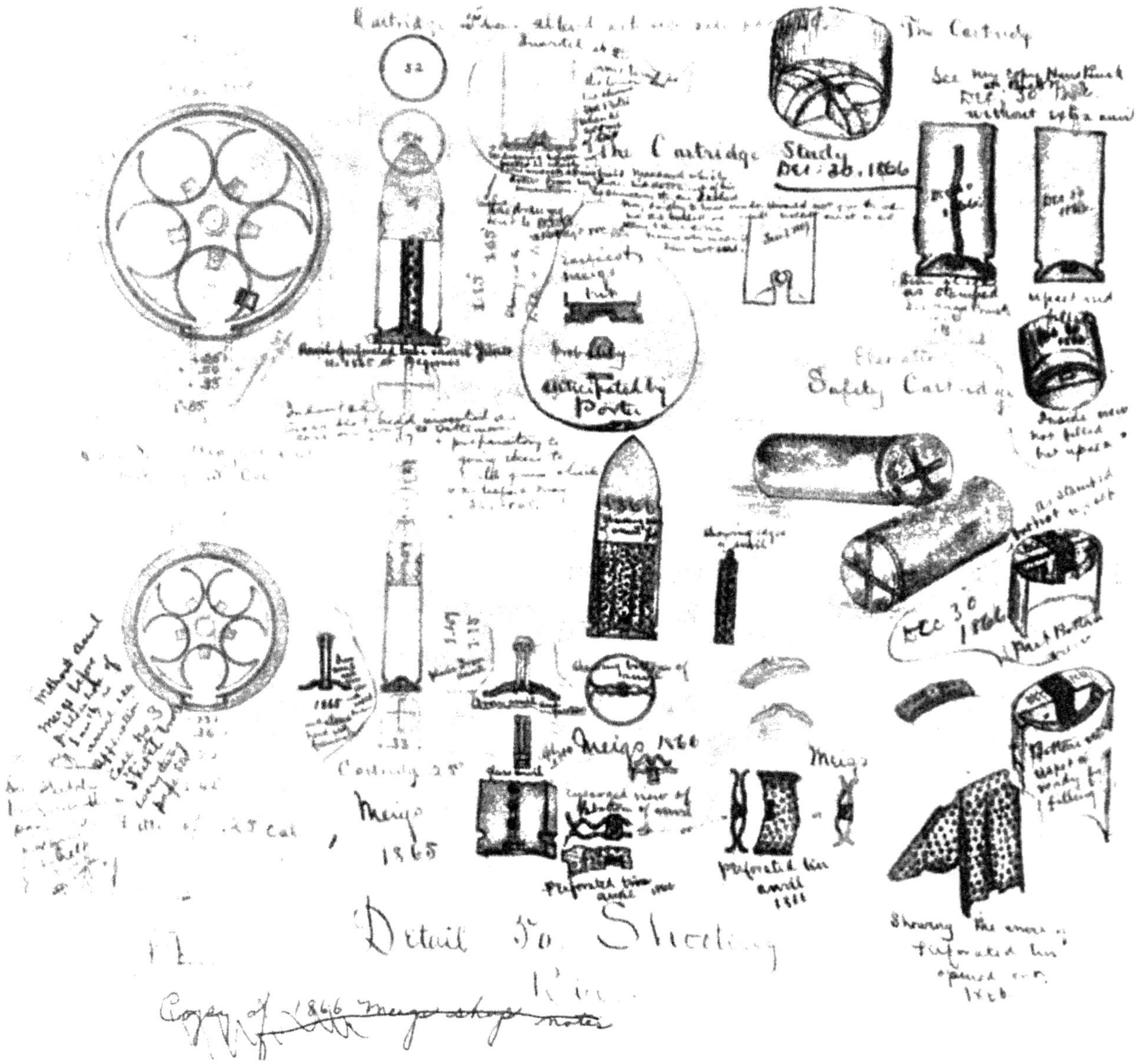

MEIGS and the U. S. Cartridge Co.

By Col. B. R. Lewis, USA

Fig. 3—J. V. Meigs (seated) with his three brothers. Shows (l. to r.) Meigs converted musket, Meigs magazine pistol, assorted Meigs cartridges, and Meigs magazine rifle. The stock of this rifle was formed by adding a cheekpiece and buttplate to the magazine tube

MANY engineers and inventors developed center-fire cartridges in the United States during the transition years 1860-1870. Their work put this country in the lead in the small-arms ammunition field during the rest of the blackpowder period. Captain Josiah V. Meigs was one of these men.

In 1868, the United States Cartridge Company was organized at Lowell, Massachusetts, to exploit the inventions and general capability of Meigs in the growing ammunition business. C. K. Farmer was Superintendent and D. W. C. Farrington was Treasurer. The company was well financed and installed tooling intended to produce 160,000 cartridges per day. The plant made the first solid-head center-fire cartridges used by the U. S. Army—a type which was very extensively adopted both here and abroad.

Good but short-lived

The Henry and Spencer rifles, as used during the Civil War, were forerunners of a long series of tubular magazine firearms. When center-fire ignition was introduced, there was much concern about accidental firing of a cartridge in the magazine by striking its primer against the point of the bullet of the cartridge behind it. To solve this problem, Meigs produced a distinctive cartridge having inside priming, which was simple, sure-fire, waterproof, strong, and required a minimum space inside the case. This was really a good method of priming, but about the time it was perfected, the Army and other users became intrigued with the reloading possibilities of the new center-fire ammunition. They insisted on that capability. As the Meigs cartridge could not be reloaded, it was soon dropped from the United States Cartridge Company line, though it could be made as cheaply as rimfire types which met no sales resistance from the reloading viewpoint.

Ammunition rare

Because it was manufactured for so short a time, the Meigs-primed ammunition is quite rare today. In 1872 the United States Cartridge Company issued an illustration of the various calibers of ammunition it was then producing. It pictured four items with the Meigs primer (Fig. 1). One (A) is cut away and shows the construction. The other three Meigs cartridges are for relatively unknown rifles. Cartridge B is for the Bethel Burton magazine rifle. This cartridge was made without a head in the usual sense to avoid the ruptures experienced with heads formed in the way then customary—by folding. This Burton series was dropped soon after the illustration appeared. Cartridge C is one of a series made for the Meigs magazine rifle, another short-lived venture. Cartridge D is for the Meigs breech-loader submitted for various State and Federal trials in 1866-1868.

Meigs' breech-loader was covered by his patent No. 54934, issued May 22, 1866. Its principal feature was a vertical breechblock actuated by moving the trigger guard back and forth (Fig. 2).

Fig. 3 shows J. V. Meigs with his three brothers, who helped with his activities.

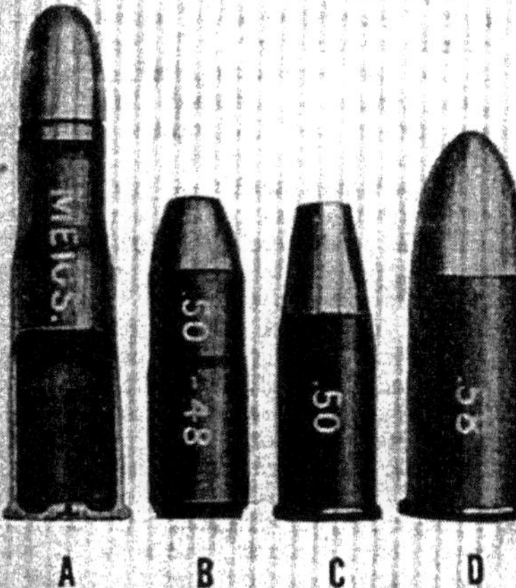

Fig. 1—A) .45 Meigs primed 1-31/32" case; B) .50-48 Burton headless; C) .50 Meigs magazine rifle cartridge; D) .58 Meigs rifle

Fig. 2—Patent drawing of Meigs breech-loader

The Meigs magazine rifle (Fig. 4) combined the trigger guard movement with a tubular magazine which formed the buttstock. Only two screws were removed to disassemble the gun, which was quite simple and compact. The Meigs magazine rifle could be fired 50 times in 19 seconds. A test rifle was fired 38,000 times.

A self-lubricating bullet

Figs. 5 and 6 show cartridges made for use with the Meigs rifle. Some used the special Meigs primers, others the more conventional type patented by Farrington. Most of them used Meigs' self-lubricating bullet. This had a base cavity filled with grease which, when the bullet was fired, was forced through holes leading to a grease groove. This idea was picked up by Smith & Wesson years later and has usually been attributed to them.

I was never able to find either an example of the Burton rifle made for the headless cartridge or any printed details of its construction. It was mentioned in general terms in the 1872 edition of Norton's *American Breechloading Small Arms*. It was there stated that Mr. Burton was applying for patents and that further information would be published soon. No patent was recorded, however, and no later reference is known to me.

How the extractor worked is a mystery. Some of the early Bergmann pistols—20 years later—used headless grooveless cartridges extracted only by residual pressure. But there is nothing to indicate that the Burton rifle had other than mechanical extraction. Fig. 7 shows a series of Burton cartridges. They appear to have evolved in three steps, starting with a rounded base, inside-primed version, and ending with the final form of Meigs priming and the United States Cartridge Company headstamp.

In addition to the special types of cartridges here mentioned, the United States Cartridge Company made the usual rimfire sizes as well as a series of solid-head recapping cartridges to meet the demand for reloadable ammunition. The firm did a world-wide business for many years, using the familiar "US" headstamp on the rimfire line, and "USCCo" or "USC" on others. The Meigs-primed items often had an interesting marking consisting of "JM" in monogram and "Pat'd" in the center, with "U. S. Cartridge Company Lowell, Mass." around the edge (Fig. 8).

This illustration gives an idea of the appearance of the heads of various types that have been mentioned in this discussion. ◆ ◆ ◆

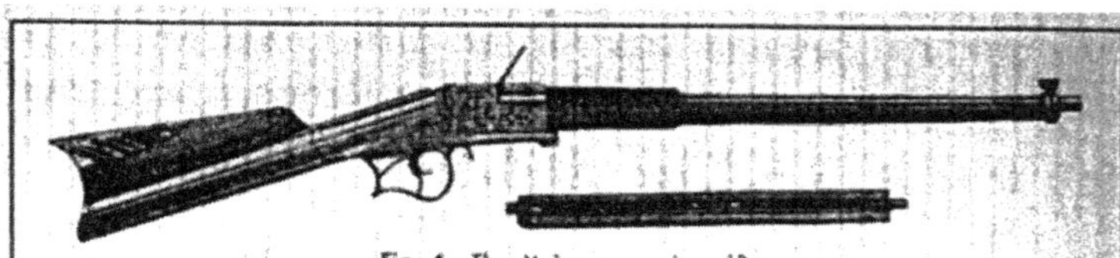

Fig. 4—The Meigs magazine rifle

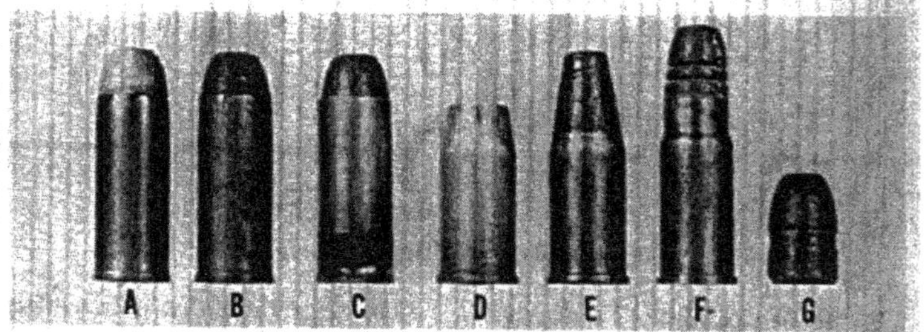

Fig. 5—Meigs cartridges for magazine rifle. A) Cal. .50-50-300, wood-bullet dummy, 1¼" case; B) Same, but conventional (Farrington) primer and Meigs self-lubricating bullet; C) Same, with Meigs primer; D) Cal. .50 blank, Farrington primer; E) Cal. .50 short, with another typical Meigs bullet; F) Cal. .44-50, solid bullet and Farrington primer, 1¼" case; G) 300-gr. cal. .50 Meigs self-lubricating bullet

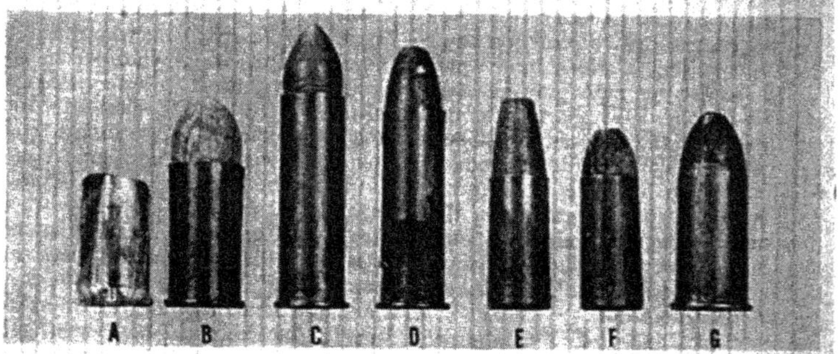

Fig. 6—Meigs cartridges: A) Tin draw-piece for cal. .58 musket cartridge, Meigs primer; B) Cal. .63 rifled musket, 1¼" case, Meigs primer; C) Cal. .52, 1-27/32" case, Meigs primer; D) Cal. .50-70 Government, 1¾" case; Meigs primer; E) Cal. .50 with patented "X" impressed anvil; F) Cal. .50 rimless; G) Cal. .58 Madrid (.577), 1¼" case

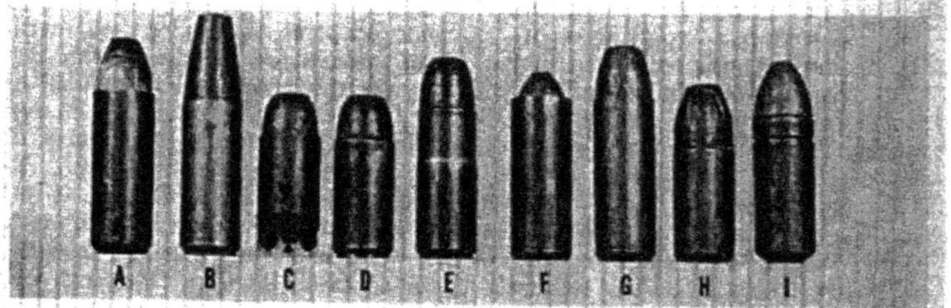

Fig. 7—Burton headless cartridges: A) Cal. .50 wood-bullet dummy; B) Cal. .50 with Meigs bullet and primer, "JM Pat'd" headstamp; C) Cal. .50 short, with Meigs bullet and variant primer; D) Same, with usual Meigs primer and headstamp; E) Cal. .50-48, with Meigs primer; F) Cal. .50 with wad-cutter type bullet, earlier head shape; G) Same, with Meigs bullet; H) Same, but short case; I) Cal. .50 with rounded base

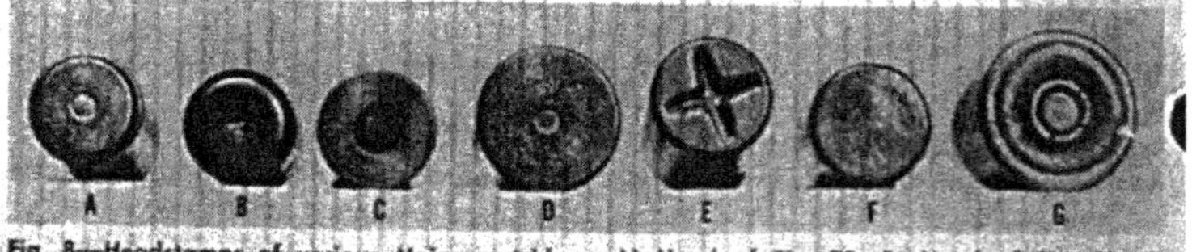

Fig. 8—Headstamps of various Meigs cartridges: A) Head of Fig. 7B; B) Head of Fig. 7F; C) Head of Fig. 5B; D) Head of Fig. 6D; E) Head of Fig. 6E; F) Head of Fig. 6F; G) Head of Fig. 6G

.58
.22
U.S.C.Co.
BLANK
.22
.43
.56-.50
.30
.42
.56-52
.32
.50
.56-46
.32
MEIGS
.44
.41
.433
.44
.44
.58
.44
.44
.58
.38
.50 -48
.58
.38
.50
.58

MEIGS CARTRIDGES

In 1868, the United States Cartridge Co. was formed in Lowell, Massachusetts to develop the inventions and to produce the ammunition of Captain Josiah V. Meigs. During the period of 1860 to 1870, great strides were being made in the ammunition business and the stockholders who formed the Company recognized Meigs as being a very talented inventor and outstanding leader in his field. D.W.C. Farrington was elected Treasurer and C.K. Farnum was appointed the Plant Superintendent. Later Farrington designed the priming system commonly referred to as the Farrington Primer. The plant was tooled with the latest cartridge making machinery capable of producing 160,000 cartridges a day. That was no small achievement in those days!

Captain Josiah V. Meigs' has provided cartridge collectors with some of the rarest cartridges ever seen. These rarities run the gamut from .44 caliber to the .58 caliber used in the Meigs breech-loader rifle. The .58 Meigs breech-loader was submitted for consideration by the United States Army Ordnance Testing Board, but unfortunately the Army became interested in the re-loading capabilities of other cartridges - a feature not possible with the Meigs. Therefore, the .58 Meigs rifle and cartridge was by-passed and soon the ammunition was dropped from manufacture.

In addition to the regular line of rim-fire and center fire ammunition, the first solid head center-fire cartridges were made and used by the Army. This ammunition enjoyed great popularity here and abroad.

Great concern was expressed over the danger of accidental explosions from self-contained cartridges and especially so, when repeating arms were invented which utilized tubes beneath the barrel or in the stock as magazines. The fear of loading ten to twelve cartridges point to base was of no small concern to the shooter. Meigs became totally engrossed in developing a truly safe cartridge, and, after many tries, succeeded in perfecting an inside-primed center-fire cartridge which was simple, waterproof, sure-fire and strong. Tests were conducted on these rounds to prove their ruggedness and that they were truly safe. However, the round could not be re-loaded and the military desired one that could. Subsequently the round was removed from production, which contributed to Meigs going out of business.

A partial listing of Meigs cartridges is as follows:

MEIGS CARTRIDGES:

.50 Rimless
.50 Rimless-"X" impressed anvil
.50-70 Govt. Meigs Primer
.52 1 27/32"Case Meigs Primer
.58 Madrid
.58 Musket-tin case
.63 Rifle Musket

MEIGS CARTRIDGES-MAGAZINE RIFLE:

.44-50 1 1/4" case
.50 Short tapered bullet
.50-50 Farrington Primed
.50-50 Meigs Primer

In addition to the named class of cartridges, the Company manufactured a solid head recapping cartridge to meet a demand for ammunition which could be reloaded. These Cartridges were test fired with 120 grains of powder, a shock force of 99.5 Ft. lbs., as applied on the head of one cartridge, and through, the point of its bullet to the cap, and head of another held securely in a die which is fitted. This is ample proof of its safety from such forces as are produced by the sudden action of bolt and hinged breech-guns.

The trials of the Meigs cartridge, shows that even steel square pointed bullets (the point being no larger than the fulminate cavity) would be eminently safe to carry in a magazine gun, when piled one on another, yet would be so sensitive as to require but a very slight blow to ignite it.

In fact, the Meigs' cartridge can only be ignited when struck in the proper place by proper direction, and with the right kind of point, as will be readily seen in the engravings.

In this system no cap protrudes (as there is none) to render the cartridge dangerous.

The protrusion of cap peculiar to all outside capped cartridges has been obviated in the solid-headed recapping cartridge, by a peculiarity of manufacture, as well as in the manner of making the cap. This cartridge is free from a projection peculiar to all cartridges of the cap class. The uncovering of the anvil or displacement of the fulminate has gradually worked out from underneath the anvil.

In this cartridge accidental firing cannot occur, as the fulminate, is fixed, and must remain so, until properly struck, it has all been expanded into gas, which pierces the charge with flame, igniting it instantly and surely. It is believed that these are qualities which are uncommon and unknown in any other cartridges".

Careful examination should be made of the photographs of the cases and primer components. Only a limited amount of information is available on the Meigs cartridges, unfortunately, and, in order for the reader to better understand the system, he should study the photos carefully.

The Author understands that the Estate of Capt. Meigs originally held the collection of complete cartridges, cutaways and bullets (along with other items). Because of the high asking price of the collection at the time of sale, it was decided to split the collection up among, I believe, six collectors. Col. Lewis obtained the loaded rounds that illustrated the fine article he wrote for the American Rifleman in June 1955.

It was one of the author's greatest disappointments not to have been able to convince Col. Lewis to have those collectors contribute information about their individual portion of the collection to a complete publication . Regretfully, Col. Lewis was not in the best of health at the time, and the idea was set aside for another day.

Back in 1970 I was fortunate to acquire a fairly large collection of Meigs inventory of what I would describe as "bits and pieces" he used in creating this primer. This material resulted from Meigs' experiments with designs of his various safety primers. Meigs devoted a lot of his efforts towards constructing a safe weapon and safe and secure ammunition. He wanted the shooter to be able to handle and operate the gun accident free.

It is unfortunate that I only have access to a limited amount of his "shop notes" that offer some explanation of the various specimens he experimented with.

1. Two of the Meigs experimental center fire cartridges utilizing the .58 Musket rim-fire case. The dimensions of these cases and that of the standard musket cartridge are:

   Case Length: 1.2" Rim: .710" Head: .636"

   The case on the left has an iron block fused (in some unknown manner) to the base of the shell. Perhaps the block originally contained the priming compound. The specimen on the right has a raised piece of copper containing the priming compound. The raised area is not "bumped" from the base of the case, but was bent into that shape and attached in some manner to the inside of the case of the cartridge.

2. These are the bases of the two cartridges above. The cartridge on the right has the primer riveted over from the separate raised piece of copper previously described. It would appear this priming method would be most difficult to manufacture with most exacting and a very critical type of construction.

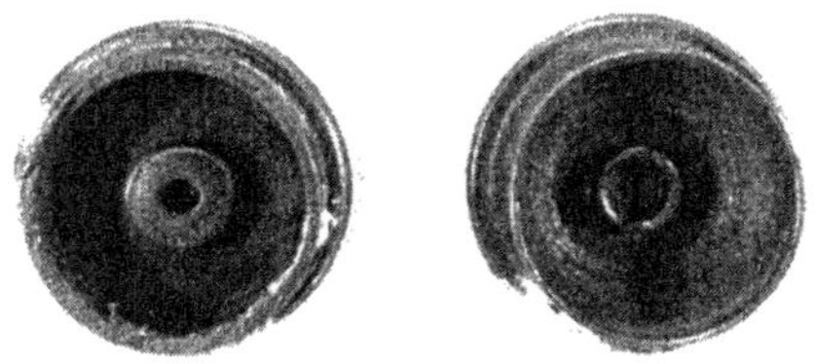

3. .58 Musket rim-fire case used by Meigs to test the durability of the case when fired. Note that a cut was made completely around the case up under the rim. All these cases, pieces, and priming components were part of an exhibit of Meigs' experiments. The hole in the case was likely a means of fastening the case to the display board.

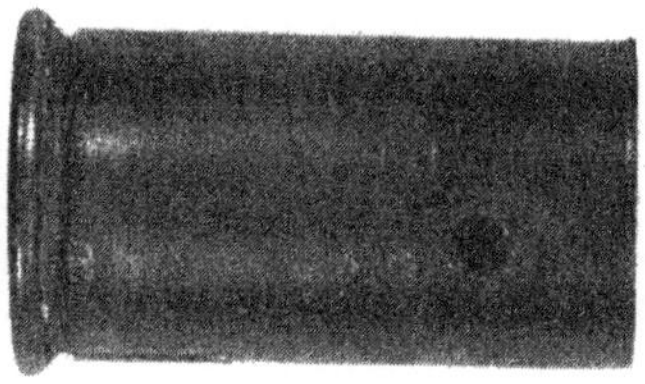

4. The base of the shell above. The .58 CAL head-stamp cases were utilized for other experiments as well. These were the Bar Anvil and Front Ignition experimental cartridges mentioned elsewhere in the text. The standard rim-fire musket round with the 58 CAL head-stamp is slightly different and is only obvious when the two are compared side by side.

5. The base of a Meigs "Solid Head Safety Primed" .58 caliber cartridge. The rim measures .656"; slightly smaller than the standard .58 rim-fire round.

In the dished base are the raised letters:

* U. S. CARTRIDGE CO. *

JM

P A T D

LOWELL, MASS.

6. A cutaway of a priming system used in the center-fire .58 Repeating Rifle. Rim diameter is .710". The center piece of metal is brass. This section holds the priming compound and the entire unit is riveted to the head of the case.

7. The reverse side of the primer system mentioned above. It shows how the piece of brass metal containing the priming compound is riveted to the base. The piece is secured by peening over the ends to the main body of the base. This experiment is identified by the markings X22 and an F scratched onto the metal

8. Three .58 cases (not head-stamped) showing the interior of the priming method being used.

9. Seven methods of priming the Meigs .58 center-fire cartridge.

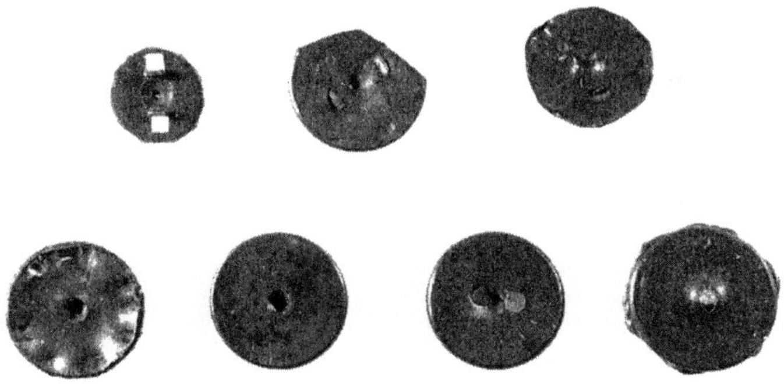

10. Reverse side of above bases.

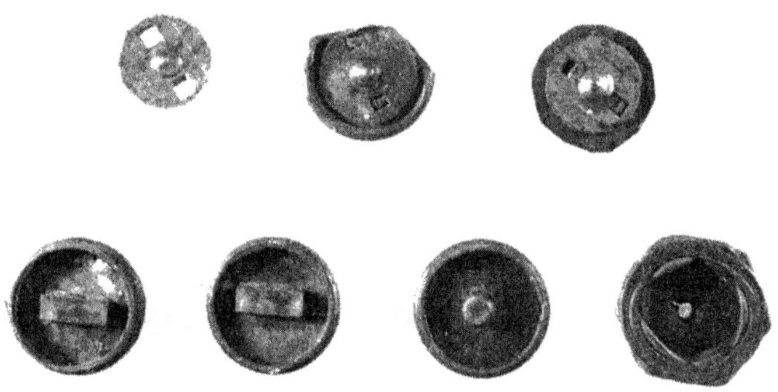

11. Photo taken from an original United States Cartridge Company advertising pamphlet showing the punishment a Meigs Solid Head Safety Cartridge can take without the round exploding. Observe how the heads shown here resemble the base pictures shown in this section.

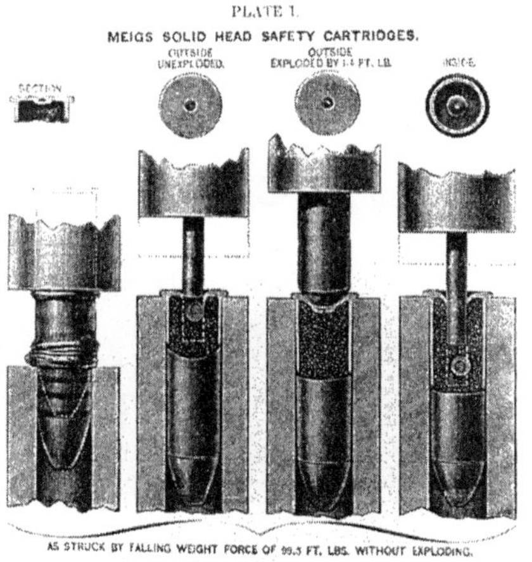

12. Four unidentified .58 caliber rounds that appeared on a very early United States Cartridge Company broadside. It is assumed that these are all center-fire cartridges. The rounds shown on the left side of this broadside depicted rim-fire cartridges only. On the rim-fire side a .58 musket round was shown. Could it be that this .58 had a U. S. raised head-stamp as was the case with the other specimens shown on that side?

13. A drawing in the Meigs patent of October 21, 1862, #36721 illustrates the breech loading weapon in detail

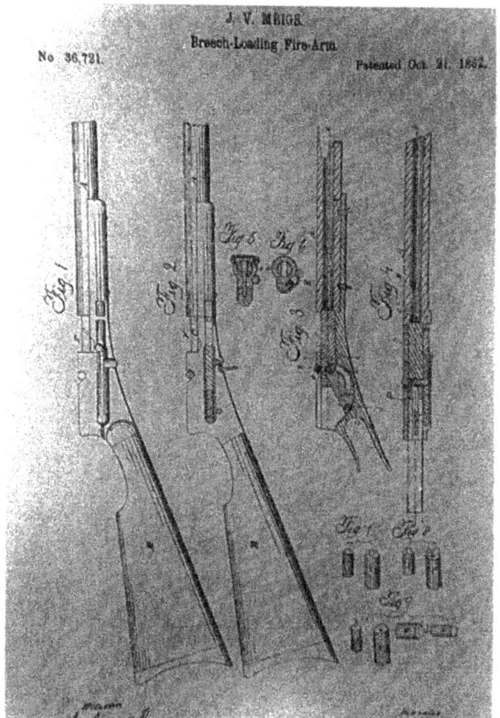

14. Experimental cartridge.

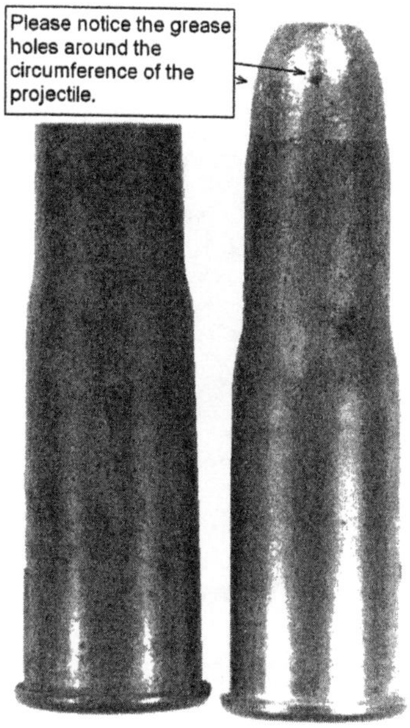

15. Self lubricating lever action rifle cartridge.

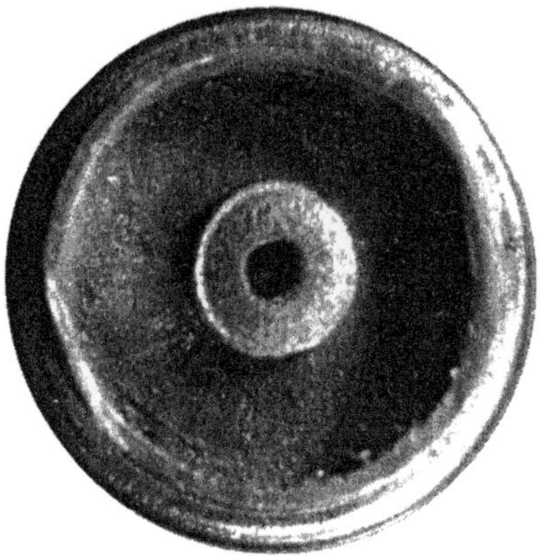

16. Cut off base of experimental primer flash hole.

17. Experimental offset primer hole.

18. Cutaway of .58 base.

19. Unknown rifle cartridge.

20. Cutaway of .58 CF.

21. .50 Blank

22. Unknown experimental primer.

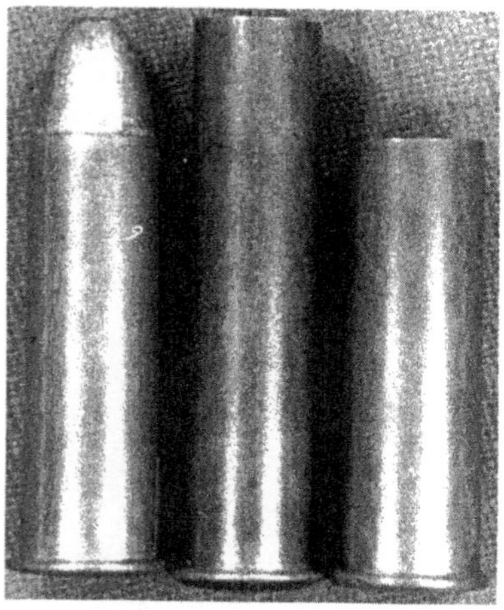

23. Meigs experimental cases of unknown caliber probably fifty.

24. .50 Short

25. Burton case.

26. .50 Caliber rifle.

27. Experimental primer types.

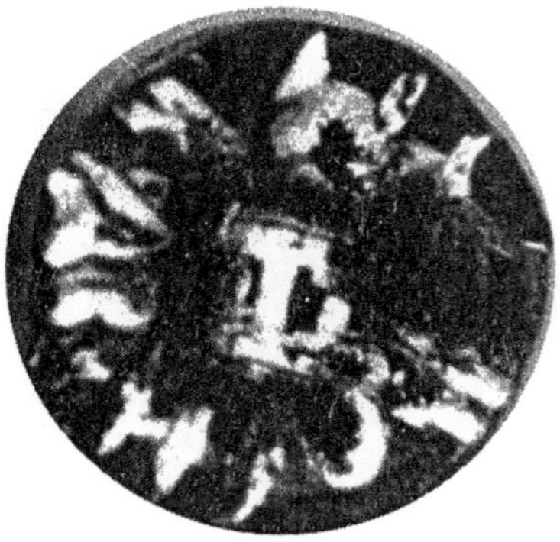

28. Head of .32 R F shell that was struck repeatedly to test its durability. The "L" is for C.D. Leet.

29. Meigs cross primer.

30. Experimental primer #6 on a .58 Roberts case.

31. Other end of Burton shell pictured above.

When I first started cartridge collecting about forty years ago, Meigs cartridge specimens were few and far between. The most common Meigs round were the .50 - 50 – 300 loaded with a flat nose lead bullet and tapered case mouth blank loading. At that time I didn't even know what a Farrington Primer was.

During the '60's I was able to buy a collection of center fire cartridges. Among the collection of 200 rounds were a large number of Meigs rounds from .22 rim fires to a lot of experimental larger rounds, shells, bits and pieces of cases, etc, he experimented with. Many of these rounds were of brass or copper case material. While it was great to acquire this material, much of it remains unidentifiable.

Luck was with me again some years later when I had an opportunity to acquire a collection of cartridges from a very old gentleman in Rhode Island who was disposing of his entire collection of guns and ammunition. The recipient of his collection was his grandson. The young man was getting ready to head for college and the grandfather wanted the young man to dispose of his collection to use for the tuition costs. The boy's father knew of my knowledge of guns and ammunition, approached me to see if I would be interested in inventorying and pricing the guns in exchange for acquiring the many rounds of ammunition the old fellow had acquired while working as a machinist for 45 years at Smith & Wesson. I wasn't too sure what I was getting into, but during the number of years he was employed at Smith & Wesson, the grandfather had acquired perhaps 100 boxes of cartridges more or less, I quickly agreed to the deal. The following night I visited his house. I was speechless to put it mildly. He took me in to his den where the stuff was laid out on the floor. In an 8 x10 room there was a tremendous array of Smith & Wesson experimental revolvers, target hand-guns, and ammunition, piled box upon box, you name it. It seems that the old fellow had it with little or no knowledge of what it was worth as collector's ammunition.

With every surprise comes a disappointment. On the first night I had spotted two quite beat-up cardboard ammunition boxes marked U.M.C. 32 L S & W. But on my return the second night the boxes I wanted so badly were gone. I inquired as to their whereabouts. My buddy said that he didn't think that I would want that old junk so he disposed of it. My heart sank but his next words were that the ammunition is right over there in that tin box. Sure enough, about 75 rounds of 1.1" long inside lubricated experimental rounds were there!

During the 1970's I was able to purchase from the Meigs Estate a copy of the Meigs story, written by Col. Berk R. Lewis on the history of Capt. V. Meigs of the U. S. Cartridge Company of Lowell, Massachusetts which appeared as a copyrighted article in the American Rifleman magazine.

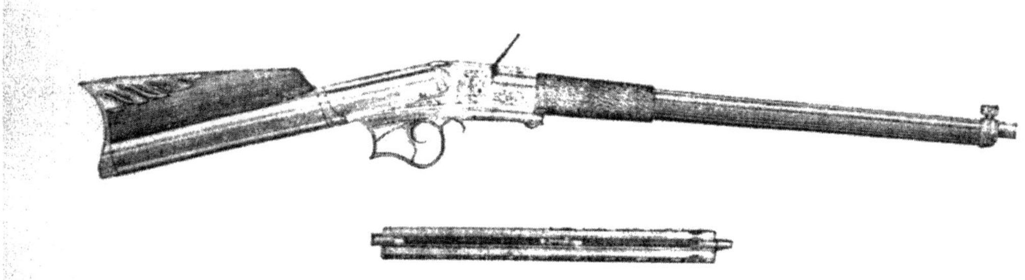

Meigs' Magazine Arm

J.V. Meigs' "50 Shooter Rifle or Shotgun" patented in 1866 was constructed around a bronze receiver with a multiple tube magazine arrangement in the buttstock which held 50 cartridges. The action was operated by drawing the trigger guard assembly to the rear and returning it to its forward position. This accomplished extraction and movement of a cartridge from the magazine into the chamber.

In his engineering notebook, Meigs described his invention as follows: "...It can be loaded and fired from the shoulder...50 shots without taking down...After placing the gun to the shoulder it requires but two motions to load, fire, and extract the cartridge shell...It has comparatively but few pieces. The main pieces being but nine and the whole number including every composite part is 42...The loss of ammunition is entirely avoided..The main points of the gun are these– 1. It is compact.–2. It has fifty shots.–3. Its weight is judiciously placed.–4. It is safe.–5. Its cartridge is safely handled or shipped.–6. Its cartridge can be safely inserted in the barrel.–7. It requires no cartridge box.–8. Its balance is perfect.–9. It weighs less than the ordinary rifle and cartridge box.–10. It requires but two motions to load and fire."

Meigs' efforts to sell his repeater to military authorities met with no success. He was way ahead of his time, as so many inventors in arms and ammunition development have been.

5.4.93A. *J.V. MEIGS.* *May 22, 1866.*

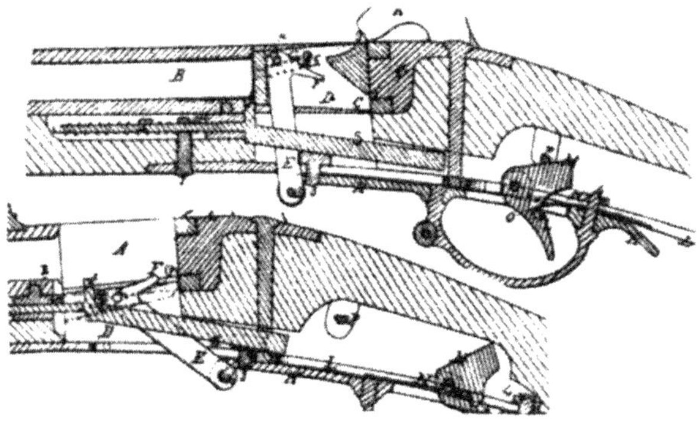

Keep in mind that the article is at least forty years old and the text and the pictures may not be of the best quality.

The following shop notes were written in a question and answer format numbering from one through seventy six and were titled "Meigs Specimen Catalog " dealing with the various experiments and thoughts of his efforts relating to designing safety primers and his thoughts on how he could improve the primer design and how to improve and correct the problems he had encountered.

I have copied his poorly written and sometimes cryptic handwriting verbatim although a number of the words are not legible. Despite these vacancies the notes make for interesting reading, so I wish the reader the best of luck in trying to understand what Meigs was writing about.

MEIGS SHOP NOTES 1866

1.
CAUSE: This shell was struck with a firing pin the point of which was too large.

EFFECT: It is not readily or certainly exploded by so blunt a point. Too much force required of the main spring.

REMEDY: The shell must be struck by a point.

2.
CAUSE: This shell is struck by a flat firing pin which is too broad. No 3 is nearer the pin but that is rather blunt for a spring right in strength.

This pin can be driven too far unless with an expensive head.

REMEDY: Round pin with shoulder

3.
CAUSE: Firing pin strikes right too broad – spring too strong

REMEDY: reduce the spring and firing pin of it's point

4.
CAUSE: The firing pin in this case was round but having too much through it penetrated the barrel and upset it and so was filed to be ½ round, and in this state failed to ignite the shells.

REMEDY: The proper throw only given and the proper point of the shell to be struck.

5.
CAUSE: Bulge on side – extractor bad one – this shell was fired with the increased charge at the.........trials and was the only un-ruptured shell from any gun so tried firing pin right for flat one with just strength of spring.

REMEDY A new extractor on form.

6.
CAUSE: No firing pin The block has a projection on its face (Q) and when struck by the hammer at (A) is driven forward and mashes the rim of the shell so as to cause the explosion. In this shell it is too narrow and strikes too high up and far in.

REMEDY: Wider firing point striking lower down

7.
The same as 6. But is made less blunt. It gave better results and warranted a new block – properly made.

8.
CAUSE: AA' A" Firing point flat and not tapered New block Y firing point sloped striking correctly but <u>not wide enough</u>

9.
CAUSE: Firing point too narrow but correct as to position

REMEDY: Make wider and sloping toward axis of bore and vice versa

10
CAUSE: Firing point correct block too short tilting and catching in dirty or rusty breech <u>Bulged Head</u> (effect)

REMEDY: Clock shorter and more rounded at ends

11
CAUSE: Block too short – Extractor takes too small a hold block dumps & upsets recessed end of barrel so that the expanded head of shell, can't be drawn out by any ordinary force. <u>Metal head</u> – <u>partly pulled off</u> (effect)

CORRECTION: Shorter block – give extractor more hook. Lessen fiction of extractor.

12
CAUSE: <u>Ruptured shell</u> – Extractor recess in end of barrel too deep as in dotted lines at A should be

REMEDY: As at A solid lines – cut as in dotted lines blew out with 120 grains of powder – block a little too short cut as solid gave correct results.

13.
CAUSE: Point of shell filed so that it's end is not square with its axis. Then put in a vice so that it's head shall rest at an angle to one face of the vice. Upon bringing the jaws of the vice together enough to bring the ends of the shell parallel the fulminate will be exploded

REMEDY: Central fire shell will only be safe. (etc. – drawing on back)

14

CAUSE: Explosion of shell by abrasion or grinding motion of rim, has occurred in Spencer and other magazine guns. The point of the bullet taps & taper and gradually presses the rim in until an abrasion is made of some part of the head over the unaffected body of the shell at the head and an explosion answer of which no explanation can be given.

15.
CAUSE: Shell exploded in Meigs guns erroneously constructed by being pulled back suddenly against to erect a <u>tipping</u> point at A and mashing head of shell same as the one in the vice

REMEDY: Central fire shell. Rim fire give A point slope as shown by the line O------P

16.
The following was erroneous – many thousand trials show that such a thing cannot be done and the only result to become is the shells were prepared <u>..........</u> Dyer & Gun

CAUSE: Shell exploded in Meigs gun, breech open – exploded by point of extractor while <u>acting</u> as inserter – head of shell probably <u>thick</u>. This is one only case of more than 2500 shells (rim fire) so inserted/and fired in this gun, but causes the conclusion that only a central fire will be safe for rapid firing (etc., etc. & sketch on back)

17
CAUSE: Opening the breech in the gun <u>with</u> the catch in rear of the guard too quick for the powder to drive the ball out of the barrel, reaction – and shell driven back into cavity, striking upon hammer is crimped with two additional heads – (etc., etc.)

18
CAUSE: Explosion by heat in end of barrel. The gun had been fired 170 rounds with greatest rapidity (etc.)

REMEDY: Central Fire Shells should always be gauged Invent gauging machine.

Central fire shell drawing of which I showed Mr. Porter in June 1865. He said he anticipated me in No. 1 exploded by the sides of the firing pin a and X. I told him if he could show me that he anticipated me that I would aid him to procure patents. This he feared to do test he should loose his place. Then I offered by a note to take out the patents in my name and transfer them to him if he could show priority, this he did not do but gave me this shell no. 19 a few days afterward. I also picked up in an old box of fired shells No. 25 the nearest to either his or mine – See 25 my .... Would not try my shell but made my bullet

20
Meigs Central Fire Shell (This is mostly a series of drawings)

21
Meigs Central Rim Fire Shell to be used with or without tape fulminate of fore
Pat Apl For. Invent night Dec. 30, 1866

22
This shell I procured at Springfield and was said by some to be the invention of Maj. Saidly and by others to be copied from some other shell. It is an expensive one.

Brass anvil, Copper cap, Iron pin. I took this example because it had no head

23
This central fire was given me in the examiners office at the Patent Office and has no anvil except a consolidated powder one with fulminate on its end upon which the copper of shell is forced by the explosion.

24.
Martins central fire shell – so called.

Used by the Ordnance Dept. but since I was at Springfield and explained my shell A No. 20. I do not know Mr. Martins .... and he may be the original inventor but Maj. Saidly did not say that it was ever Martins when my drawings were exhibited and asked to be made but said that everything had been tried and that it would not do. My anvil filled one whole base of the shell. Then I afterwards saw Martins shell exhibited as the best central fire. My new one I believe to be better (C No 20)

25
Central fire shell of ....
Picked up the day I exhibited my drawings in a box in the office used at a printing office of Mr. Porters Office (MM) at Sp'd Mass. June 1865. He did not know whose it was. This is the nearest to mine but is not the same because the fulminate is not in mine compressed against the head of the shell. This is the only copper anvil that I ever saw—

26
Berdan Central Fire – Brass shell & copper cap.

The strange thing is that fulminate can't be exploded without being struck by a point. Reason Copper on brass

28
Rim fire pistol without head less liable to explosion than ordinary rim fire. Would be dangerous in magazine/gun Inventor unknown. Exploded by inside blow

29 Small pistol rim fire/cartridge, of no value except as for a parlor pistol. Perhaps large enough with the coming explosives for great work but bullet too small.

30
Tit central fire inventor – unknown Mr. Rollin White says it is his and shows models and tools very early made too dangerous for pistol.

31
Pistol Rim Fire

32
Meigs Central Fire – (anvil) incomplete not primed showing shape of shell for Meigs 50 shooting magazine/gun. (str case 25mm long). Exterior .... Inventors wrong Experimental

33
LaFaucheau Pin Fire 4 models French
Expensive English make – Shotgun-Pasteboard – brass copper cap.

34
English Shot gun center fire/paste board, brass, copper cap

35 – 335'
3 models. Gomez cartridges exploded & unexploded. They are made by placing a very quick detonator infallible powder between the rolls of a paper strip so as to give time for fuel burning of the explosive or otherwise there would be a burst barrel .... this I drew from Gomez – but they are using in Europe some thing of the same kind. Gomez powder will stand water/a limited time but need, I think to be put in a metallic shell so as

to be used in a magazine gun/he asks $150.00 for his invention for the U. S. 35' cartridge unexploded

36
Martin Colt Smith & Wesson Pistol. This cartridge was invented after I was in Springfield with one of my cartridges which is the same except the anvil. (rest of this is description and drawing)

37
Colts Outside capped

38
Unknown

39
Martin Anvil
.... Reinforce

40-41 Meigs Magazine pistol

42
Flobert

43
4 Piece Dangerous Fulminate

44
Meigs Experimental paper .... Pressed up at one operation Nov. 13, 1867. Placed inside copper shell for preservation this was made to try dies which were made to .... In composition .... Of fiber/ nibbe and paper.

45
Shows how Fulminate burned under potash and .... Fire

46
Shows how breech block in Meigs musket gun can be .... Down after setting fire to rim fire and have shell to explode without damage

47
Outside reinforced shell/unknown

48
Meigs Experimental 1867 – To burn powder better

49
Unknown Pistol

50
Bar Cast Reinforced bar anvil center fire fired with ………. Powder Fled around ………. No hurt done to shell/two charges two balls

51
Meigs – 3 models – center Fire – Plain and ……… trim down this is useless for it is just as good plain. This is the best cartridge yet made

52
Martin bar anvil/not reinforced

53
Gutta percha – Unknown from S.H. & G.N.Y.

54
Berdan Tinned cup

55
U.S. Cartridge Co. – Recapping for Meigs Magnum – Dummy Rollin White Tinned Cap. Cap inside

56
Volcanic Pistol Cartridge charged with powder

57 – 58
Riveted Center Fires – cups

59
Meigs Center ……….

60
Riveted Center

61
Rim Fire rivet

62-63-64-65-66-67
Simply to be put under charge of powder on stiff card

68
Riveted center

69
Head for paper cartridge – unknown

70
U. S. Cartridge Co. – Pierced cap – Cap put in wrong side out.

71
Morse – made at Pistol Shop when Star Arms Co.

72
Cup Anvil – unknown

73
Colt – Berdan

74 - 75 – 76
Meigs accelerating Cartridge. The tubes in each case penetrate the base of the bullet and are charged with fine powder. They are ignited and start (over) the ball when the main charge is ignited at its front driving it – as soon as test machine is ready. I will try these fully - great results are hoped for.

END OF NOTES

This brings to an end the story of a very talented inventor and gunsmith. There are many more cartridges he developed but they were not of .58 caliber sizes.

One important item was his invention of the self-lubricating bullet. This was a revolver cartridge with a hollow running from its base about 75% up to near the top of the round with small holes that permitted grease to run up the center of the ball, a lead plug was forced forward acting as a plunger which forced the lubricating grease out and into these small holes on the side of the bullet. This in turn left lubricating grease to coat the sides of the bore. They came in sizes .32, .38, .44 Russian and .45 Colt. A special experimental size titled .32 Extra Long in 1.1" long was also listed.

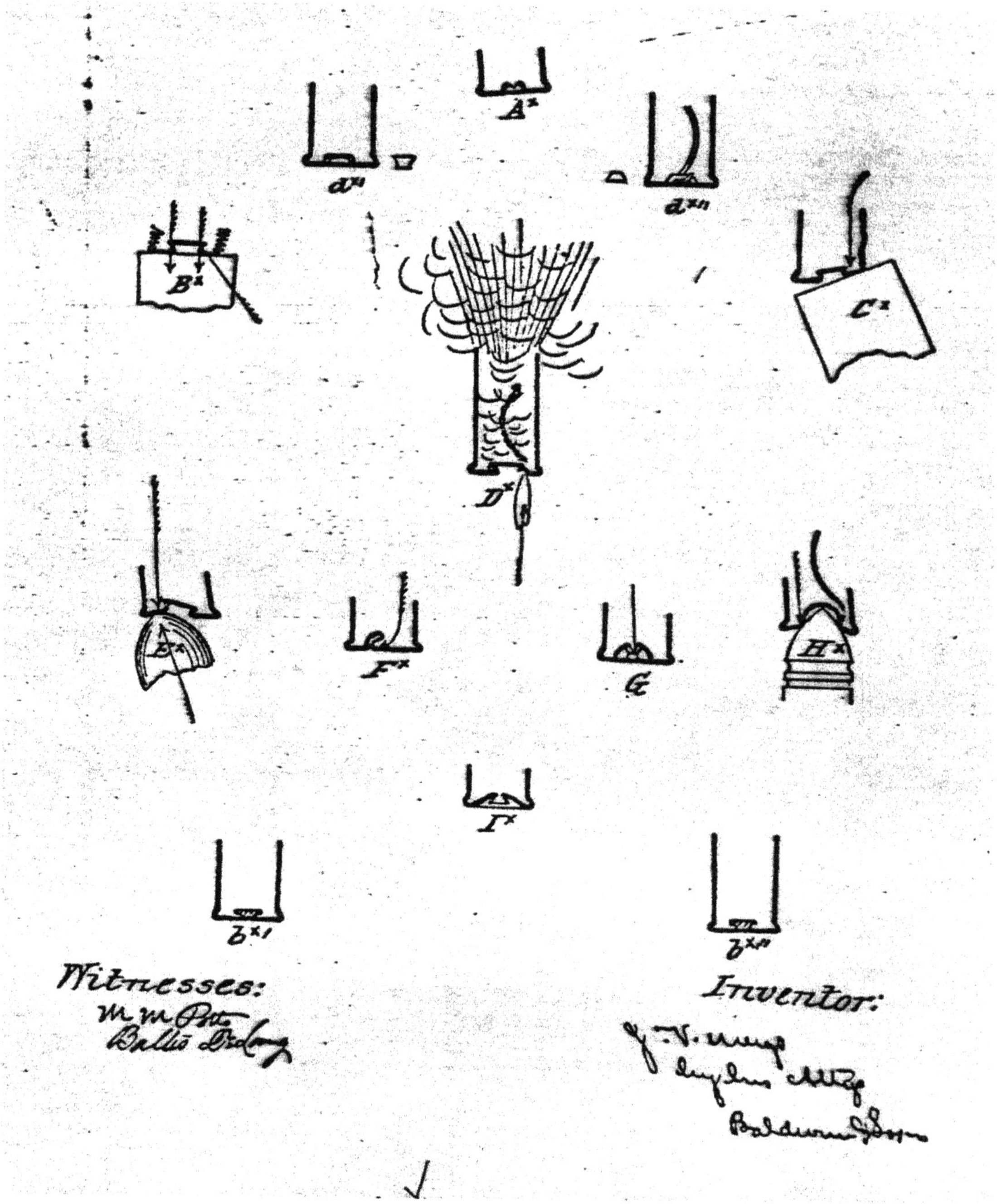
Witnesses:
Inventor:

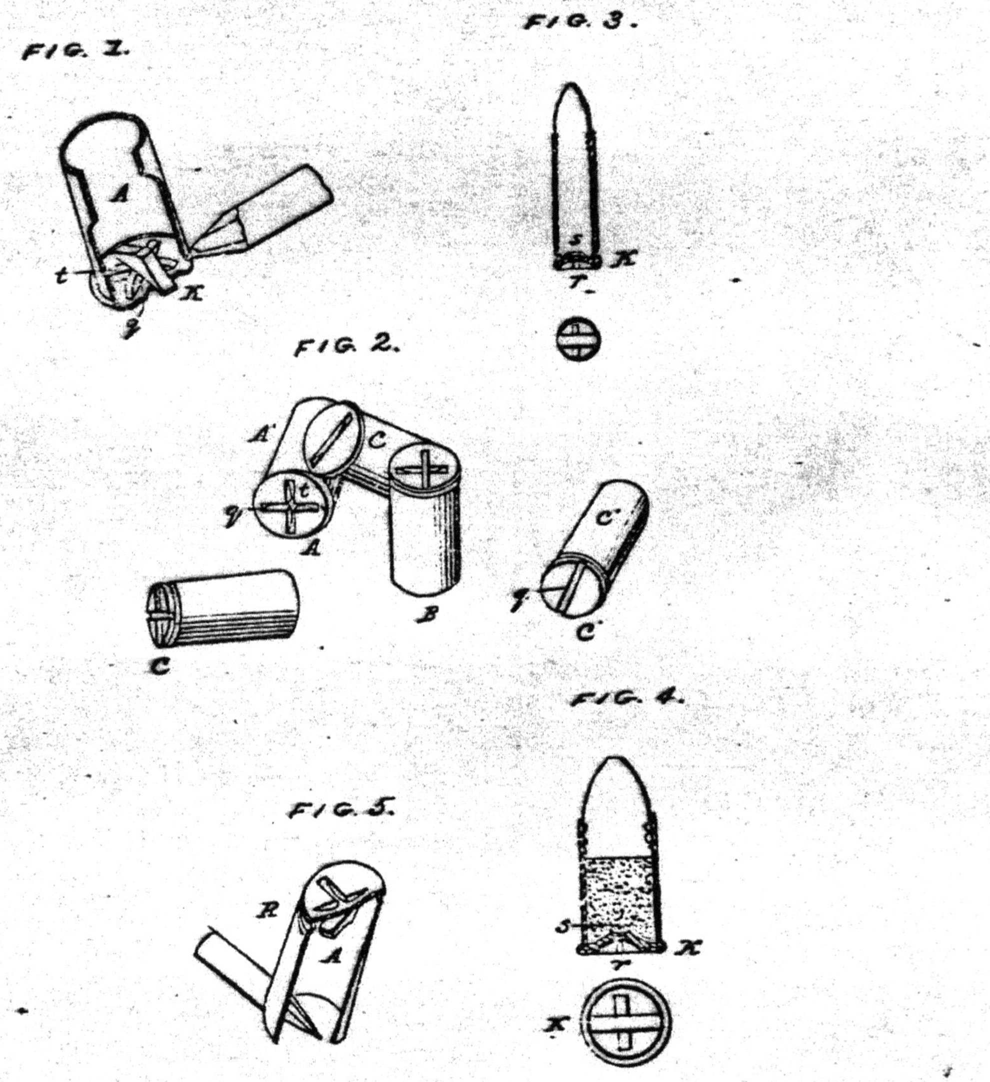

WITNESSES:

J. J. Peyton

Theodore Lang

INVENTOR.

by his Attys

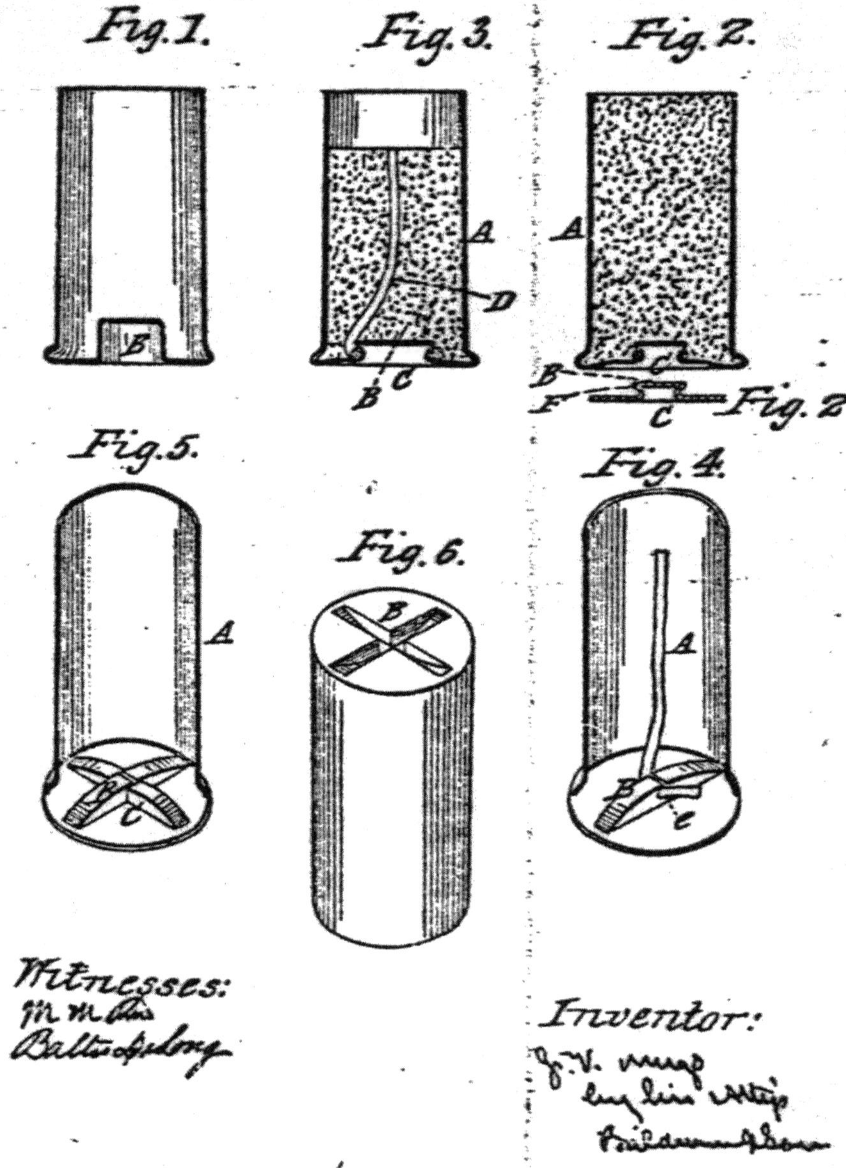
Fig. 1.
Fig. 3.
Fig. 2.
B
A
D
C
Fig. 2a.
F
Fig. 5.
Fig. 4.
Fig. 6.
Witnesses:
Inventor:

TITLE: .58 MADRID CARBINE-MEIGS COPPER

HEADSTAMP: Plain

DIMENSIONS:

Rim:	.744"
Head:	.651"
Neck:	.606"
Bullet:	.565"
Case:	1.115"

OTHER DETAILS:

Bullet Grains:	450
Powder Grains:	40
Bullet:	Tapered
Case Material:	Copper
Total Grains:	589

OVERALL: 1.650"

A variant round.

The total weight is sixty-three grains less due to the different shape and size of the projectile. All other features of this Meigs cartridge remain the same except for the lack of the neck crimp that is found on the other round.

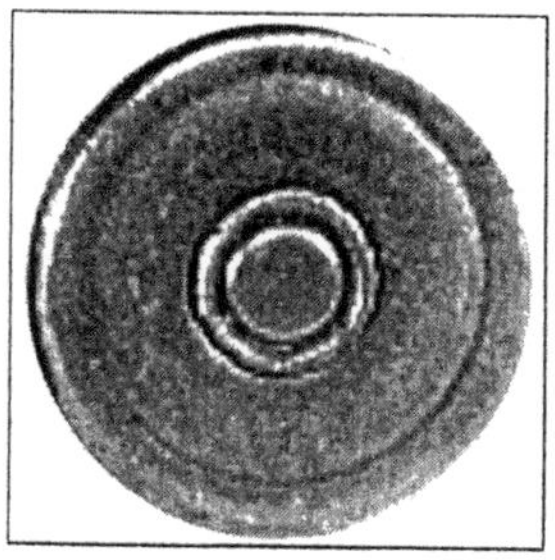

TITLE: .58 MADRID CARBINE-MEIGS BRASS

HEADSTAMP: Plain

DIMENSIONS:

Rim:	.737"
Head:	.645"
Neck:	.605"
Bullet:	.575"
Case:	1.225"

OTHER DETAILS:

Bullet Grains:	500
Powder Grains:	60
Bullet:	Tapered
Case Material:	Brass
Total Grains:	652

OVERALL: 1.666"

Capt. J. M. Meigs patented this cartridge manufactured for him by the United States Cartridge Company of Lowell, Massachusetts.

The odd Meigs primer is unlike any other type of primer. It is held together by a rivet which is peened over the primer pocket.

It is possible that the bullet in this Carbine round is of this type?

Although a Madrid Berdan Rifle exists, it does not seem possible this carbine cartridge could be chambered in the rifle model due to considerable variances in rim and case body dimensions.

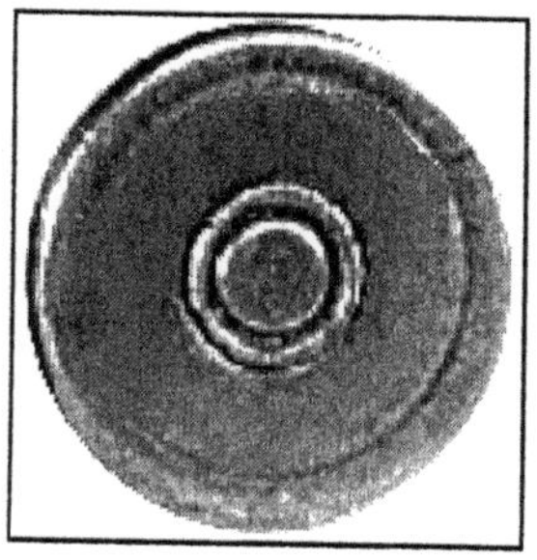

TITLE: .58 MEIGS CONVERSION CENTER FIRE

HEADSTAMP: Plain

DIMENSIONS:		OTHER DETAILS:	
Rim:	.737"	Bullet Grains:	525
Head:	.650"	Powder Grains:	60
Neck:	.620"	Bullet:	Lead
Bullet:	.610"	Case Material:	Copper
Case:	1.250"	Total Grains:	?

OVERALL: 1.850"

The above measurements were taken from a chamber cast of the actual rifle submitted to the 1863 trials.

No cartridge specimen was available for examination.

MILBANK

This is a copy of the report on the Milbank Primed cartridges as appears in METALLIC CARTRIDGES, AS MANUFACTURED AND TESTED AT THE FRANKFORD ARSENAL, PHILADELPHIA, PA. - 1860-1873 (Plate XXXV):

Remarks:

A number of the above cartridges, Milbanks Patent (May 1870) manufactured by the Winchester Repeating Arms Co., New Haven, Conn. Labeled solid head, central fire, reloading cartridges, were tested at this Arsenal, April 1873.

The head is claimed to be a solid one, which is presumed to mean one folded closely at 0, similar to the Dutch carbine cartridge. As it does not fulfil the conditions of a solid head it is rated under Class 1. The cartridge case metal is quite thick, about .05". The flanges have a greater variation in thickness than any other cartridge tested here from .062" to .079" and are variable in thickness on the same flange at different points, showing bad work at an important point.

The primer is made somewhat like a rim-fire case, having a recess in the centre holding the priming, partly in the folded rim and at the bottom.

The composition is of a dark color resembling that in the Eley cap and is covered by a paper wad, the open end being slightly closed to facilitate its insertion into the pocket of the case.

It was claimed to be superior in the following points: First: Certainty of fire; Second: Non escape of gas at primer; Third: Impervious to moisture; and Forth: Facility in reloading.

Comparative performance of Winchester and service ammunition-object of trial-certainty of fire, of both cartridges with the Springfield and Remington guns. Recapitulation-Winchester ammunition, Springfield gun-fired in all 352 rounds with the following results: 70 required 2 blows to explode; 3 failed with 2 blows to explode. The primers were in a number of instances discolored by leaking gas. Winchester ammunition, Remington gun-Fired in all 250 rounds with the following results: 14 cartridges required 2 blows to explode; 7 cartridges failed to explode on 2 blows.

Service ammunition, Springfield gun-fired in all 200 rounds with the following result: No failures of any kind except 1 case burst at flange.

Using a Springfield gun with both cartridges, 100 rounds each of Winchester and service ammunition that were in water for five, (5) days, were fired with the following results;

WINCHESTER AMMUNITION	SERVICE AMMUNITION
70 required two blows 11 hung fire, 2 on 1st blow 14 on 2nd blow 3 cartridges failed on 2nd blow	No failure of any kind-all exploded promptly on 1st blow

TITLE: .58 CONVERTED INSIDE PRIMED

HEADSTAMP: Plain

DIMENSIONS:

Rim:	.747"
Head:	.666"
Neck:	.645"
Bullet:	.612"
Case:	1.222"

OTHER DETAILS:

Bullet Grains:	530
Powder Grains:	40
Bullet:	Tapered Point
Case Material:	Copper
Total Grains:	675

OVERALL: 1.700"

In no way is the title meant to be an official identification.

This does show the ingenuity a shooter will go to in order to be able to fire his weapon. The case has obviously been fired as the usual crimps have blown out. The head has been cut into in order to insert a shotgun type primer for ignition.

There is nothing really new or novel about this type of alteration as man has been known to go to any lengths to reload and reuse his "brass". The Indian found he could reuse the fired shells by drilling out a so-called primer hole, using either a match head or a percussion cap for ignition. It became the Indian Squaw's duty to gather up the cases on the battlefield, assist in capping and molding bullets in a stone mould and reloading the shells.

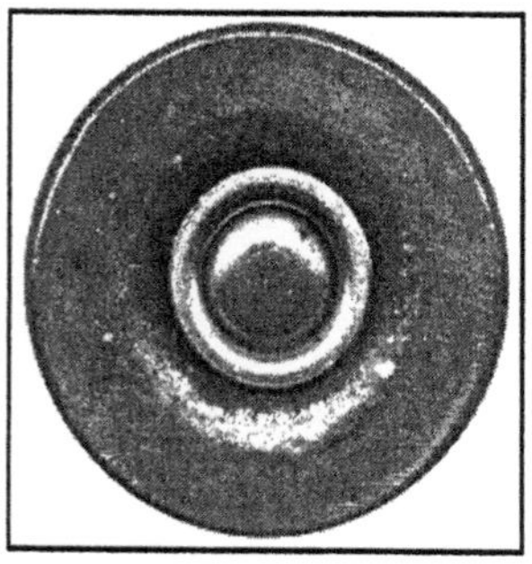

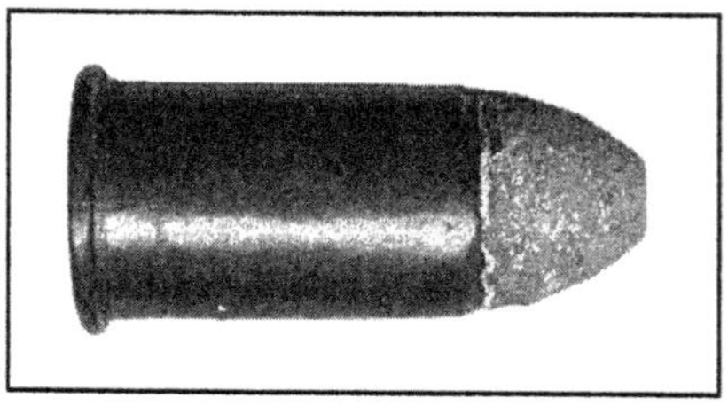

TITLE: .58 REMINGTON CARBINE-BRASS

HEADSTAMP: Plain

DIMENSIONS:

Rim: .739"
Head: .655"
Neck: .622"
Bullet: .592"
Case: 1.250"

OTHER DETAILS:

Bullet Grains: 525
Powder Grains: 40
Bullet: Smooth, Tapered, Flat Nose
Case Material: Brass
Total Grains: 673

OVERALL: 1.745"

This is a Remington Carbine round manufactured by Winchester Repeating Arms Company. The base is very smooth and contains a small brass primer. The actual case-length is a little shorter (1.214") than the standard length of 1.25" for these rounds.

The bullet is distinctive. It has no grease grooves as compared with the more common U. M. C. cartridge.

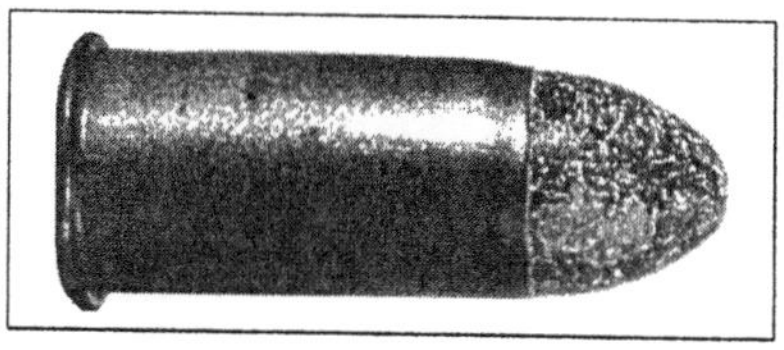

TITLE: 14.5 WANZL

HEADSTAMP: Plain

DIMENSIONS:

Rim:	.687"
Head:	.600"
Neck:	.590"
Bullet:	.566"
Case:	1.300"

OTHER DETAILS:

Bullet Grains:	425
Powder Grains:	-0-
Bullet:	Lead, One Knurl
Case Material:	Brass
Total Grains:	557

OVERALL: 1.859"

This specimen defied identification for many years but has been now identified as the Austrian Wanzl breechloader rifle cartridge.

It would appear that this is a salesman's sample display round rather than a board dummy. Board dummies have uneven raised edges around the holes instead of the four smooth holes drilled in the side of this case. The size of the round would require only two holes, instead of four, to attach to an advertising board.

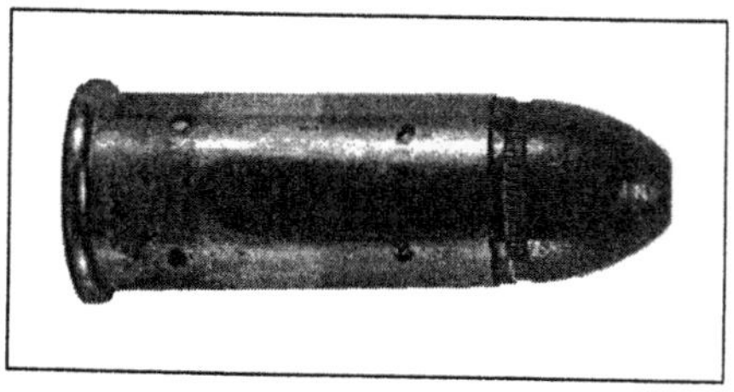

TITLE: 14.5mm AUSTRIAN WANZL

HEADSTAMP: G R ***

DIMENSIONS:		OTHER DETAILS:	
Rim:	.673"	Bullet Grains:	458
Head:	.613"	Powder Grains:	68
Neck:	.591"	Bullet:	Lead, Two Grooves
Bullet:	.565"	Case Material:	Brass A base
Case:	1.300"	Total Grains:	680

OVERALL: 1.961"

This is the center fire version of the rim fire round. The round was used in converted rifles utilizing the breech-loading action of the Vienna Gunmaker, J. Wanzl, circa 1870.

The head-stamp is that of the ammunition manufacturer, G. Roth of Vienna, Austria.

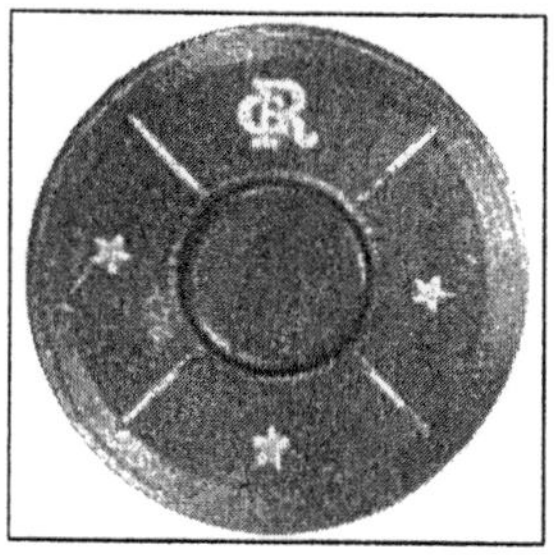

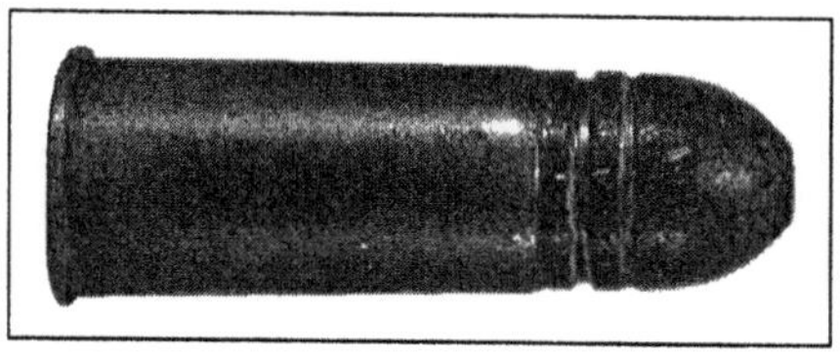

TITLE: .577 MODEL 70 SERBIAN – PEABODY

HEADSTAMP: G r on Primer

DIMENSIONS:		OTHER DETAILS:	
Rim:	.747"	Bullet Grains:	?
Head:	..650"	Powder Grains:	?
Neck:	.410"	Bullet:	?
Bullet:	.592	Case Material:	Brass
Case:	1.321"	Total Grains:	?

OVERALL: 1.970"

A very early drawn brass case manufactured by Gevelot of France. Dimensions are close to the .58 Roberts American centerfire cartridge with a raised ring head.

No speciman was available.

TITLE: 14.8mm SERBIAN PEABODY

HEADSTAMP: Plain

DIMENSIONS:

Rim:	.736"
Head:	.641"
Neck:	.585"
Bullet:	.558"
Case:	1.326"

OTHER DETAILS:

Bullet Grains:	?
Powder Grains:	?
Bullet:	Lead, pointed
Case Material:	Brass
Total Grains:	700

OVERALL: 1.730"

This is a U.M.C. specimen made for the 14mm (.58) Serbian Peabody converted musket.

After the Civil War, the Providence Tool Company of Rhode Island took over the task of obtaining contracts to convert the great number of left over muzzle-loaders to a center fire system. The Serbian model was one of those conversions.

The Providence Tool Company took all their ammunition problems to U.M.C. and the two worked hand-in hand in adapting ammunition to their products in both rim and center fire systems.

U.M.C. had much to gain from such a relationship not only with the Providence Tool Co. but with all the weapon manufacturers. The end of the Civil War caused a great decrease in the demand for ammunition and without contracts to supply cartridges for these converted muzzle-loaders many of the firms were heading for bankruptcy.

TITLE: .58 SERBIAN - AUSTRIAN CARBINE

HEADSTAMP:

DIMENSIONS:		OTHER DETAILS:	
Rim:	.649"	Bullet Grains:	?
Head:	..569"	Powder Grains:	?
Neck:	.550"	Bullet:	?
Bullet:	?	Case Material:	Brass
Case:	1.340"	Total Grains:	?

OVERALL: 1.780"

The above dimensions were taken from a description of the cartridge that was on an old U.M.C. board.

The case length corresponds to the .58 Roberts BreechLoader but other case dimensions differ greatly. It is not unreasonable to assume that the round is for a Roberts because the weapon was widely made and distributed and the ammunition was manufactured by U.M.C.

No specimen was available for examination.

TITLE: .58 GATLING CENTER FIRE

HEADSTAMP: Plain

DIMENSIONS:

Rim:	.710"
Head:	.615"
Neck:	.615"
Bullet:	.600"
Case:	1.356"

OTHER DETAILS:

Bullet Grains:	565
Powder Grains:	86
Bullet:	Lead, pointed
Case Material:	Brass
Total Grains:	778

OVERALL: 2.130"

As developments in the design of Gatling guns went from percussion cap ignition and then to rim-fire, it was logical a large caliber center-fire weapon would be next. The .58 center-fire followed. In effect, it was a copy of its rim-fire brother discussed and pictured elsewhere in the book. The bullet in both rounds appears to be identical and the total grains of the cartridge is within twenty grains. This small difference is accounted for by the heavier case material in the center-fire version.

It would appear that U.M.C. was the only manufacturer of this ammunition.

Interestingly enough, the 1889 Hartley and Graham catalog lists and pictures the Gatling center-fire and also states it was used in the Snyder Rifles-Turkish Model. Selling price in those days was a mere four cents a round.

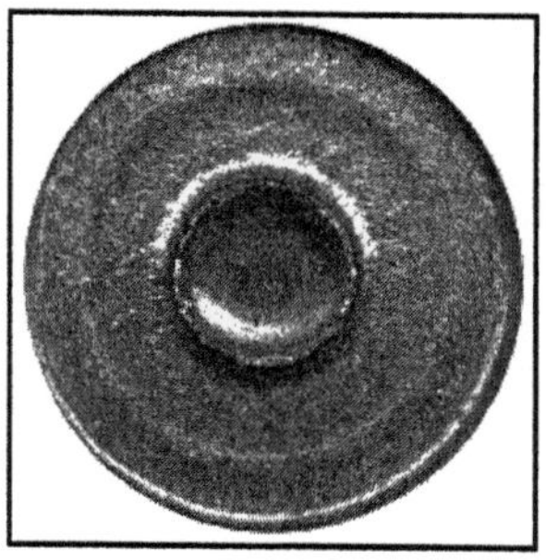

THE 15MM HANDGUN CARTRIDGE

Here is a scarce assortment of .58 caliber cartridges from a relatively unknown European manufacturer. Gevelot manufactured a variety of sporting ammunition especially in handgun sizes. I was not privy to any information regarding the manufacturing of these .58 caliber products.

No. 1) is a 15 mm dummy cartridge. No. 2) is a standard specimen with intertwined GG logo by Gevelot. No. 3) is one of their GG 15 mm head stamp with a copper battery ringed protected primer. No. 4) is a very scarce head stamped Gaupillat – Paris of the early variety with a copper battery ringed protected primer. No. 5) is a brass dummy with GG 15 mm markings. Nos. 6 & 7) are also dummy rounds with similar head stamps with a peculiar group of holes on the side of the cases.

In forty years of cartridge collecting this is the first time I had ever encountered specimens like Nos. 6 & 7 with this pattern of holes on the side of each cartridge.

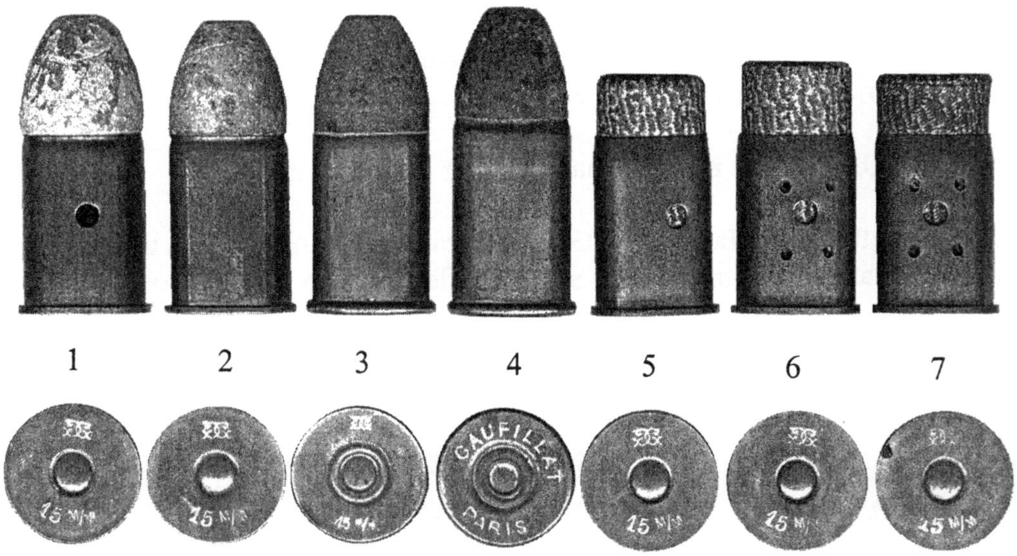

TITLE: 15 x 34 SERBIAN PEABODY
CANNISTER LOADING

HEADSTAMP: Plain

DIMENSIONS:

Rim: .735"
Head: .646"
Neck: .617"
Bullet: .562"
Case: 1.360"

OTHER DETAILS:

Bullet Grains: ?
Powder Grains: ?
Bullet: Lead, blunt, pointed
Case Material: Brass
Total Grains: 503

OVERALL: 1.709"

This is an extremely rare loading of the Serbian Peabody cartridge. X-rays indicate that the pointed lead bullet sits on top of two rows of two smaller round balls.

The reader will note that this specimen originated from the White &Munhall collection as evidenced by the inscribed number 8475 on the side of the case.

The light total grains should be discounted as the round was either deactivated by drilling a hole in the case to remove the powder, or it might be an authentic dummy round.

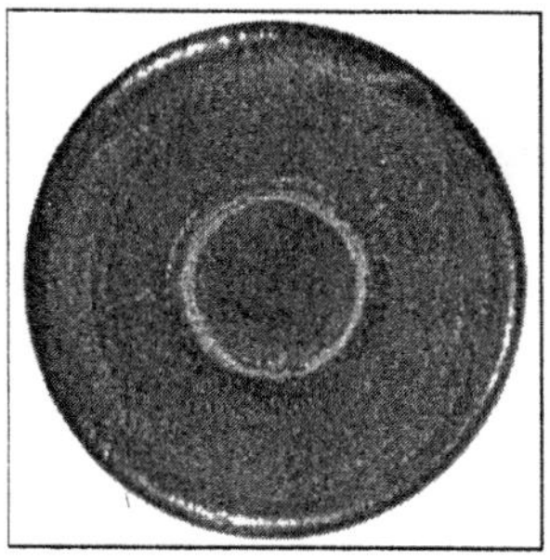

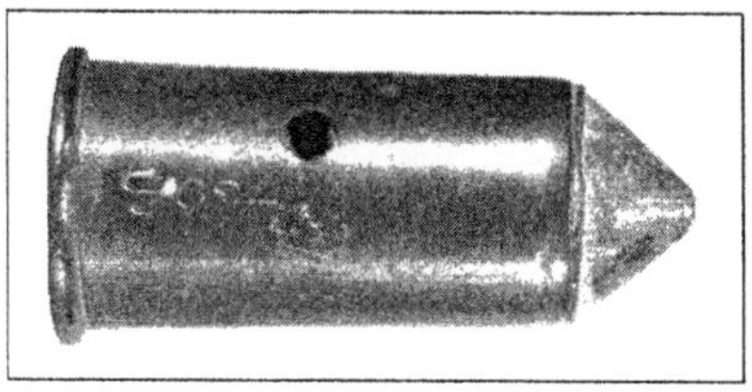

TITLE: 15 x 34 R mm

HEADSTAMP: G on primer

DIMENSIONS:

Rim:	.748"
Head:	.649"
Neck:	.623"
Bullet:	.598"
Case:	1.360"

OTHER DETAILS:

Bullet Grains:	?
Powder Grains:	?
Bullet:	Lead, Blunt Point
Case Material:	Brass
Total Grains:	664

OVERALL: 1.821"

The round presently defies identification.

The G on the copper primer would indicate that the ammunition was made by Gevelot of France.

Note the most unusual configuration of the lead projectile. A similar specimen appears in Hoyem's Vol. II CFR #208b.

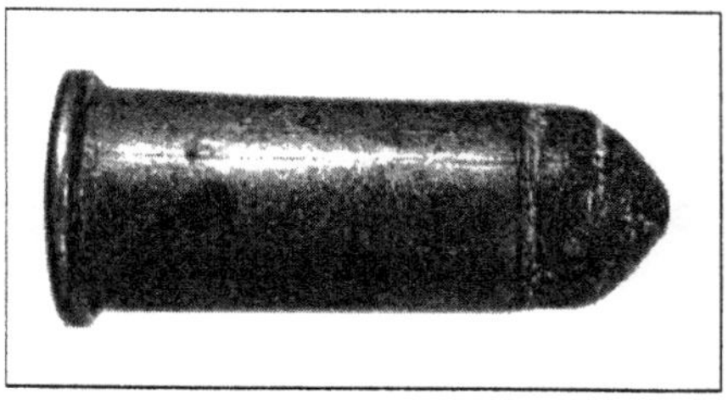

TITLE: 15.2 x 40Rmm

HEADSTAMP: Plain

DIMENSIONS:

Rim:	.722"
Head:	.660"
Neck:	.655"
Bullet:	.589"
Case:	1.586"

OTHER DETAILS:

Bullet Grains:	?
Powder Grains:	?
Bullet:	Pointed, Patched
Case Material:	Brass
Total Grains:	751

OVERALL: 2.205"

This is an unidentified round that appears to be of European origin.

The beveled rim is extremely thin (.052") and contains a flush brass primer of .256" diameter.

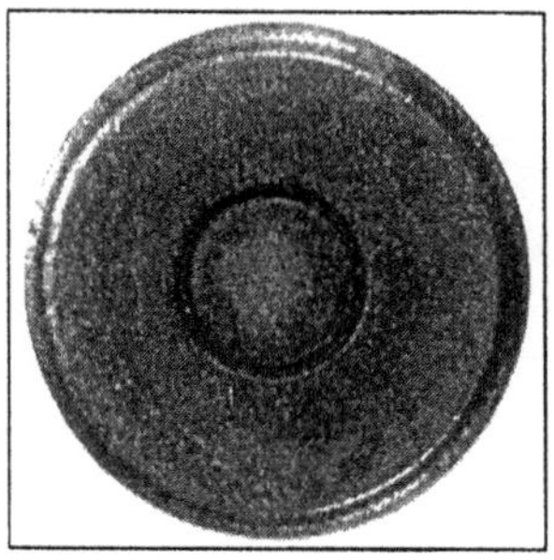

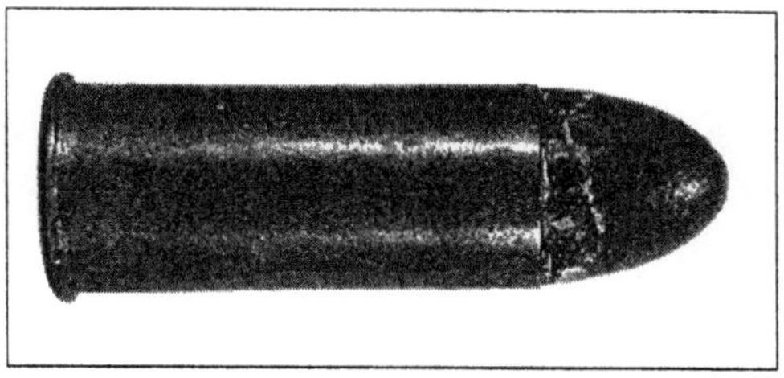

TITLE: 15mm x 48Rmm

HEADSTAMP: Plain

DIMENSIONS:		OTHER DETAILS:	
Rim:	.761"	Bullet Grains:	550 (?)
Head:	.668"	Powder Grains:	70-80 (?)
Neck:	.626"	Bullet:	Ribbed, One Groove
Bullet:	.606"	Case Material:	Brass
Case:	1.868"	Total Grains:	945

OVERALL: 2.602"

This cartridge is of unknown origin, but it is most likely of European manufacture as evidenced by the heavily raised ringed head construction. The case comes very close to being of "A" base type configuration.

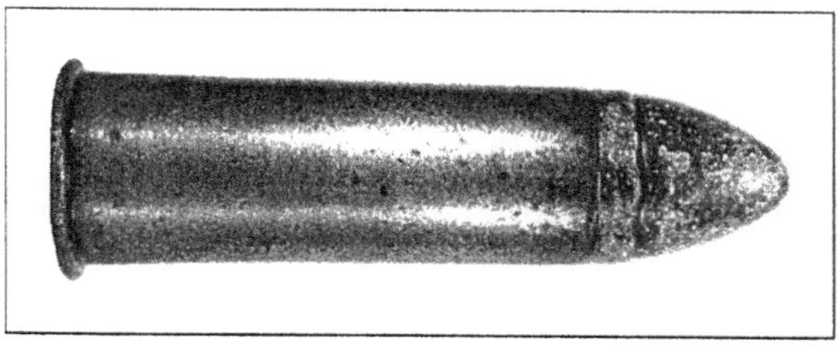

TITLE: 15mm x 49Rmm

HEADSTAMP: /GR/*/*/*

DIMENSIONS:		OTHER DETAILS:	
Rim:	.722"	Bullet Grains:	?
Head:	.647"	Powder Grains:	?
Neck:	.609"	Bullet:	Tapered/Blunt
Bullet:	.556"	Case Material:	Brass
Case:	1.940"	Total Grains:	692

OVERALL: 2.411"

It might be possible that this round was intended to be used in a Snider conversion but due to the smaller rim size, the author has declined to list it under the Snider category.

The cartridge was manufactured by G. Roth of Austria, and the number 12214 inscribed on the side of the case indicates it came from the White & Munhall collection.

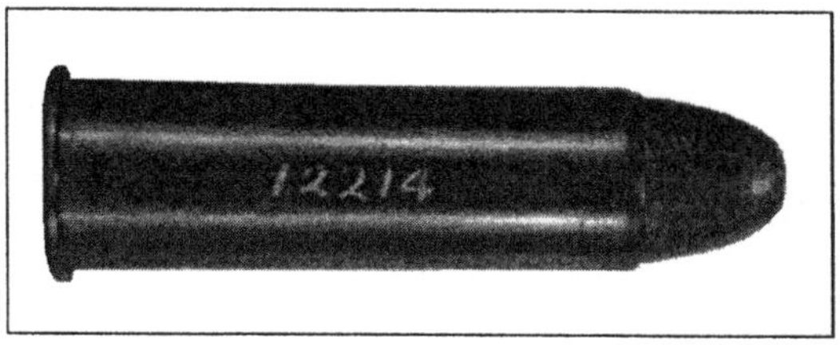

TITLE: 15mm x 49Rmm

HEADSTAMP: /GR/*/*/*

DIMENSIONS:

Rim: .716"
Head: .647"
Neck: .602"
Bullet: .594"
Case: 1.943"

OTHER DETAILS:

Bullet Grains: ?
Powder Grains: ?
Bullet: Lead, Flat Nose
Case Material: Brass
Total Grains: 692

OVERALL: 2.302"

While possibly another Snider round, the small rim diameter of .716" (instead of .735" to .740" would tend to rule out that this cartridge was made for a British Snider conversion. Possibly it was made for some similar conversion originating in Austria where the round was made.

The extremely flat nose bullet is unique in that it is head-stamped!

The "Crown" and letters "H D" intertwined have been impressed into the nose of the projectile. This marking can be found on Waenzl cartridges.

The numbers 8467 have been coded on the side of the case indicating it was originally in the White & Munhall collection.

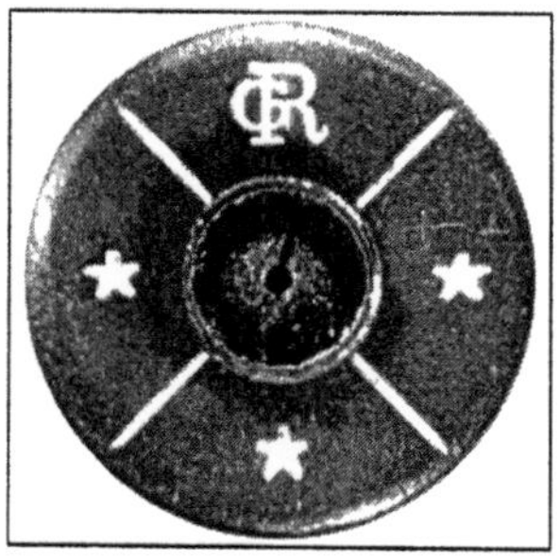

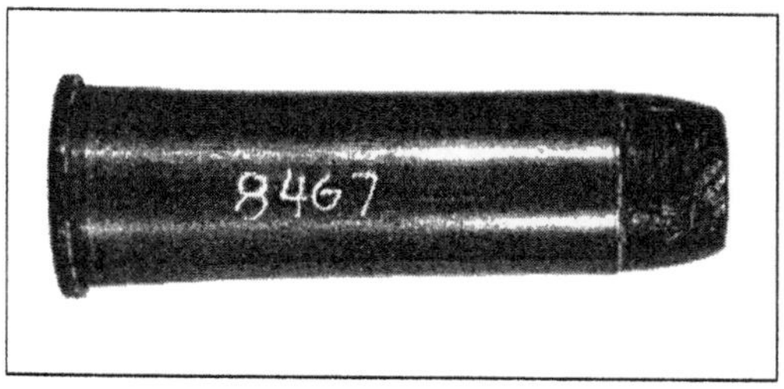

TITLE: 15mm x 49Rmm

HEADSTAMP: Plain

DIMENSIONS:

Rim:	.765"
Head:	.651"
Neck:	.645"
Bullet:	.615"
Case:	1.943"

OTHER DETAILS:

Bullet Grains:	550 (?)
Powder Grains:	70 (?)
Bullet:	Ribbed-Flat Nose
Case Material:	Brass
Total Grains:	872

OVERALL: 2.468"

The brass cased British Snider compares favorably with the 15mm x 49Rmm round but because of the "A" base construction of this cartridge, it is doubtful if it was ever intended for the Snider rifle.

It is more likely that the cartridge was meant for an early made converted muzzle-loader. Note the very large rim diameter and neck diameter of .765" and .645" respectively.

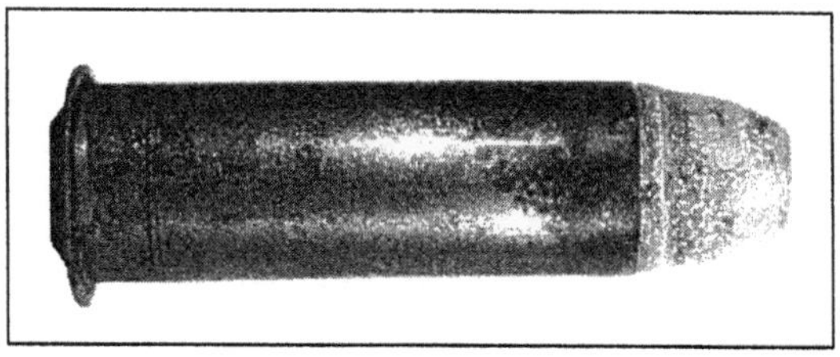

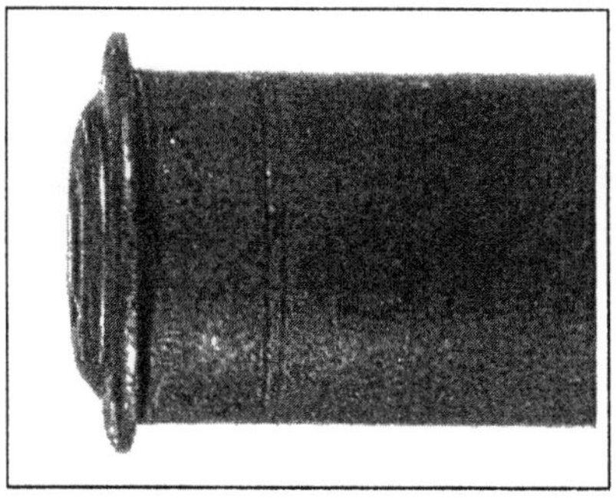

TITLE: .58 DANGERFIELD & LEFEVER

HEADSTAMP: Plain

DIMENSIONS:

Rim: .704"
Head: .635"
Neck: .635"
Bullet: .615"
Case: 2.135"

OTHER DETAILS:

Bullet Grains: ?
Powder Grains: ?
Bullet: Flat Nose - Heavy Band
Case Material: Brass
Total Grains: 942

OVERALL: 2.861

This round is for a comparatively unknown weapon patented on September 3, 1872.

The arm was supplied with both rifled and smooth bore barrels; hence the ammunition, while generally found as simply primed empties, will occasionally be found loaded. By supplying the loading tools with each weapon, the shooter then had his choice of loading shot shells or molding his own bullets to put into the cases for use in the rifle barrel.

All the cartridge cases I have examined have been made by U.M.C. and constructed with a reinforced head and three hole Berdan primer. There is a great similarity between these cases and the common 2 1/8" Winchester or U.M.C. case. The empty cases weigh 164 grains, the U.M.C. brass weighs 140 grains. The heaviest case is the W.R.A. Co. which weighs 196 grains.

It is the author's opinion that Winchester might have copied their shot shell case from the D & L weapons set because they did not offer the .58 Shot Shell case to the public until 1881. The patent for the D & L was issued late in 1872.

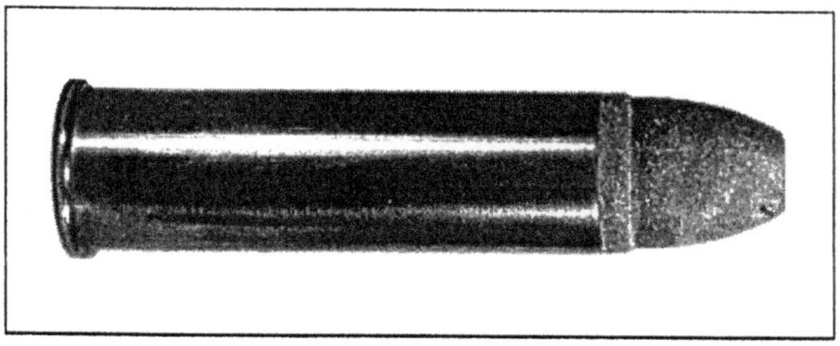

TITLE: 15mm x 57Rmm BOARD DUMMY

HEADSTAMP: Plain

DIMENSIONS:

Rim: .740"
Head: .660"
Neck: .637"
Bullet: .606"
Case: 2.217"

OTHER DETAILS:

Bullet Grains:	?
Powder Grains:	?
Bullet:	Flat Nose
Case Material:	Brass
Total Grains:	N. A.

OVERALL: 2.698"

This specimen came out of the White & Munhall collection as evidenced by the file # X359 scratched on the side of the case.

Unfortunately, their reference file gives no clue as to whether or not this or many of their other cartridges were run-of-the-mill variety or of an early experimental nature.

There are two holes drilled in the side of the case which leads me to believe the round was originally for display purposes.

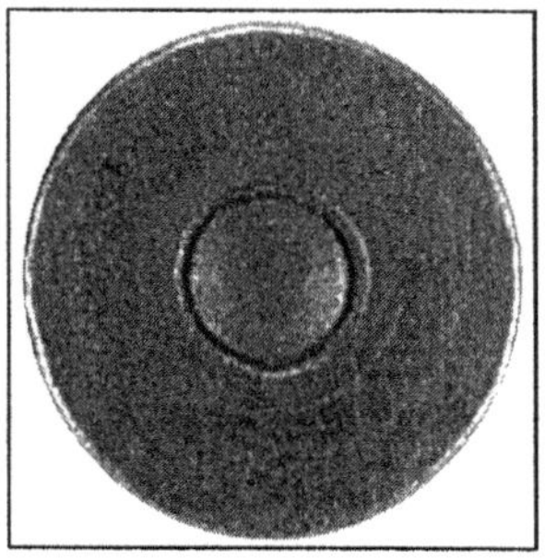

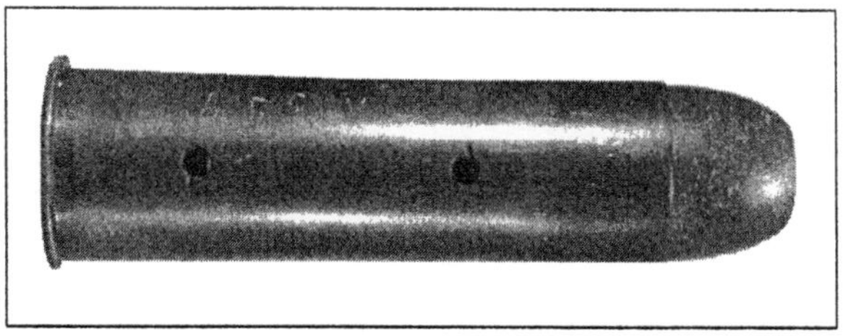

TITLE: .577 – THREE INCH

HEADSTAMP: B.E.L.L. .577 BASE N. E.

DIMENSIONS:		OTHER DETAILS:	
Rim:	.745"	Bullet Grains:	-0-
Head:	..674	Powder Grains:	-0-
Neck:	.605"	Bullet:	Jacketed
Bullet:	.592"	Case Material:	Brass
Case:	3.000"	Total Grains:	

OVERALL: 3.682"

A modern day three-inch case length.

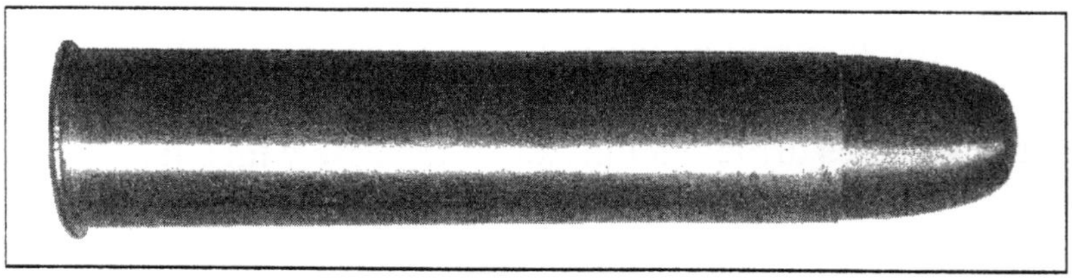

TITLE: 577 TYRANNOSAUR

HEADSTAMP: A-SQUARE 577 TYRANNOSAUR

DIMENSIONS:

Rim: .686"
Head: .689"
Neck: .611"
Bullet: .588"
Case: 3.000"

OTHER DETAILS:

Bullet Grains:
Powder Grains:
Bullet: Jacketed, Round Nose
Case Material: Rmls. Brass, Necked
Total Grains:

OVERALL: 3.690"

A modern day 577 cartridge.

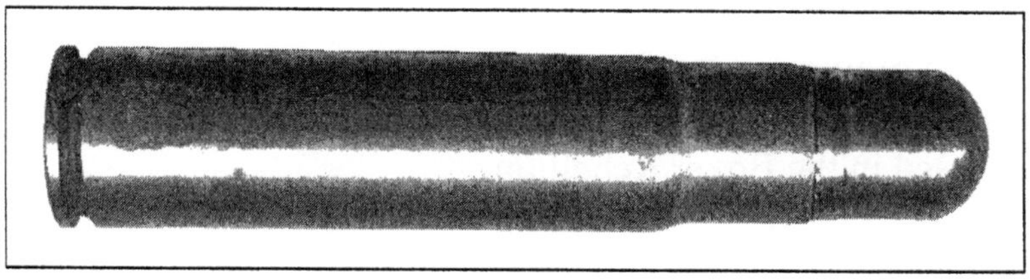

TITLE: .60 CHINESE JINGAL

HEADSTAMP: CHINESE MARKINGS

DIMENSIONS:		OTHER DETAILS:	
Rim:	.907"	Bullet Grains:	1.420
Head:	.817"	Powder Grains:	?
Neck:	.641"	Bullet:	Lead, Tapered
Bullet:	.585"	Case Material:	Brass
Case:	3.637"	Total Grains:	N. A.

OVERALL: 4.600"

The word "Jingal" has a literal translation of "two man matchlock" weapon. The action required to handle this long case would require a large arm. One can readily understand the need of two men to shoot it. It is likely that these guns were used as wall or rampart guns. The bolt action gun and its companion round date back to the early 1890s. They were used in the Boxer Rebellion.

The head-stamps vary from cartridge to cartridge. It is the opinion of the author that the place of manufacture was West Peiyang, China.

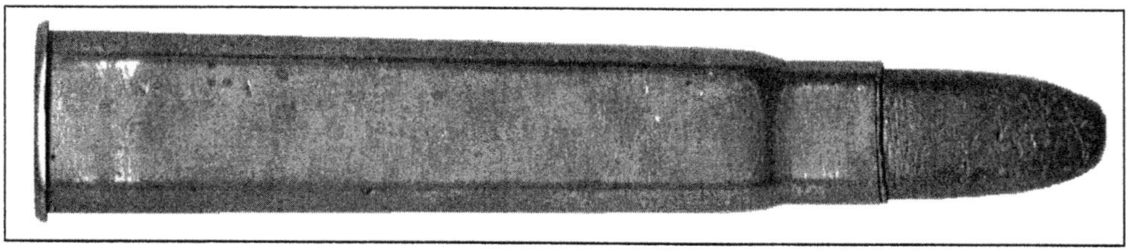

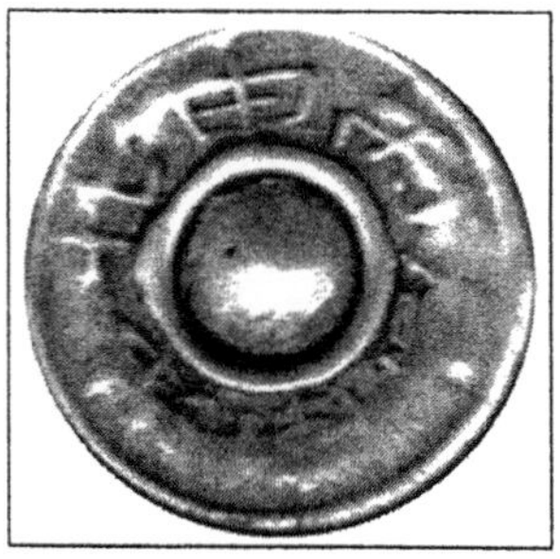

MORSE CARTRIDGES (HITCHCOCK & MUZZY)

George Washington Morse is a name to be forever remembered in the annals of firearm and cartridge invention and manufacture. As is frequently the case, his contributions to firearm and cartridge design were not appreciated until after his death (in 1888).

He can lay claim to a number of firsts: his .55 caliber rifle was the first breech-loader to use a metallic cartridge ever made at Springfield Armory; the gun had an inside hammer, a rebounding floating firing pin and a spring loaded double hook extractor-all features years ahead of their time.

George Morse was a nephew of the famous inventor of the telegraph, Professor Samuel F. B. Morse; hence inventiveness was part of his background.

Morse left his home in South Carolina in the mid 1850s to seek out Hitchcock & Muzzy, manufacturers of iron and steel rifle and shot barrels, located in the Merrifield Building, at the corner of Union and Exchange Streets in Worcester, Massachusetts.

A diligent examination of Worcester City directories revealed that Morse resided at Prescott Street in 1853 and at 4 Grove Street. His occupation was listed as a machinist. Histories of George Morse would indicate that, in Baton Rouge, Louisiana, he first obtained his patent papers #15995 and #15996 on October 28, 1856. He then came North to find someone to make his rifle and cartridges. Not so. From the dates of residency, it would appear that the experimentation work was done in Worcester and then he went to Louisiana to obtain the patent papers. What was his reason for not having the papers prepared in the North? Did he feel that war was inevitable even at this early date (1855-1856) and that being a Southern sympathizer he wished to register his invention in the South? Perhaps he felt he would be better protected and the South would benefit from his arms? No one knows.

George Morse remained in Worcester and Lowell, Massachusetts only a short time and then moved to Springfield, Massachusetts to be close to the activities surrounding the manufacture of his arms and cartridge cases. In 1861 he lived in Washington, D. C. at 185K Street.

Work had been started on his rifles at Harpers Ferry Armory for only a short time when in March 1861, the machinery, along with Morse, was transferred to the Tennessee Armory where he became Superintendent of Works. The approach of Northern troops required another move, this time to Atlanta. And, finally, with the inevitable fall of the South, George Morse lived in Columbia, South Carolina, where he formed the Morse Arms Company. He operated with the remnants of the machinery that had been captured from Harpers Ferry until 1865. It is said that he was able to produce enough of his Confederate Morse breech-loading, brass framed carbines to equip a company of Confederate soldiers.

At the close of the Civil War, he returned to live in Washington, D. C.. He soon brought suit for infringement of his patents relative to the alteration of muzzle-loading system and infringement of his patents on cartridge making. He sought damages of $5.00 for each of 130,000 rifles from the Government and Winchester as joint defendants. After much delay, the suit was found in his favor. It was not until after his death on March 28, 1888 that a cash settlement was made to his widow.

Perhaps because of his sympathy with the South little recognition was given to him. His death notice in the New York Tribune simply stated:

> "Washington, March 28, Col. G. W. Morse, the inventor of the Morse cartridge and breech-loading gun, and a nephew of Professor Morse, who invented the telegraph, died at his home in this city today. His funeral will take place Saturday."

Thus, the passing of a prolific inventor whose ideas were unappreciated and years ahead of his time. It is only in reviewing his work that we realize just how much he had to offer. Had not the Civil War interfered, who can tell what further developments in arms and ammunition might have resulted?

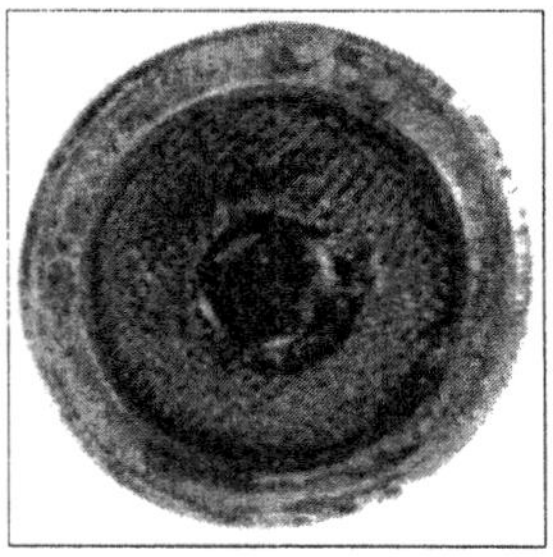

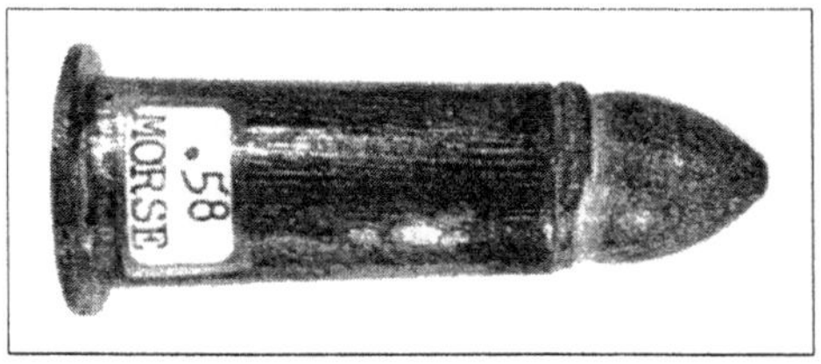

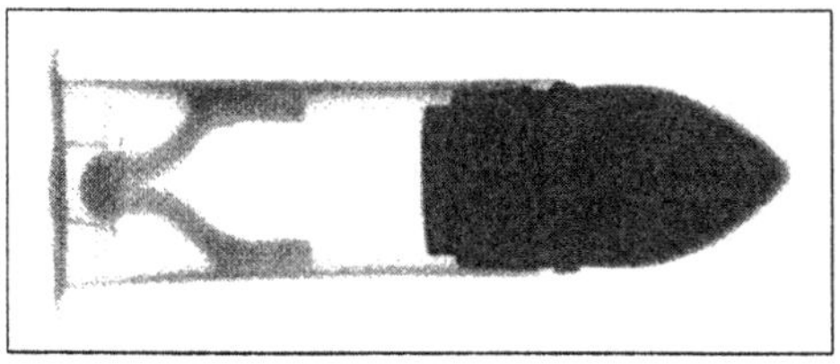

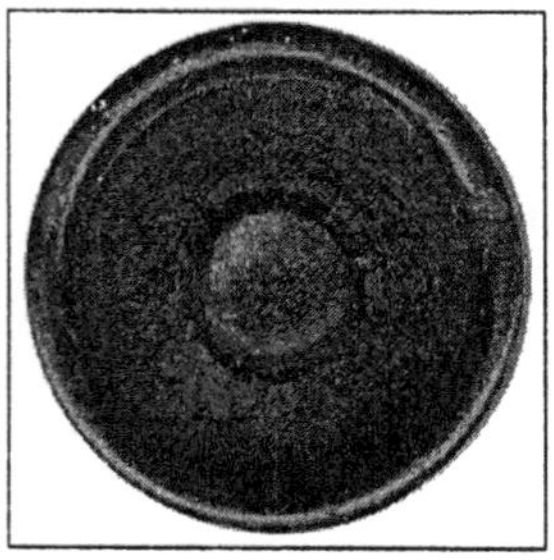

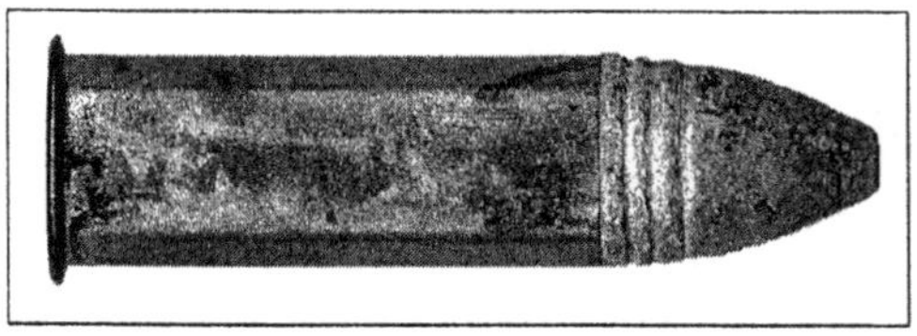

UNITED STATES PATENT OFFICE.

GEORGE W. MORSE, OF BATON ROUGE, LOUISIANA.

IMPROVEMENT IN CARTRIDGES.

Specification forming part of Letters Patent No. 15,996, dated October 28, 1856.

To all whom it may concern:

Be it known that I, GEORGE W. MORSE, of Baton Rouge, Louisiana, have invented a new and useful breech and prime-sealing (capable of being readily reprimed) and automatically-retractible cartridge shell or case for breech-loading fire-arms or cannon, light enough for transportation, yet strong enough to resist the blow of the hammer; and I do hereby declare the following to be a full, clear, and exact description of the same, reference being had to the annexed drawings, made part of this specification, and lettered to correspond therewith, similar letters referring to similar parts in the several figures.

The nature of my invention consists in providing a soft-metal cartridge-case with a priming apparatus, whereby I seal the breech and vent of a breech-loading gun at firing as effectually as if the breech were solid metal around and back of the bore, leaving no escape for gas except at the muzzle. My cartridge-case is also capable of being automatically withdrawn, whether fired or not fired, and is so constructed that sufficient resistance is made to the blow of the hammer so as always to effect ignition if the primer does not fail, and if this does fail, it can be readily removed or replaced. My cartridge-case is also capable of being practically useful for breech-loading cannon as well as small-arms.

Figures 1, 2, 3, 4, and 5 are central and longitudinal sections, in isometrical perspective of the cartridge-case and prime-sealers.

a is the cylindrical part of the shell; *f*, the expansible portion of the after part, and made of sufficient substance not to burst and leak at the breech-joint; *l*, the unchangeable part for automatic retraction; *h*, the outer part of the after part of the shell, being a head-piece of hard material, affording the unchangeable part, as at *l*, in some instances, or set in the soft material as in Fig. 2. *c* is a copper eyelet, by which a thin cylinder may be attached to the hard head-piece *b*, as in Fig. 1, the substance at *f* being furnished by a leaden ring or a wad, as in Fig. 5. *d* is the touch-hole. In Fig. 5 the cylinder *a* is made to clasp the piece *b*; but the reverse of this may be used, the piece *b* being grooved to receive the cylinder *a*. In Fig. 3 the soft material is attached to the hard material by being forced through holes *g* in the piece *b*. In Fig. 4 the soft material is cast around the hard material, consisting of one or more pieces of tin. In Fig. 4 the primer is shown in position, with the priming between it and the hard material *b*, the copper disk *e* being larger in diameter than the percussion-pin driven against it by the hammer, and the edge of the disk having a rim to fasten it to the soft material, so that if one primer fail another can be set in the same place and fixed there. *e*, Fig. 2, has the same characteristics, and *e*, Fig. 1, is made to clasp the head-piece *b* when *b* forms the part for automatic retraction. Fig. 3 shows the copper disk *e*, made to adhere to the outer side of a disk, *k*, of paper, of the size of *h*. On the inner side of *k* the priming *i* is fixed, and the disk *k* is pasted or glued to *e*, where it clasps the piece *b*. This method may be used with all the other figures.

When the gun is fired the part *f* swells out and seals the breech-joint; the part *b* resists the hammer-blow; the disk *e* or *k*, together with the after part of the cartridge-case, is driven back against the breech-piece, and, being broader than the percussion-pin, or the aperture for the passage of the pin through the sliding breech-piece, seals around it, so that the more forcibly the cartridge recoils the more securely is the vent sealed, there being no vent-hole through the copper disk. Finally, the part *l* remains unchanged, whether the cartridge be fired or not fired, and allows the cartridge-case to be automatically withdrawn from the gun. If the primer fail to fire, another can readily replace the one failing.

Having thus fully, clearly, and exactly described and represented the nature and scope of my invention, I would state that I am aware that a metallic cartridge-holder of elastic metal has been invented and used for preventing the escape of gas through the joints of breech-loading guns; but in this case the gas was found to escape at the vent, and also between the case and the bore of the gun.

I am also aware that a cartridge has been made so as to contain the powder in the cavity of the projectile, and to have the powder covered by a cylindrical cap of metal, which contained the percussion priming, and fitted over the cylindrical part of the projectile, while a small orifice in this cap permitted the

G. W. MORSE.

Cartridge-Case.

No. 15,996 Patented Oct. 28, 1856.

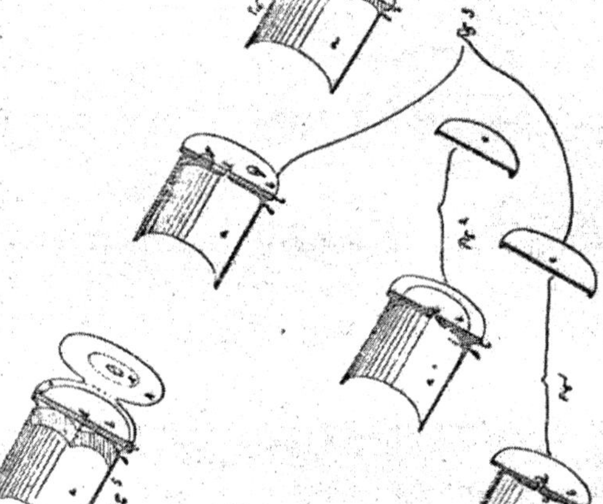

UNITED STATES PATENT OFFICE.

GEORGE W. MORSE, OF BATON ROUGE, LOUISIANA.

IMPROVED CARTRIDGE-CASE.

Specification forming part of Letters Patent No. 20,214, dated May 11, 1858.

To all whom it may concern:

Be it known that I, GEORGE W. MORSE, of Baton Rouge, in the State of Louisiana, have invented an Improvement in Cartridges or Cartridge-Cases for Breech-Loading Fire-Arms; and I do hereby declare that the same is fully described and represented in the following specification and the accompanying drawings, of which—

Figure 1 is an external or side view of a metallic cartridge-case constructed on my improved plan. Fig. 2 is a rear end view of the same. Figure 3 is a longitudinal section of it.

In these drawings, A denotes the tubular case or body of the cartridge, such case being formed open at both ends and with an annular shoulder, *a a*, extending around its rear end, as shown in the drawing. B is an expansible head, which is made of thin metal, and somewhat in the shape of a cup, and so as to slide and fit into the rear end of the tubular case A. Within the said case A there is arranged a furcated tige, C, which consists of a wire bent, as shown in Fig. 3, and having the two extremities of its prongs *b b* soldered or otherwise properly fastened to the internal surface of the case A. The middle part or nipple of the tige may be formed so as to receive a percussion-cap, as the nipple of a gun-lock usually receives the same.

Preparatory to loading the cartridge a percussion-cap is to be placed on the tige, or, instead thereof, the inner surface of the expansible head may be coated with percussion-powder or paste. This having been done, the expansible head B is to be inserted within the case, and the latter filled with a charge of powder and shot or ball, as circumstances may require. After this has been accomplished the cartridge is to be inserted within the open breech of a gun-barrel in such a manner that the shoulder on the case or body of the cartridge may rest against the rear end of the barrel. If we suppose the gun to be furnished with a hammer arranged so as to strike on the expansible head or against it when the trigger of the gun-lock is pulled backward, a part of such head will be driven against the tige, and so as to cause the percussion-powder to be exploded and fire the load of the cartridge, the lateral expansion of the head B, within the body A, being sufficient to prevent the smoke and flame of the discharge from escaping back between them, and also to expand the rear part of the body A, so as to perfectly fit the bore of the gun.

By constructing the tige and applying it to the body in the manner as described, its prongs serve as braces, or operate somewhat like an arch to hold the tige in position, and still allow of the lateral expansion of the case A, while they maintain it within the case to much better advantage than it would be by a straight bar extending across the case and having a tige projecting from its middle part.

I do not claim a cartridge made so that the whole body of the same may expand under the force of an explosion, as such is the ordinary mode of making a cartridge for the well-known saloon rifle or pistol.

Nor do I claim a cartridge-case having an expansible head or closer separate from the body, and constructed and applied to the body, so as to expand laterally during the explosion of the cartridge and make a close joint around the sides of the said head or closer.

Nor do I claim a furcated tige, applied within a case or body of a projectile, as shown in Riera's British patent of May 6, 1853.

Nor do I claim the application of a tige and a movable head to a cartridge, as exhibited in in the French patent of Chaudun, described in Brevet d'Invention, new series, Vol. 13, and represented in Plate 2 (Figs. 24 and 29) of said volume.

My invention differs from all these, inasmuch as its expansible head or closer is not made to encompass or encircle the outer shell of a cartridge-case, but is arranged within the same, and so that it may slide freely toward and away from the tige without any obstruction from the rear end of the cartridge, whereas as expansible heads have heretofore been applied, they have been arranged so as to abut against the rear end of the cartridge-case and encompass the said rear end, serving to arrest their forward movement.

Furthermore, my improved cartridge-case is formed with a shoulder, *a a*, to abut against the end of the gun and to hold the body of the cartridge within and prevent it from being discharged from the gun when an explosion of its charge takes place.

My arrangement of the cartridge-closer B,

G. W. MORSE.

Cartridge.

No. 20,214. Patented May 11, 1858.

UNITED STATES PATENT OFFICE.

GEORGE W. MORSE, OF BATON ROUGE, LOUISIANA.

IMPROVEMENT IN CARTRIDGES.

Specification forming part of Letters Patent No. 20,727, dated June 29, 1858.

To all whom it may concern:

Be it known that I, GEORGE W. MORSE, of Baton Rouge, in the State of Louisiana, have invented a new and Improved Mode of Manufacturing Cartridges for Breech-Loading Fire-Arms; and I do hereby declare that the following is a full and exact description thereof, reference being had to the accompanying drawings, and to the letters of reference marked thereon.

My invention relates to that class of cartridges for breech-loading fire-arms which has a metallic case, and which is inserted in the rear end of the barrel; and it consists in the mode or modes of providing the metallic case with a tige, secured in the interior of the same, as hereinafter described, and extending toward the rear of the cartridge, so as to serve as a means of exploding the percussion-cap when it is struck by the dog or hammer of the gun.

It also consists in the combination and arrangement of the percussion-cap, with a perforated disk of caoutchouc, or some other equivalent elastic material, as described, whereby the contents of the cartridge-case are well secured against moisture, the cap protected from accidental explosion during transportation, and the cartridge readily exploded by the hammer whenever it is desired to discharge the gun.

To enable others skilled in the art to make and use my invention, I will proceed to describe its construction and operation.

I construct my cartridge-case, in the usual manner, of brass, copper, tin, or any other suitable metal or alloy, open at both ends, and having a slight flange on the rear end, to facilitate the withdrawal of the cartridge from the barrel of the gun, as shown in section in Figures 1, 5, and 6. I then secure the tige in the case in one of the three different modifications of the same contrivance, as illustrated in Figs. 5, 1, and 6. In Fig. 5 the tige is formed of a strong wire, bent as shown in the figure, and soldered at the ends to the inner sides of the cartridge-case.

Fig. 2 represents the tige, as shown in Fig. 1, where it is also secured to the inner sides of the case by solder.

Fig. 7 represents a plan of the tige, which is shown in position, in section, in Fig. 6. The tige having been secured in the cartridge-case in either of the three modes above described, and illustrated in the drawings, I take the disk of caoutchouc, prepared as illustrated in Fig. 3, and carrying the percussion-cap, and insert it in the rear of the cartridge-case, pressing it firmly into its place until the inner portion of the percussion-cap shall reach, or nearly reach, the end of the tige. The cartridge-case is then ready to receive the powder and ball, which are inserted in the usual manner of loading cartridge-cases.

It will be observed that the percussion-cap does not extend entirely through the perforated disk, but that the body of the disk projects beyond it, thereby protecting the cap from accidental blows during transportation, and obviating one of the greatest objections to self-primed cartridges.

Having thus fully described and shown the character of my invention, I will now proceed to state what I claim as new, and for which I ask Letters Patent.

I claim—

1. The tige secured in the cartridge-case, in either of the modes above described, and all equivalents thereto, for the purpose above mentioned.

2. The combination and arrangement of the percussion-cap and perforated disk, as described, and for the purpose above mentioned, and any and all equivalents thereto.

In testimony whereof I have signed my name before two subscribing witnesses.

GEO. W. MORSE.

Witnesses:

N. CALLAN,
A. HERBERT.

G. W. MORSE.

Cartridge.

No. 20,727. Patented June 29, 1858.

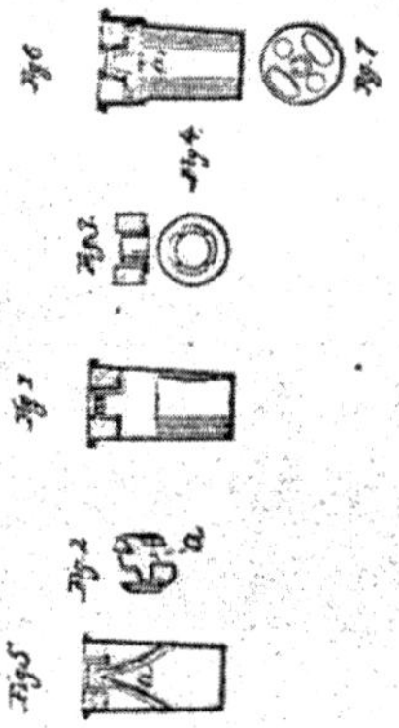

HITCHCOCK & MUZZY

A most elusive firearms manufacturer!

It is known that George Morse came from South Carolina, supposedly seeking Hitchcock & Muzzy to manufacture his newly patented rifle and ammunition. I qualified the statement by using the phrase "supposedly seeking" Hitchcock & Muzzy; after much diligent searching, little information could be uncovered regarding the history of this firm. Therefore, it is most difficult to believe that they would have been so popular and well known for Morse to have knowledge of them as far away as South Carolina.

An exhaustive search of reliable information in Worcester, Massachusetts revealed only their small ad in the City Directories, which indicated they did business in the Merrifield Building on the corner of Union and Exchange Streets. The ads stated simply that they specialized in iron and steel rifle and shot barrels. These advertisements appeared in City Directories for the years 1854 through 1857.

As previously pointed out, George W. Morse is listed in the 1853 Worcester directory as a mechanic residing at Prescott Street, therefore, it would appear logical that Morse first perfected his invention with the assistance of Hitchcock & Muzzy, and when he felt satisfied with the gun and related cartridge, returned to Baton Rouge, Louisiana to secure the necessary patent papers.

One cannot help but wonder what part Ethan Allen's firm might also have played in the development of Morse's firearm. The Merrifield Building was a complex containing other manufacturing firms. Two of the tenants in this building were Frank Wesson and Ethan Allen.

Muzzy entered into a contract with George Morse in 1858 to manufacture 100 cased sets containing one stock and action, three interchangeable barrels, cartridges and loading accessories. There was a .54 caliber carbine; a .69 caliber (16 gauge) shot barrel, and a .50 caliber rifle. All components utilized the same case rim diameter but had different sized washers for Morse's Type I cartridges (patent #20727, dated June 29, 1858). These weapons are not military items but commercial firearms for the sportsman.

The manufacture of shoe cutlery was an important part of Worcester's economy as early as the 1830s. Therefore, a number of skilled machinists were working in the Worcester area and it was a short step from being employed in this business to switching to the manufacture of firearms. By 1850, about the time when Morse appeared on the scene, an array of Hitchcocks, Muzzys, Stowes and Careys all appear in the local directory as mechanics employed in or operating shoe manufacturing businesses. These individuals connected with the firearms manufacturer also had relatives in the shoe manufacturing business.

A list of those individuals is as follows:

NAME	TITLE	DATES
Carey, George	Armorer	1854
Carey, Joseph W.	Armorer	1855-1862
Carey, William A.	Armorer (H & M Co.	1857-1859
Carey, L. H.	Armorer (A & W)	1860
Hitchcock, Dexter	Armorer	1851-1859
Hitchcock, David	Boots, Shoes	1860-1869
Hitchcock, Jeremiah Evarts	Armorer (H & M Co.)	1857-1862
Hitchcock and Muzzy	Barrel Manufacturer	1854-1862
Morse, George W.	Machinist	1853-1854
Muzzy, Edwin A.	Shoe Manufacturer	1857-1862
Muzzy, Nathan M.	Armorer (H & M Co.)	1851-1862
Muzzy, Nathan (father of Nathan M. Muzzy)	Overseer of the Poor	1853
Paine, J. P.	Armorer (H & M Co.)	1857
Stowe, Addison	Armorer	1852-1855
Stowe, Luther & Co.	Shoe Manufacturer	1857-1862

Confusion surrounds Hitchcock & Muzzy: the year 1857 was the last date their ad appeared in the city directory yet various names in the list indicated that they were employed as late as 1862 by that firm. This is borne out by the record showing that Nathan M. Muzzy was still connected with Hitchcock & Muzzy until 1862. An examination of the documents from 1858 through 1863 failed to show any mention of the name Hitchcock & Muzzy. Nathan M. Muzzy was again to appear in business with the firm of Muzzy & Trumbull at 45 South Canal Street, Chicago for the period from 1882 through 1884. After that date, Nathan Muzzy remained in business by doing gunsmithing himself until 1889.

One cannot help but wonder how Morse happened upon the firm of Hitchcock & Muzzy. It was a little known firm, when one considers such names as Allen & Wheelock and Frank Wesson resided in the same city and manufactured firearms in the same building. Why Muzzy and not the others? Allen & Wheelock certainly had the expertise in cartridge making to be one of the first successful producers of reliable self-contained ammunition. In another chapter, we will find that Hitchcock & Muzzy could not properly "draw" a cartridge case, but had to resort to casting their (Morse's) bronze cases.

It is some source of amazement that so little is known of this firm that contributed in so many ways to the pioneering of arms and ammunition just prior to the Civil War. It just does not make sense that someone, somewhere did not write a history of the company and its involvement with such an important firearms inventor. How is it possible for such information to elude us? Even an examination of all the early Scientific American issues dating back to 1855 failed to make mention of the firm or the inventor, George W. Morse.

HITCHCOCK, MUZZY & CO.,

MANUFACTURERS OF

CAST STEEL AND IRON

Rifle and Shot Gun Barrels,

Merrifield's Buildings, Cor. Union & Exchange Sts.,

WORCESTER, MASS.

Cast Steel and Iron Target and Hunting Rifle Barrels,

OF EVERY DESCRIPTION:

CAST STEEL AND IRON SHOT BARRELS,

FOR BOTH SINGLE AND DOUBLE GUNS:

Fine Twist Rifle and Shot Barrels:

Heavy large bore Shot Barrels,

FOR DUCK GUNS AND SEA SHOOTING:

RIFLE TELESCOPES OF A SEPERIOR QUALITY.

Thankful for the former patronage of our customers, we beg leave to present the following inducements for its continuation:—

The superiority of well made Steel Barrels over Iron, for durability, force in shooting, and fineness of finish, is now well established. These advantages are fully secured only by the use of the best material, and by scientific treatment in its preparation and manufacture into barrels.

The STEEL we use is manufactured to our order of the best stock, expressly for barrels. Having made sure of the material, we have peculiar advantages for annealing it to the proper condition for working, without, at the same time, destroying its life; a point which every good Rifle-maker knows to be of vital importance. Constant and careful practical experiments, under various circumstances, combined with many years' previous experience in both Barrel and Rifle Manufactories, have enabled us effectually to secure that perfectly even temper, which best fits it to be wrought into the fine shooting rifle.

OUR IRON barrels are also made of the clearest and strongest Iron—made and imported to our order, sound and free from greys; *and we will warrant them* SUPERIOR *to any other Iron barrels made in the country.*

We have spared no pains to secure the permanent services of the most experienced and skillful borers and straighteners, so as to make the shooting qualities of our barrels perfectly reliable.

Every process in the manufacture is constantly under our immediate supervision, and subject to our personal inspection;—so that deficiency in any part is at once noticed and remedied.

Our department for Rifling and Muzzling nice Target Barrels, is under the direction of an OLD and EXPERIENCED RIFLE-MAKER, whose long practice in making and shooting the Target Rifle, has enabled him to ascertain and apply the GAIN TWIST most appropriate to every variety of bore in order to the most effective shooting at long distances.

Our method of boring Shot Gun barrels to improve their shooting, has been tested by many of the leading sportsmen, and has proved highly satisfactory. We have also received the voluntary testimony of the most celebrated Riflemakers in various sections of the country, as to the superiority of our barrels.

THE FOLLOWING IS A

LIST OF OUR CASH PRICES:

CAST STEEL BARRELS, of usual proportions,	$3.50	each.
" " " weighing over 6 lbs.,	.40	per lb. will be added.
IRON BARRELS, WELDED, of usual proportions,	2 25	each.
" " " weighing from 8 to 12 lbs.,	.10	per lb. will be added.
" " " " " 12 to 18 lbs.,	.25	" " "
" " " " " over 18 lbs.,	.40	" " "
" " Drilled from solid bar of Hammered Iron, 10 to 12 lbs.,	3.00	each.
Iron Barrels, drilled from solid bar, weighing over 12 lbs.,	.25	per lb. will be added.
Twisted, plain, of usual proportions,	4.00	each.
" " weighing over 6 lbs.,	.40	per lb. will be added.
Matched Barrels, for double guns, Cast Steel,	8.00	per pair.
" " " " " Plain Iron,	4.50	
" " " " " Twisted,	8.00	
Heavy Barrels for sea shooting, from 20 lbs. upwards,	.50	pr lb.
Rifling Steel Target Barrels,	5.00	each.
" Iron " "	2.00	"
Patent Loading Muzzles,	3.00	"

☞ For putting on cone seat, grooving for rod, extra length, or otherwise unusual proportions, a corresponding additional charge will be made.

Telescopes with Cast Steel Tubes drilled out of the Solid Bar, $25.00.

Telescopes with welded Iron Tubes, $20.00.

Telescopes with Brass Tubes, $18.00.

Putting in Patent Breeches, fitting on Breakoffs, Finishing up Cone Seats, fitting up double Rifle and Shot Barrels, and putting on Sights of all descriptions, done to order, in the best manner.

HITCHCOCK, MUZZY & CO.

WORCESTER, July, 1856.

180 ADVERTISEMENTS.

HITCHCOCK & MUZZY,

MANUFACTURERS OF

IRON AND STEEL

RIFLE AND SHOT BARRELS,

MERRIFIELD'S BUILDING,

(EXCHANGE STREET,)

WORCESTER.

BORING, RIFLING, AND MUZZLING DONE TO ORDER.

ADVERTISEMENTS. 191

HITCHCOCK & MUZZY,

MANUFACTURERS OF

IRON and STEEL

RIFLE BARRELS,

Corner of Union and Exchange Sts.,

Worcester, Mass.

TITLE: .58 ROBERTS BREECHLOADER

HEADSTAMP: Plain

DIMENSIONS:

Rim:	.724"
Head:	.646"
Neck:	.618"
Bullet:	.605"
Case:	1.344"

OTHER DETAILS:

Bullet Grains:	480
Powder Grains:	60-70
Bullet:	Tapered, Pointed
Case Material:	Brass
Total Grains:	687

OVERALL: 2.037"

The head and primer construction resembles the early U.M.C. manufactured Berdan cartridges; and quite likely; this specimen was made by them. The large copper primer is deeply recessed in the base.

The bullet lacks any grease grooves and is much longer and more pointed than other identifiable Roberts rounds.

The numbers 14104 and 2266 inscribed on the case indicate the round was in the White and Munhall collection.

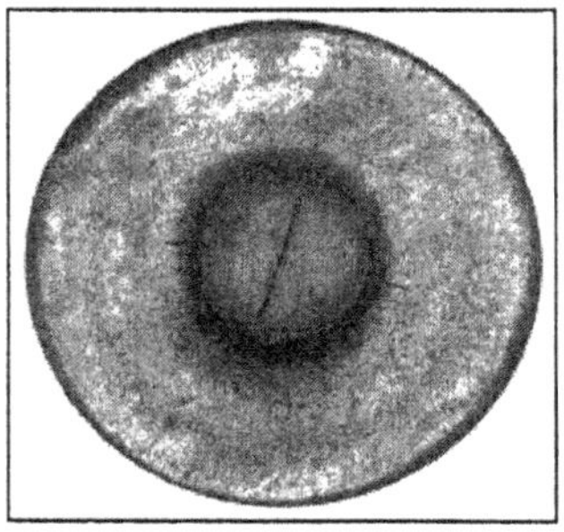

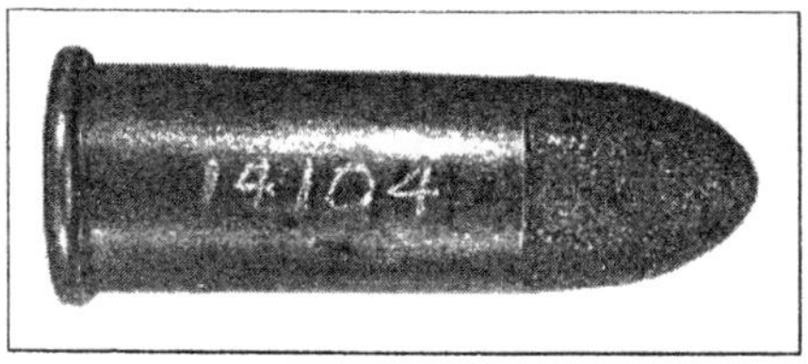

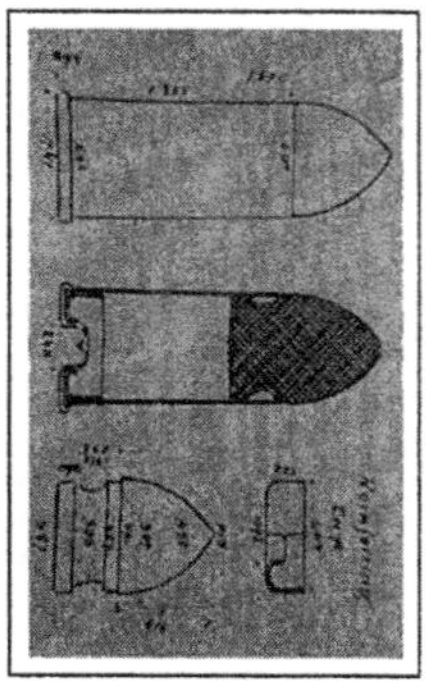

TITLE: .58 ROBERTS BREECHLOADER

HEADSTAMP: Plain

DIMENSIONS:

Rim:	.743"
Head:	.658"
Neck:	.605"
Bullet:	.603"
Case:	1.345"

OTHER DETAILS:

	WRA	UMC	RemUmc
Bullet Grains:	480	527	620
Powder Grains:	60-70		
Bullet:	Pointed		
Case Material:	Brass		
Total Grains:	760		

OVERALL: 1.906"

These cartridges were manufactured under the Roberts Patents of 3/20/1866, 9/29/1868 and the S. W. Woods Patents of 4/1/1862 and the 4/2/1872. It was noted in the pictures of the box labels. The ammunition was used in altered Springfield and Enfield Rifles, and, while the breech system was recognized as successful and reliable, it was never accepted by U. S. Ordnance for issuance to its troops. The State of New York purchased a supply of these converted arms for their own militia's use during the late 1860s.

Credit should be given to Roberts for designing not only a reliable breech system, but designing the first tapered style case. It was done either with the purpose of making the case easier to extract after firing and/or a means of securing the reinforcing cup that was inserted into the base of the cartridge.

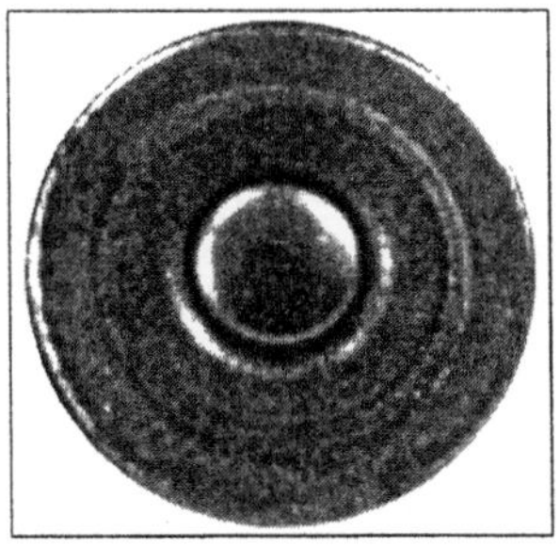

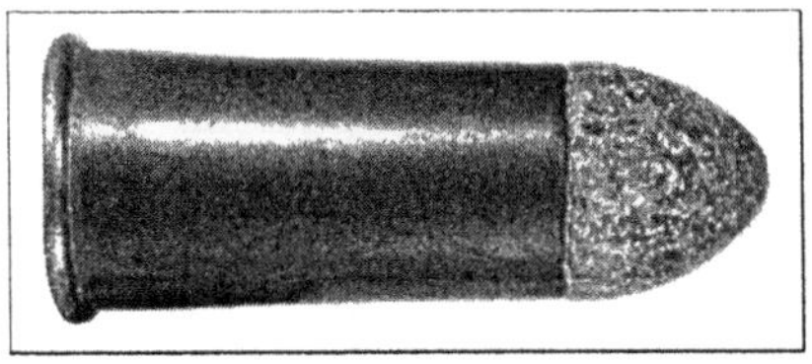

TITLE: .58 ROBERTS-BLANK

HEADSTAMP: Plain

DIMENSIONS:

Rim:	.733"
Head:	.660"
Neck:	.618"
Bullet:	----
Case:	1.375"

OTHER DETAILS:

Bullet Grains:	----
Powder Grains:	60
Bullet:	----
Case Material:	Brass
Total Grains:	250

OVERALL: 1.375"

A blank round for use in the Roberts Breech Loading Rifle.

The mouth has the common roll crimp with a white wad closing the powder.

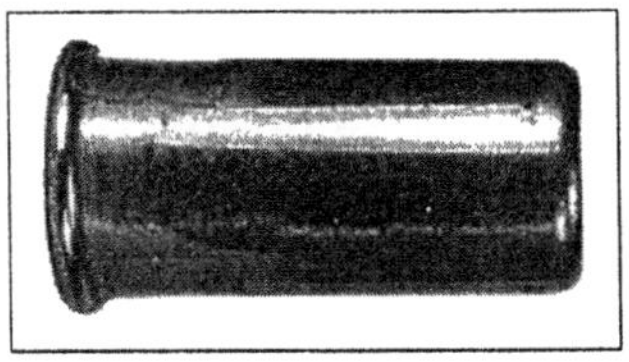

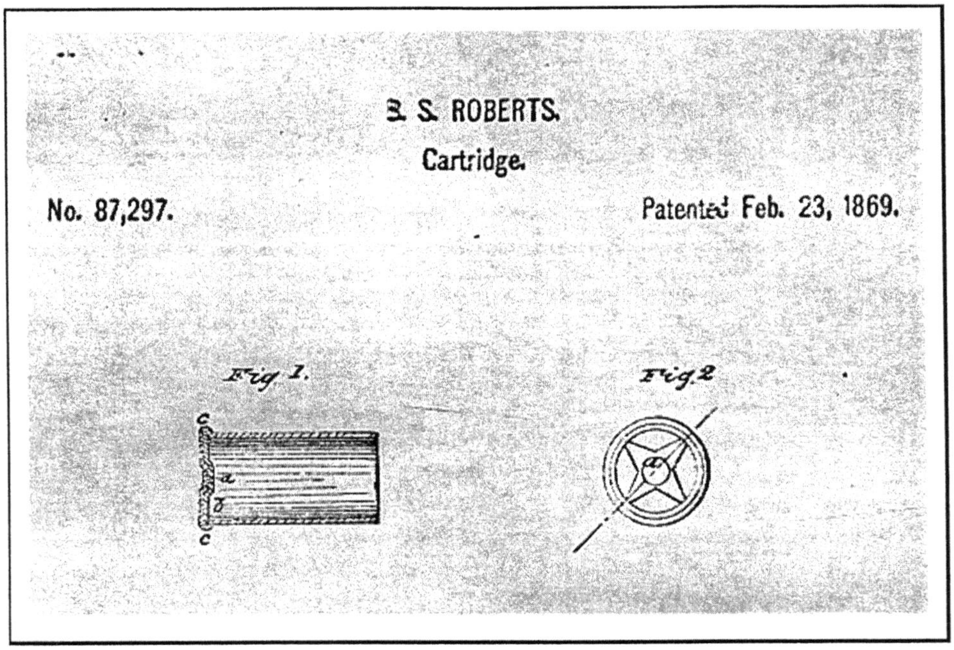

B. S. ROBERTS.
Cartridge.

No. 87,297. Patented Feb. 23, 1869.

TITLE: .61 ROBERTS

HEADSTAMP: Plain

DIMENSIONS:		OTHER DETAILS:	
Rim:	.778"	Bullet Grains:	?
Head:	.679"	Powder Grains:	80-85
Neck:	.625"	Bullet:	Lead
Bullet:	.576"	Case Material:	Brass
Case:	1.640"	Total Grains:	811

OVERALL: 1.958" (with projectile)

The shop records of a Winchester engineer by the name of John Gardner illustrated a drawing of a .65 General Roberts cartridge. Reference is made in the text to an unknown .61 caliber Roberts rifle. Quite possibly the two are related.

The brass case is of two piece construction, a reinforcing cup with a large brass, five hole Berdan primer. This rare round is easily confused with the .58 Roberts or Berdan Musket cartridge. The .61 caliber has a larger rim and a slightly longer case. These characteristics would prevent the round from chambering in any other rifle.

An old U.M.C. cartridge board specimen of very similar dimensions refers to a .58 Madrid and/or Japanese round. Perhaps they are all interchangeable.

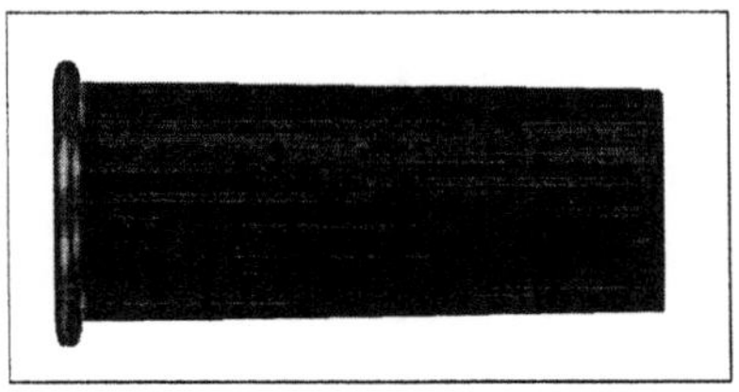

TITLE: .58 WINCHESTER - SHOT SHELL

HEADSTAMP: WINCHESTER .58

DIMENSIONS:	
Rim:	.704"
Head:	.635"
Neck:	.633"
Bullet:	------
Case:	2.125"& 2.500"

OTHER DETAILS:	
Bullet Grains:	m.t.
Powder Grains:	m.t.
Bullet:	m.t.
Case Material:	Brass
Total Grains:	206

OVERALL: 2.125"& 2.500"

The first listing of this round was in the W. R. A. Co. catalog of 1881 under the section listing brass shot shells. They were sold only as primed empties and were intended for use in the old .58 converted muskets. They were last listed for sale in the 1919 catalog.

Old advertisements stated that the shorter case length (2.125") was for the "rifle bullet". No explanation was given as to what the reference meant. Was it a particular rifle chambered for this shorter case or simply a lead round ball loading instead of shot? Except for the case length, the dimensions of both case sizes are identical.

Below left is two distinct sizes of head stamp lettering. In this instance the 2.500" cartridge has the smaller lettering of the two.

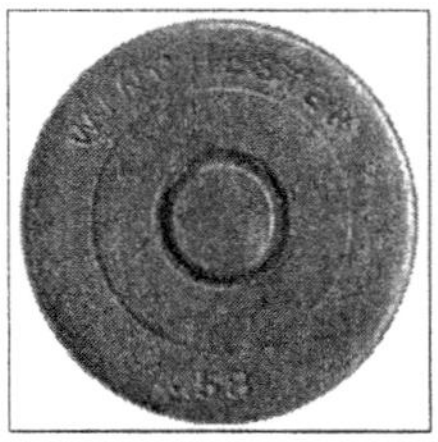

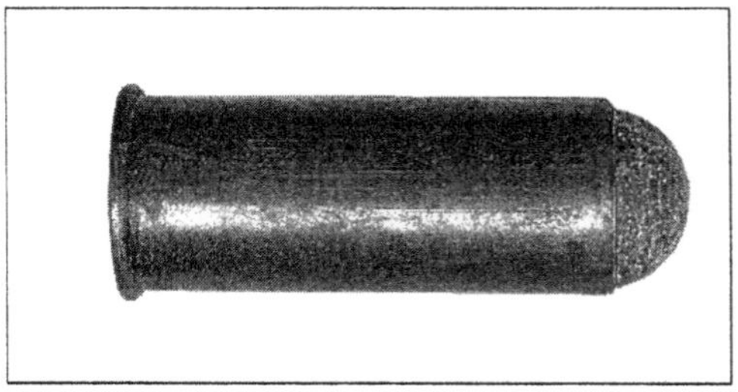

This Shot shell has been altered to accommodate a lead ball projectile. (Overall: 2.0")

TITLE: .58 SHOT SHELL

HEADSTAMP: Plain

DIMENSIONS:		OTHER DETAILS:	
Rim:	.715"	Bullet Grains:	-----
Head:	.636"	Powder Grains:	-----
Neck:	.636"	Bullet:	-----
Bullet:	-----	Case Material:	Brass
Case:	2.130"	Total Grains:	140

OVERALL: 2.130"

This round is a product of U.M.C.. It varies slightly from the better known W. R. A. Co. specimen. The primed empty case is much lighter (it weighs only 140 grains versus 206 grains of the Winchester). The weight difference is obviously due to its inner construction.

The primed empty has a three hole Berdan primer pocket PLUS a center hole.

As in the case of the Winchesters, these U.M.C.'s were sold only as empties and the purchaser was expected to load his own.

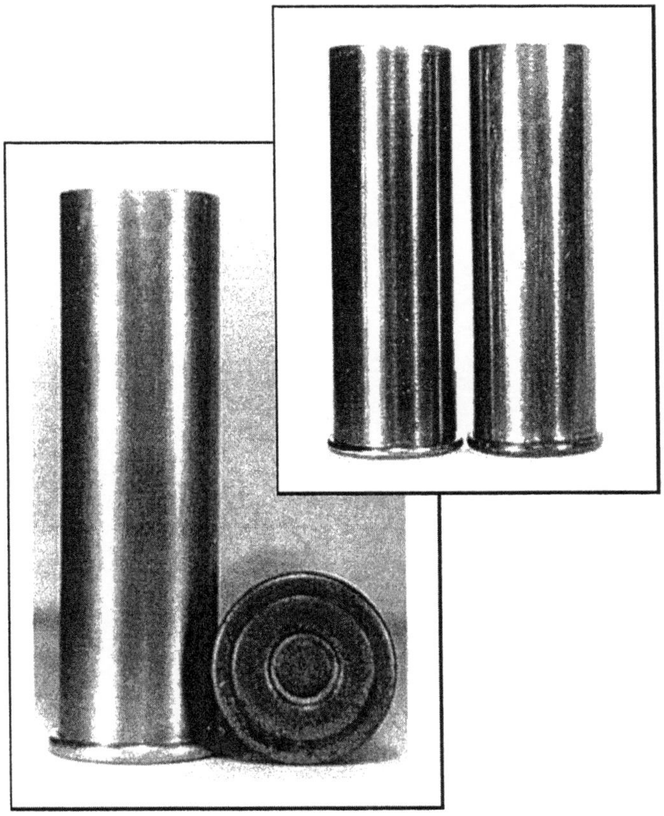

II

Experimental

THE BERGEN PATENT FRONT IGNITION SYSTEM

During the late 1860's and early 1870's there was much experimentation with priming systems in cartridges. Ignition from the base of the shell was the technique generally used. Ammunition makers were concerned about the powder loss (through the barrel) typical of base ignition. They felt that some of the powder was being forced out of the barrel in an unburned state when the cartridge was fired thereby losing the benefit of the entire charge.

Experiments were tried using the front ignition system. This method was tried in the .58 caliber cartridges with limited success.

On May 17, 1864, Cornelius J. Bergen of Brooklyn, New York was granted patent #42815 for a system to adapt a front ignition system to the .58 caliber musket cartridge. The patent was assigned to the Moore's Patent Fire Arms Co. of Kent Avenue and Hawes Street in Brooklyn. This firm changed its name to the National Arms Company on March 9, 1865.

The patent describes the round as a double metallic copper cartridge case with channels extending from the fulminating material to the front portion of powder. The copper cup had a rounded base and corrugated sides. The cup was inserted into a normal primed rim-fire case. In the only specimen known to exist, the case was manufactured by Joseph Goldmark evidenced by the J. G. head stamp. After inserting the powder charge into the cup, the bullet was loaded and the case crimped.

Despite x-ray examination of other J. G. head stamped rounds, this is the only specimen known to exist.

An excellent article on the Bergen Patent Front Ignition system by George Kass, appeared in the August 1975 issue of the Arms Gazette. Due to the efforts of James Tillinghast, John Hintlian, Jerry Teesdale and Dwight Demeritt this information and the pictures of the patent and fired case were made available for study.

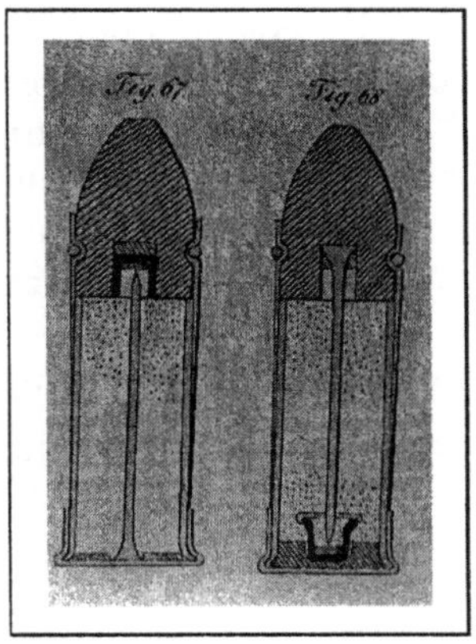
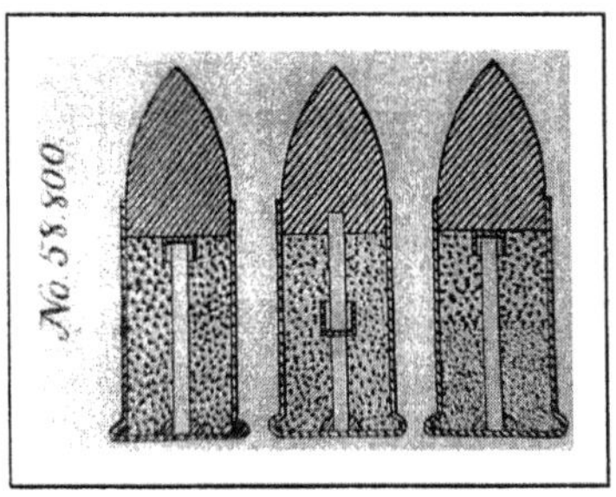

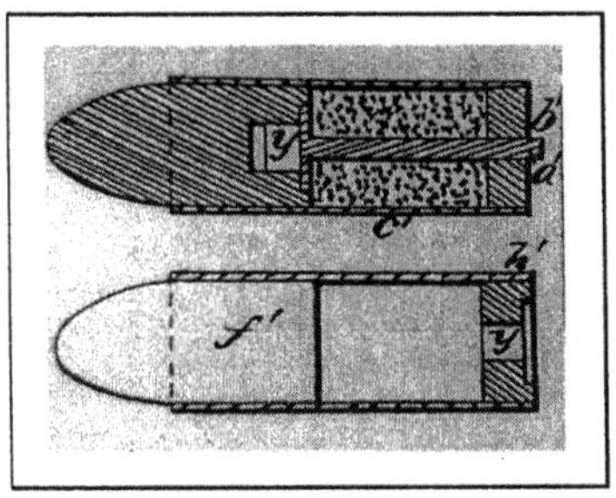
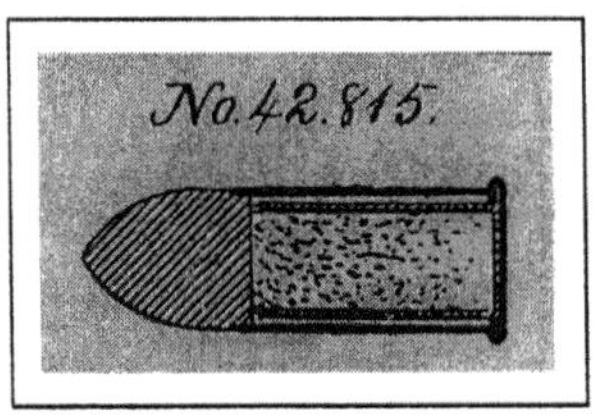

CORLISS EXPERIMENTAL

In a never ending search to find a better means of exploding the propelling charge of black powder in cartridges, experiments were made in America and Europe with placing the primer in the base of the bullet instead of the rim of the shell.

Variations of these experiments were also tried. For example the primer was set in the center of the needle or rod; this was a combination of a primer in the base of the bullet and a primer in the conventional position. The primer cap was also fixed on the front of the rod instead of on the base of the bullet. These ideas combined with the use of two types of powder charges, one slow-burning charge and a charge of faster burning powder near the rim.

These patent variations will be discussed in greater detail later in this chapter.

The Corliss Front Ignition Experimental was developed at the Frankford Arsenal. In Major T. J. Treadwell's report of METALLIC CARTRIDGES, tested at Frankford Arsenal, the tests on the Corliss were described as follows:

> "The Corliss Needle Cartridge for Springfield B. L. Rifle. The needle was attached at the base to a disc crimped into the flange of the case. The point was made to conform exactly with that of the service firing pin. The fulminate was held in a copper capsule in the base of the bullet open to the rear and covered with tin foil. In Consequence of the extremely limited longitudinal motion which the needle experiences from the indentation of the cartridge case, the bullet was carefully gauged in the loader to bring the fulminate almost in contact with the needle point. Notwithstanding the close gauge employed, the first 25 cartridges failed entirely to explode in the gun. On examination, the fulminate was found to have been forced into the bullet instead of being crushed by the needle point. 30 more cartridges were made with a needle flat point fronted and square at its apex like the arbor of a watch; which it was thought would effectually explode the fulminate. Out of 18, 5 of them exploded. Average velocity of service 1320'; needle 1.100. As was to be expected, the needle cartridge gives a higher velocity, but in uniformity is much below the service. A target of 20 shots, were attempted to be fired; 13 shots were secured in comparison with service. Practice at 500 yards. Mean absolute deviation, Service 1'856; needle 2'169; corrected angle of sight, service 1 21'11", needle 1 21'41". Fouling from 14 shots, service c grs.; needle 5 grs. Of the above 13 needle cartridges, 1 cartridge exploded on the 1st blow, 1 on 2nd blow, 6 on 3rd, 2 on 4th, and 1 on 5th blow. Five needle cartridges also failed entirely to explode. The same want of uniformity that was exhibited by the velocities is here also in the mean absolute deviation. The fabrication of these cartridges was extremely troublesome and attended with no little danger of premature explosion."

Obviously, the failure of the cartridge was the critical distance required between the end of the needle and the fulminate in the cup. If the distance was excessive, the needle would be unable to reach the cup and explode the charge. If the needle was too close, the force from the firing pin would simply force the cup into the bullet causing a misfire.

One of the many difficulties in producing good reliable ammunition of these times-1865 to 1870, was the failure of the propelling charge to be totally consumed before the bullet left the barrel. Unburned powder residue was always in evidence. Generally, this was attributable to the poor quality and the lack of uniformity of black powder then being manufactured.. Primer construction was reliable and the burning qualities of the powder in the cap was no better than the charge. Ordnance thought that by burning the powder in the front part of the shell, the entire propelling charge would be consumed before the projectile left the barrel. We will look at some who set the primer cup in the center of the propelling charge, while others tried a combination of fast and slow burning powder.

The Corliss experimental round was based upon the .58 caliber rim-fire case. The dimensions are:

CASE LENGTH	1.211"
RIM:	.705"
BODY:	.624"
GRAINS:	106 (case only)
HEAD STAMP:	.58 CAL.

The head-stamp has four tick marks around the letters .58 CAL. the regular .58 rim-fire rounds with this head-stamp that I have examined generally lack these tick marks, but the one rim-fire round I do have, is only slightly different from this one. Another .58 caliber Bar Anvil has these identical tick marks as the Corliss. There are two possible explanations of these marks. Experimental loadings may have been marked in this manner. Certain loading machines were constructed differently, and simply had a different means of gripping the head of the shell possibly leaving these lines on the bottom of the case.

It would be nearly impossible to identify a loaded and unfired Corliss round except through the use of a magnet. Fortunately, this is a fired case and the needle can be easily seen inside the case. The needle is .450" from the edge of the case mouth or conversely, three-quarters of an inch in length

As rare and unusual as this ignition appears to be, a through examination of American and European patents reveals that others thought of the front ignition system and constructed variations of it. It is also highly likely some even predated the Frankford Arsenal Corliss experiment.

Other patent variations of the front and middle ignition systems are:

AMERICAN PATENTS:

September 15, 1857, #18217, L. Wells:

> Hard metal tube forms shell. Cap on anvil extending back from bullet, hole in leather wad closing rear of the tube.

August 18, 1868, #81283, J. Merlett

> No description. Depicted is a center-fire cartridge with a cap in the base of the bullet.

May 6, 1873, #138679, Matt and Gardiner

> Insulated wire extending from rubber disk in base of shell nearly to fulminate in base of bullet, so that cartridge may be fired by ELECTRICITY.

May 17, 1864, #42815, CC. J. Bergen

> This is a very rare American experimental cartridge using a .58 Rim-Fire case. This .58 has been referred to many times as the "Springfield Case". As was the situation with the Corliss Front Ignition cartridge, the Bergen round was found loaded in a .58 Rim-Fire case with the head-stamp J. G. (Jacob Goldmark).
>
> In order to ignite the front of the propelling charge, two tubes on opposite sides of the case extended nearly to the base of the bullet. The tubes were crimped in place by the flange or rim of the case where the priming compound was placed. When the firing pin struck the primer, the flame was transmitted up the two tubes. The intention of the patent was to ignite the charge from the front by the flame emerging from the two tubes.

The inventor hoped that by using this method, the entire powder charge would be consumed.

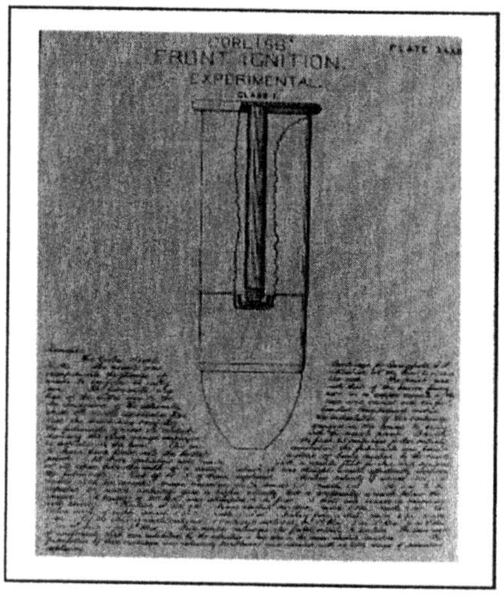

GEORGE A FITCH'S IGNITION SYSTEM

George A. Fitch was granted patent papers for a novel means of igniting the powder in a cartridge case at various points, that is to say, at the front or front and rear combined, or in the center. An accelerating charge of powder could also be utilized. When the charge is ignited at the rear, that charge burns first. When the ball exits the barrel before the powder is consumed a portion of the powder is blown out unburned. In such cases a smaller portion of the powder will be utilized and the ball will be projected with decreased power.

A pin is secured by solder within the case. This pin will be of such a length that a cap can fit at the base of the case and rest against the base of the projectile. When the hammer strikes the base of the case where the pin is attached, the pin will be driven forward and ignite the fulminate of the cap which, being at the forward position will ignite the fulminate of the cap and, at that point start burning from the front backward. In this position none of the powder can escape from the barrel without being burned. When loading the powder in the case, two types of powder are utilized. The powder first loaded is coarse or slow burning to start the bullet gradually thus permitting the bullet to take the rifling without slipping. As the ball nears the muzzle, the finer powder at the rear takes afire and burns faster thus following up the bullet with an accelerating force, besides utilizing the entire quantity of powder used.

His patent papers picture five rounds which with some modifications can accomplish the same effect within the .58 caliber cartridge.

No cartridge specimen was available.

GEORGE A. FITCH OF KALAMZOO, MICHIGAN. IMPROVEMENT IN PRIMING CARTRIDGES. #58800, OCTOBER 10, 1866

The Fitch patent has four plans in his improvement in priming cartridges, namely:

> #1. The conventional "front ignition" similar to the Corliss except the rod was to be attached to the base by solder. The Corliss was attached to a disk which was secured in place by the crimp of the rim or flange of the shell.
>
> #2. Similar to #1, but in addition to the primer cup in the base of the bullet, fulminate was also placed in the rim as in a regular rim-fire cartridge. This would require two firing pins. One to hit the edge of the rim and one to hit a blow in the center of the cartridge.
>
> #3. This plan was to have the charge ignited in the center. In this method the rod was made in two pieces. One section was attached to the base of the case and the other section secured to the base of the bullet. In the center of the bottom rod a priming cup was fashioned on the rod, and when struck (center-fire) by the firing pin, the cup was forced against the top rod, thereby igniting the fulminate. By this manner, the propelling

charge would be set off in the center assuring that the entire propelling charge would be totally consumed.

#4. This modification of #3 provides for the rod to be cup shaped to hold the fulminate, thereby, doing away with the necessity of having to use a separate priming cup which could be easily displaced.

The patent also points out that in all cases, the rod could be constructed as a cup and do away with the problem of placing a separate cup on the rod.

Undoubtedly, the safety of any four of these systems concerned the inventor. He points out in the text that he makes the base concave so that accidentally hitting the stem and prematurely exploding the cartridge would be avoided.

Experiments with this front ignition system continued at Frankford Arsenal utilizing the .50-70 Springfield round. At the 1876 Philadelphia Exposition, Frankford Arsenal exhibited specimens of these .50-70 front ignition cartridges. These were #324 and #325 appearing on one of their ammunition boards on display.

The text describes them as follows:

#324 Needle or disk and bolt, anvil. Primer in base of bullet. Disk held by fold at head. Bolt is iron and pointed like a firing pin. Disk is brass. Looks like a rim-fire cartridge. C. L. 1.768"

#325 Like last, but bolt squared up in front, like winding arbor. Frankford Arsenal. The first type tried, with a round bolt, simply flattened at the end, gave poor ignition. These cartridges can be distinguished from rim-fires by their attraction to a magnet, but they cannot be told from each other by external characteristics.

G. A. Fitch.

Cartridge.

No 58800. Patented Oct. 16. 1866.

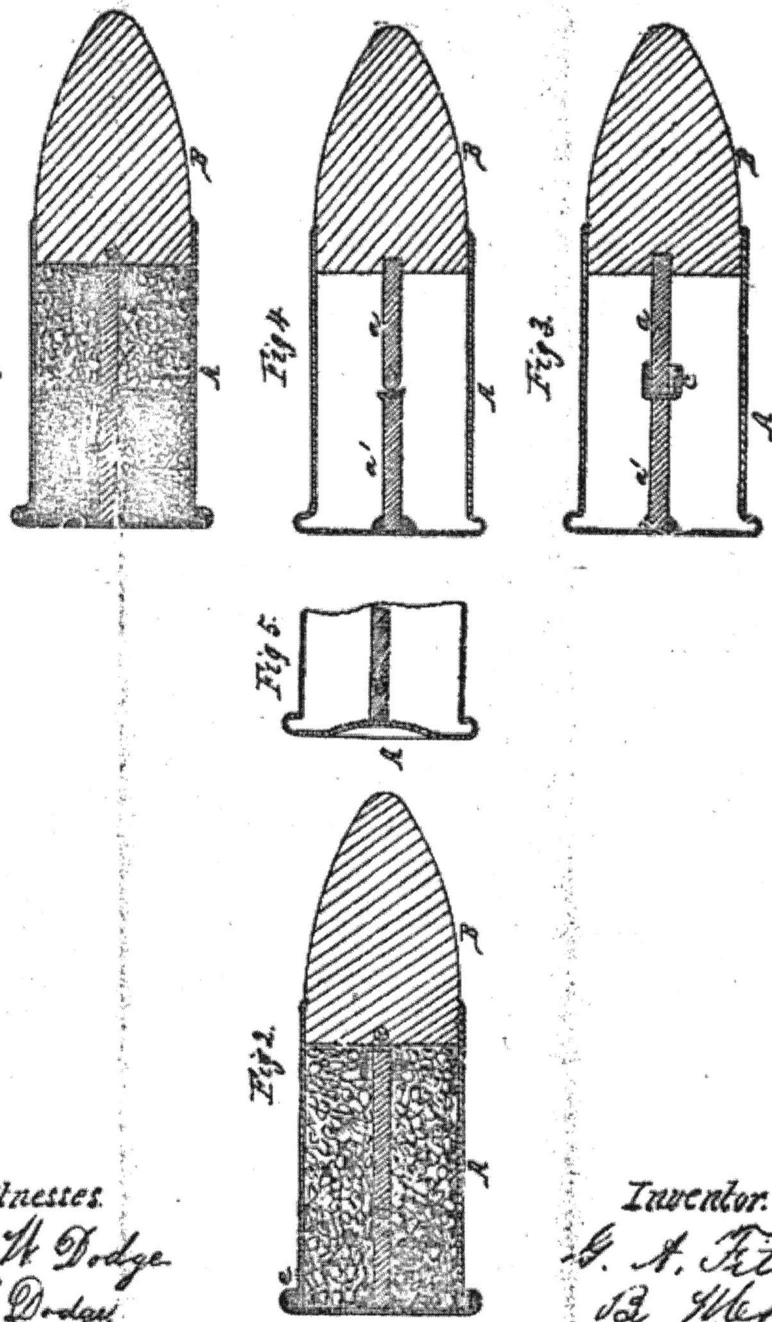

Witnesses.
W. W. Dodge
P. T. Dodge

Inventor.
G. A. Fitch
By W. W. Dodge
Atty.

UNITED STATES PATENT OFFICE.

GEORGE A. FITCH, OF KALAMAZOO, MICHIGAN.

IMPROVEMENT IN PRIMING CARTRIDGES.

Specification forming part of Letters Patent No. 58,800, dated October 16, 1866.

To all whom it may concern:

Be it known that I, GEORGE A. FITCH, of Kalamazoo, in the county of Kalamazoo and State of Michigan, have invented certain new and useful Improvements in Cartridges; and I do hereby declare that the following is a full, clear, and exact description thereof, reference being had to the accompanying drawings, making part of this specification, and to the letters of reference marked thereon, like letters indicating like parts wherever they occur.

To enable others skilled in the art to construct and use the invention, I will proceed to describe it.

My invention consists in a novel means of igniting the powder in the cartridge-case at various points—that is to say, at the front, or front and rear combined, or in the center, as may be desired—and using in combination therewith an accelerating-charge of powder, as hereinafter explained.

It is well known that but a limited quantity of powder can be burned in guns where the cartridge is ignited at the rear of the charge, and that, as the powder in such cases burns at the rear first, the ball leaves the barrel before all the powder is consumed, and consequently a portion of the powder is blown out of the gun unburned. Where the bore of the gun is small, a given quantity of powder will necessarily occupy more longitudinal space in the barrel, and will, of course, require a longer time for its consumption, and in such cases it follows that a still smaller portion of the powder will be utilized, and the ball consequently projected with a decreased power.

To remedy these difficulties I construct the cartridge as shown in the drawings, all of which are longitudinal sectional views of a cartridge, all being modifications of the same general idea.

In Figure 1, A represents the cartridge-case, and B the bullet, united in the usual manner. Within the case A, I secure a longitudinal pin or stem, *a*, which is secured to the center of the base of the case by solder, or in any other suitable manner. This stem *a* is made of such a length that there will be just room enough for a cap, *c*, between its front end and the base of the ball when the latter is inserted in place, as clearly shown in Fig. 1.

It will be seen that when the hammer strikes the base of the case A at the point where the stem *a* is attached, the latter will be driven forward and ignite the fulminate of the cap *c*, which, being at the front of the charge of powder, will inflame the latter at that point, and that, burning from the front backward, none of the powder can escape from the barrel without being burned. When using a cartridge of this style I fill it with two grades of powder, as represented, that in front being coarse or slow-burning, to start the bullet gradually, and thus permit it to take the rifling without stripping, and then, as the ball nears the muzzle, the finer powder at the rear takes fire and burns quickly, thus following up the bullet with an accelerating force, besides utilizing the entire quantity of powder used.

Fig. 2 represents a cartridge having the central stem, *a*, and the cap *c* located at the front, the same as in Fig. 1, with the addition of fulminate in the rim or flange, as in the ordinary cartridge. In firing this cartridge, the hammer of the arm must be made to strike the cartridge both at the center of the base, opposite the end of pin *a*, and on the flange *e*, as shown in Fig. 2. To accomplish this the hammer H may be provided with the points *o n*, as there represented, or it may have its face extended in a straight line from *o n*, as may be found most expedient.

It will be seen that when thus arranged the powder will be ignited simultaneously at the front and rear, and that when thus burning from both ends double the amount of powder can be burned in the same length of time that could be if burning from one end only. In this case the accelerating-charge of fine powder may be omitted, if desired, as a cartridge thus constructed and ignited will burn all the powder required before the bullet can leave the barrel, especially in long barrels, such as are used by infantry troops. In arms having short barrels, such as carbines and pistols, the accelerating-charge may be found necessary, however.

Fig. 3 represents a cartridge in which I accomplish the same object by igniting the charge at the center. In this case the central stem or pin consists of two parts, the part *a'* being secured to the base of the case A, and the other part, *a*, being secured to the rear end of the bullet B, as shown clearly in Figs. 3 and 4. A cap, c, is placed on the end of one of the parts of the stem in such a position that when the pin a′ is driven forward by the hammer the cap will be forced against the end of the stem a and exploded, thus igniting the powder at the center of the charge, from whence it will burn both forward and backward, thus insuring the burning of the entire charge.

Fig. 4 represents a slight modification of this last style, the front end of the pin a′ in this case being made cup-shaped to receive the fulminate, as shown, and thus avoid the necessity of using a cap, which would be more liable to become displaced or lost. It is obvious that this cup form of stem may be used also in the other cases, if desired.

In order to prevent the accidental explosion of the cartridge by handling or transportation, I propose to make the base concave, as shown in Fig. 5, so that there will be no danger of accidentally hitting the stem a and prematurely exploding the charge.

It is also obvious that my invention may be applied to artillery as well as to small-arms, especially in all breech-loading styles of small caliber.

Having thus described my invention, what I claim is—

1. Igniting the cartridge at the front by means of the stem a, in combination with the accelerating-charge, as shown in Fig. 1, and as herein described.

2. Igniting the charge at both front and rear, as shown in Fig. 2, and as herein set forth.

3. Providing a cartridge with the stems a and a′, when arranged to act in combination, as shown in Figs. 3 and 4, for the purpose of igniting the charge at the center, substantially as set forth.

GEO. A. FITCH.

Witnesses:
W. C. DODGE,
HENRY R. WALTON.

HENRY KELLOG

Henry Kellog, a gunsmith in New Haven, Connecticut, applied for and received two cartridge patents that are interesting and unusual. He also applied for two patents for firearms to utilize the ammunition. The calibers were approximately fifty-eight. They were truly unique rounds.

Patent #35356, dated May 20, 1862, described "a new and useful improvement" in breech-loading firearms. The novel design required the breech to have a hollow punch or piercer, arranged so when a loaded metal case was in the chamber, the action of closing the breech would force the punch to penetrate the base, or center of the head, of the metallic cartridge. The opening created in the case lined up with the percussion cone. This permitted the flame from the percussion cap to ignite the powder charge in the metallic case.

The Greene breech-loader percussion carbine had a similar piercer. It was described in the patent above eight years earlier on June 27, 1854. Henry Kellog did not claim the hollow punch. In this patent, he did claim an extractor for removing the cartridge case after the discharge.

The case dimensions were as follows:

Case Length: 1.125" Rim: .657" Head: .586" Mouth: .567"

The second cartridge patent awarded to Henry Kellog was equally unusual and interesting.

On July 15, 1862, (reissued March 27, 1877 - #7569) Kellog was granted patent #35878. The case had a projecting teat. It was cut off by the act of closing the breech lock, exposing an opening into the powder charge. As before, this permitted the percussion cap to ignite the cartridge. The teat contained no powder.

The rifle parts for this and the earlier conversions were the Model 1859 Sharps carbine. The breech mechanism operated on the original Sharps "lifter principal". When the breech closed it sheared off the base of the linen cartridge. It could also cut off the protruding teat leaving an opening with access to the propellant. The priming system was the Sharps Disk Primer devise located within the lock system of the Model 1859 Sharps.

Kellog was greatly concerned with preventing moisture from entering the case (an attempt to waterproof the case). He also wanted to avoid the serious effects created by self-priming materials used in the rim-fire cartridges. The following statement was made in his patent application:

> "to preserve the powder from the atmosphere and the deleterious and dangerous effects of fulminate usually used within the cartridge case"

An attempt was also made by Kellog to extend the life of the metallic case by coating the interior surface of the cartridge case with a solution of gun cotton in sulfuric ether. He felt this would insure more certain ignition of the entire powder charge and prevent corrosion of the metal or other harmful effects of chemical action of the powder in contact with the metallic case.

Dimensions of this cartridge are:

Case Length:	1.12"	Rim: Head:	.583"	Mouth:	.565"
Bullet:	.56"	Overall:	1.95"		

Exact details of the powder charge and the weight of the projectile were never made available. It would be reasonable to assume that they would be very close to that of the Sharps Linen Cartridge.

There is no doubt that Henry Kellog hoped to obtain a Government contract to manufacture the weapon for the military. He submitted the rifle and fifty cartridges for testing on October 8, 1864. A report was filed by 2nd Lt. Howard Stockton, Ordnance Department and submitted to Major J. G. Benton, Commanding Officer, Washington Arsenal.

No doubt the only surviving specimen of the Kellog case is the case attached to the patent card and filed with the Patent Office back in 1863.

TITLE: .58 CARBINE-THIN RIM

HEADSTAMP: Plain

DIMENSIONS:

Rim:	.708"
Head:	.660"
Neck:	.624"
Bullet:	.572"
Case:	1.187"

OTHER DETAILS:

Bullet Grains:	530
Powder Grains:	440
Bullet:	Flat Nose
Case Material:	Brass
Total Grains:	595

OVERALL: 1.671"

All attempts to identify this round have been to no avail. No clues were revealed by X-rays. Numerous examinations by advanced collectors around the country have produced no hint to its authenticity. Comparison with other .58's in the same general case size shows distinct differences and a much lighter weight in total grains.

The cartridge has been in the possession of an old time gunsmith friend of the author who upon learning of his interest in the .58 caliber field presented it to him with a smile and a "Beats me, what do you think it is?"

Could it be a variation of the tapered brass case (1.206" case) Roberts Inside Primed Cartridge?

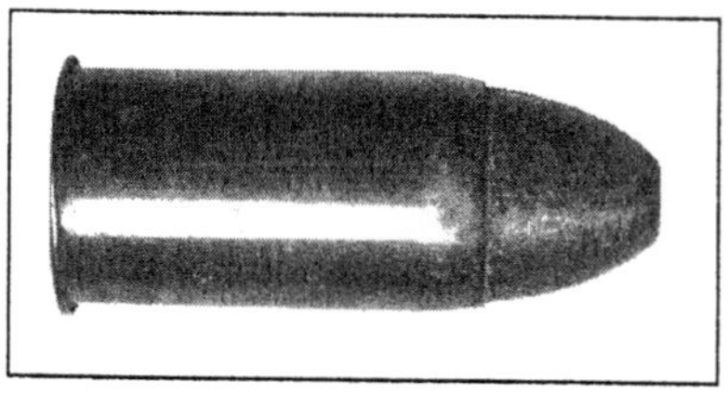

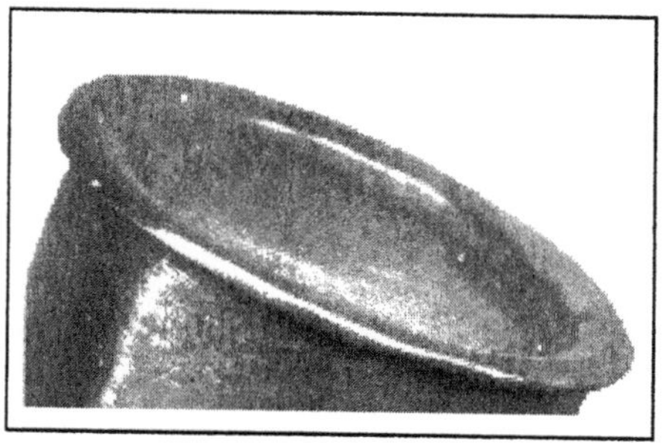

TITLE: .58 INSIDE PRIMED
DOUBLE CRIMPS

HEADSTAMP: Plain

DIMENSIONS:

Rim:	.745"
Head:	.651"
Neck:	.616"
Bullet:	.595"
Case:	1.195"

OTHER DETAILS:

Bullet Grains:	500
Powder Grains:	60
Bullet:	Lead, tapered
Case Material:	Copper
Total Grains:	700

OVERALL: 1.685"

At first glance this round appeared to be a malfunction of the crimping machine operation in assembling the round. However, after much research, it became obvious to the author that here was a distinct variation from the common, two -crimp, inside-primed ammunition.

The total grains is about correct for the two-crimp round but the overall length of 1.675" is less than the 1.75" of the regular Benet Primed cartridge. Had the bullet been simply pushed down into the case, its diameter would not then be larger than the more common Inside-Primed cartridge.

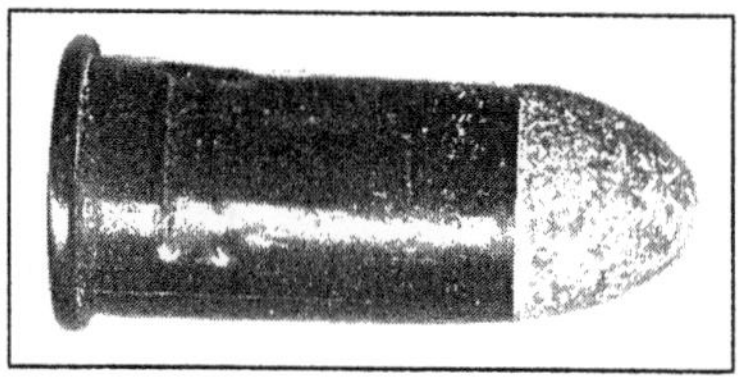

TITLE: .58 INSIDE PRIMED DOUBLE CRIMPS OFFSET

HEADSTAMP: Plain

DIMENSIONS:

Rim:	.757"
Head:	.658"
Neck:	.625"
Bullet:	.600"
Case:	1.195"

OTHER DETAILS:

Bullet Grains:	500
Powder Grains:	50-60
Bullet:	Tapered
Case Material:	Copper
Total Grains:	718

OVERALL: 1.689"

A variant of the preceding round. This specimen has the double crimps offset rather than being exactly over each other.

The total grain weight of this round is eighteen grains more due to the heavier bullet.

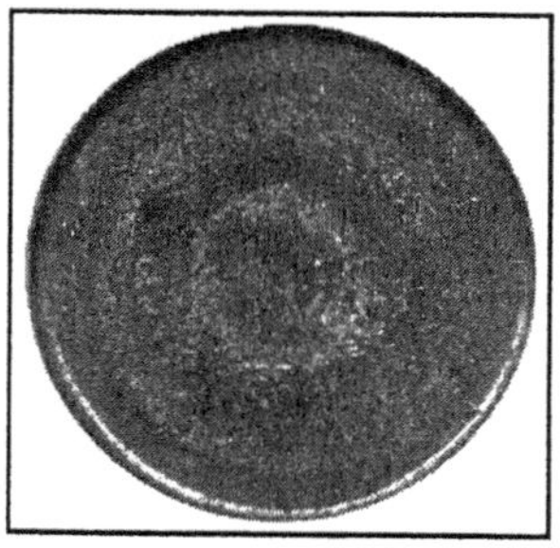

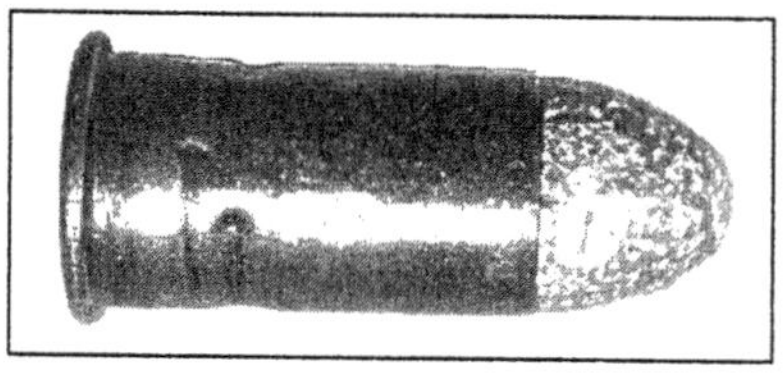

TITLE: .58 INSIDE PRIMED-TWO CRIMPS
SMOOTH BASE

HEADSTAMP: Plain

DIMENSIONS:

Rim:	.752"
Head:	.651"
Neck:	.626"
Bullet:	.585"
Case:	1.200"

OTHER DETAILS:

Bullet Grains:	475
Powder Grains:	50
Bullet:	Lead, tapered
Case Material:	Copper
Total Grains:	666

OVERALL: 1.705"

There is no question that this two-crimp Benet Inside Primed cartridge is different from the common Inside Primed round. Unfortunately, no positive identification could be made. Because only two were available for examination, it would seem likely that this one is the first type manufactured.

The bullet has a different shape; it is shorter and has a different taper from the common round. The crimp on the case neck is more rounded. An examination of the base of the cartridge reveals an extremely smooth surface, whereas the regular round has a "dimpled" appearance. Examination by X-ray revealed the primer construction to be no different than any other .58 Benet Inside Primed.

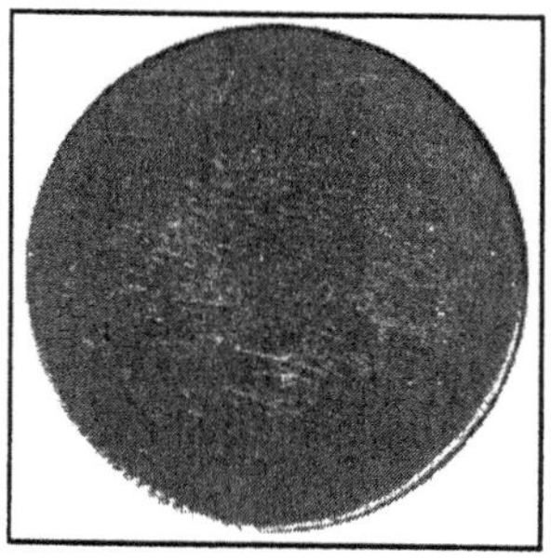

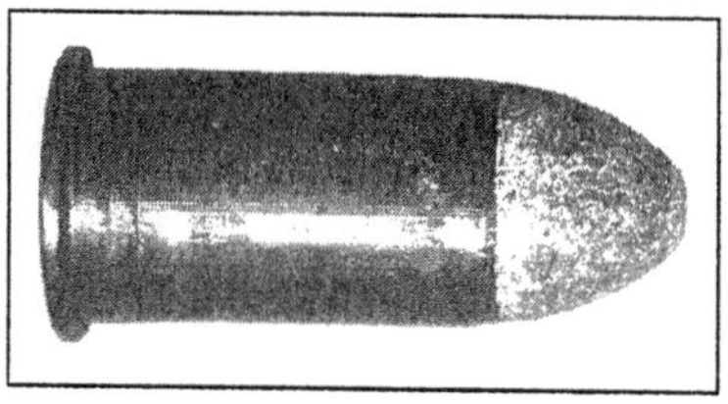

TITLE: .58 INSIDE PRIMED
THREE CRIMPS

HEADSTAMP: Plain

DIMENSIONS:

Rim:	.750"
Head:	.655"
Neck:	.628"
Bullet:	.595"
Case:	1.203"

OTHER DETAILS:

Bullet Grains:	500
Powder Grains:	50-60
Bullet:	Pointed
Case Material:	Copper
Total Grains:	690

OVERALL: 1.766"

This cartridge came from a box by E. Remington & Sons with the label indicating they were manufactured "FOR ROBERTS RIFLE".

The three crimp version was another early attempt to create a successful center-fire ignition system. The purpose of the crimps was to hold the reinforcing cup containing the primer mixture against the head of base of the cartridge.

Due to the scarcity of the three crimp cartridges, it would appear these were truly experimental in nature. The three crimp procedure was shortly replaced by the common two crimp style Benet Inside Primed variety.

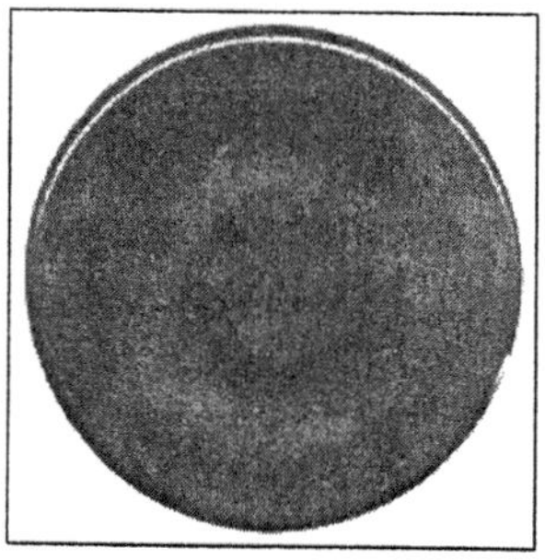

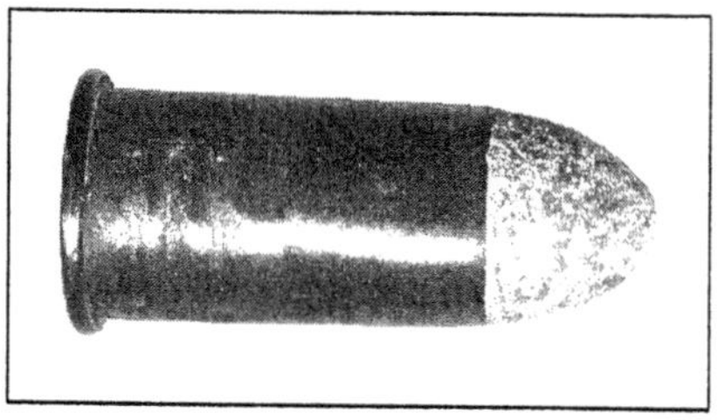

TITLE: ROBERTS INSIDE PRIMED-BRASS CASE
NO CRIMPS

HEADSTAMP: Plain

DIMENSIONS:		OTHER DETAILS:	
Rim:	.740"	Bullet Grains:	450-500
Head:	.654"	Powder Grains:	50-60
Neck:	.628"	Bullet:	Pointed
Bullet:	.597"	Case Material:	Brass
Case:	1.210"	Total Grains:	671

OVERALL: 1.800"

The Brass cased Roberts inside primed without crimps is a very rare variation of the more commonly encountered specimen which used a copper case.

Dimensions and weights of the two are approximately the same; however there is a difference of Nearly 20% between rim diameters. The brass cased cartridge rim is .740" vs. .753" of the copper cased cartridge. It would appear reasonable to question whether both rounds would successfully chamber in the same weapon.

Tolerances were rather flexible in the 1860s and while one might stretch a point and accept a five- percent variation; twenty percent seems excessive.

If the reader will refer to the Roberts inside primed cartridge with the common copper case, he will discover how the round was constructed. There is no reason to repeat it here.

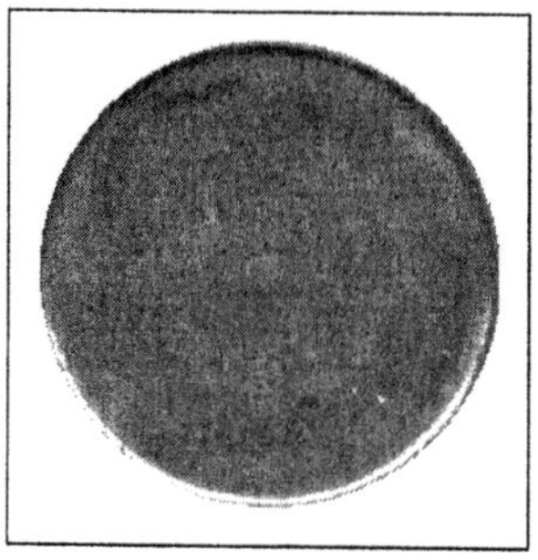

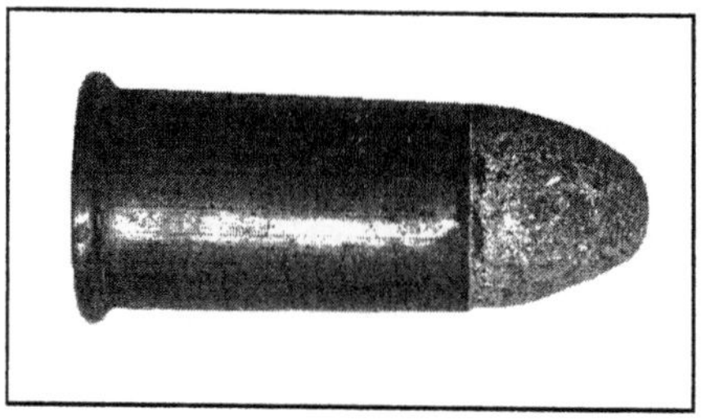

TITLE: ROBERTS INSIDE PRIMED-COPPER CASE
NO CRIMPS

HEADSTAMP: Plain

DIMENSIONS:		OTHER DETAILS:	
Rim:	.753"	Bullet Grains:	450-500
Head:	.655"	Powder Grains:	50-60
Neck:	.628"	Bullet:	Pointed
Bullet:	.570"	Case Material:	Copper
Case:	1.219"	Total Grains:	675

OVERALL: 1.625"

The Roberts inside primed cartridge was the forerunner of the two and three crimp variety.

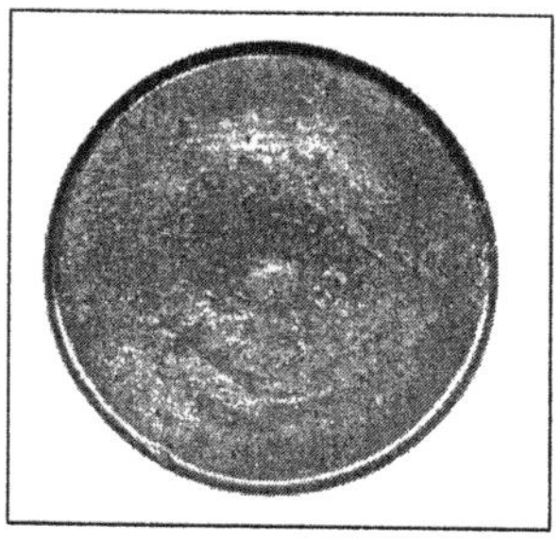

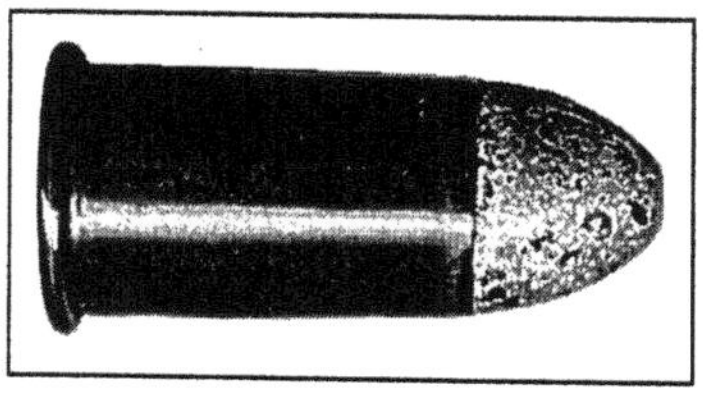

If you examine the picture of the fired case, you will observe that these inside primed rounds contained a reinforcing cup which were all similar in shape and approximately the same size; when it appeared in either a tapered or a crimped case. The two vents must have been punched through its base by a blunt instrument as the holes are always rough and of an irregular pattern. They were never round and smooth. The fired casing pictured here was found near an old fort outside of Tucson, Arizona; reasonable proof that they were placed in service and issued to troops.

These cartridges originated with the rim-fire system using as a base, .58 caliber cases and were made to be used in converted muskets.

To prime the shell, the fulminate appears to be impregnated onto a small square of nitrated paper. The fulminate was attached to the exact center of the base of the shell. Next the vented reinforcing cup was set snugly onto the base over the priming compound. The propelling charge of sixty grains of black powder was then loaded; and finally, the 450-500 grain lead bullet was secured in the case.

To secure the reinforcing cup in place, the case was tapered towards the bullet. In this manner, the cup was tightly retained in position.

Roberts always advocated a tapered case in all his cartridges to enable the fired casing to be easily extracted. He was probably the first person to use this idea.

Extensive testing and field service indicated many instances where the taper was not reliable enough to hold the cup in place. Rough handling would jar the cup out of position resulting in misfires. The distance between the base of the cartridge case, fulminate, and/or the cup was critical and if not tightly held against each other, the round would fail to fire.

The logical conclusion was that the reinforcing cup had to be held firmly in the case, and the proper way to do this was by crimping. Hence, we find evidence of three crimp, and finally the two crimp method, which proved to be satisfactory.

Manufacture of this ammunition was performed at the National Armory during the year 1865.

TITLE: BENET'S INSIDE PRIMED
TWO CRIMPS

HEADSTAMP: Plain

DIMENSIONS:

Rim:	.749"
Head:	.655"
Neck:	.631"
Bullet:	.588"
Case:	1.219"

OTHER DETAILS:

Bullet Grains:	500
Powder Grains:	50-6-
Bullet:	Conical
Case Material:	Copper
Total Grains:	690

OVERALL: 1.750"

Earlier attempts at securing the reinforcing cup in the base of the cartridge proved the three crimp rounds to be a more reliable cartridge than the Bar Anvil priming method. Having had success with this idea the next step was to eliminate one of the three crimps. Because metal as soft as copper is easily weakened from being "worked", i.e. Bent or folded, it was felt the case would be strengthened by the elimination of one crimp.

Col. Benet redesigned the case and used only two crimps. The crimps were positioned directly on opposite sides of the case, approximately seven millimeters up from the rim and they were thirteen millimeters in length. Minor variations from these measurements will be found, but they are not considered significant.

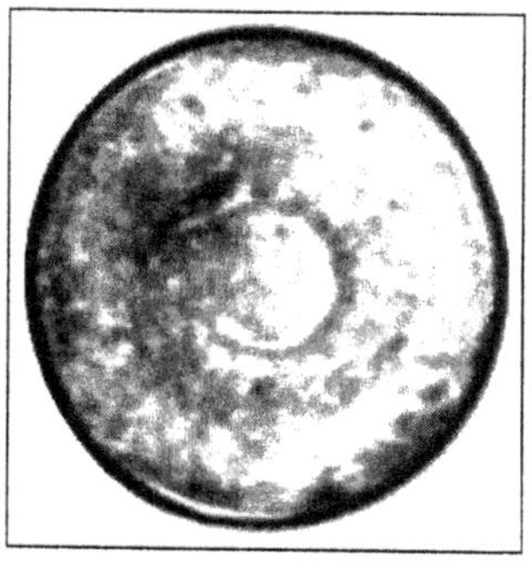

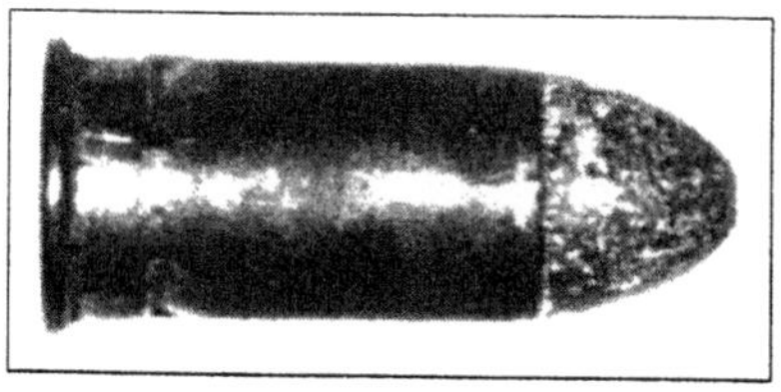

A detailed description of the Benet Inside Primed two crimp cartridge appears in the report "METALLIC CARTRIDGES", manufactured at Frankford Arsenal, Philadelphia, Pa. The following is quoted from that publication:

> The cup anvil for service cartridges was introduced and successfully applied by Co. S. V. Benet, Comdg. Frankford Arsenal, March 1868. It was made from tinned sheet iron as an experiment; the crown was indented to hold the priming, with vents for ignition. To keep and support it in place the copper was crimped under its edges. Its superiority over the bar anvil is as follows: - 1st. The cup anvil being cylindrical in form furnished more surface at its edges to crimp under, making it more rigid and permanent and greatly increasing the certainty of ignition. 2nd. The crimping and forcing the copper from the sides of the case at a greater distance from the flange considerable lessened the strain at or near the fold, giving a greater strength at head and a smaller percentage at fold. 3rd. The impossibility of getting the priming up side down, as was sometimes the case with the bar anvil.
>
> The weak point of the cup anvil cartridge is at the fold, it not being supported in the gun as other parts of the case, depends on its own inherent strength; the binding of the metal causes stretching on the outside and a closing on inside at fold, and occasional fracturing and demoralizing according to the quality of metal.
>
> Sheet metal is not free from such defects as seams, flaws, etc. which, not appearing on the surface, escape inspection, developing only when the cartridge is fired.
>
> Col. Benet experimented with solid heads which gave sufficient strength, but their manufacture was difficult and expensive.

Despite the fact that the inside primed cartridge was difficult to manufacture and expensive, a great number of cartridges were made. Specimens are very common in .50-.70, .45-.70, and in the caliber .44 and .45 Revolver sizes

One feature all of these cartridges had in common was the cup style. Simply a cup of either iron or copper with two vent holes punched into the base. The holes would appear to have been made with a nail or some other equally unscientific type tool as the holes are always very rough and ragged.

Below is a scarce example of an experimental brass cased specimen.

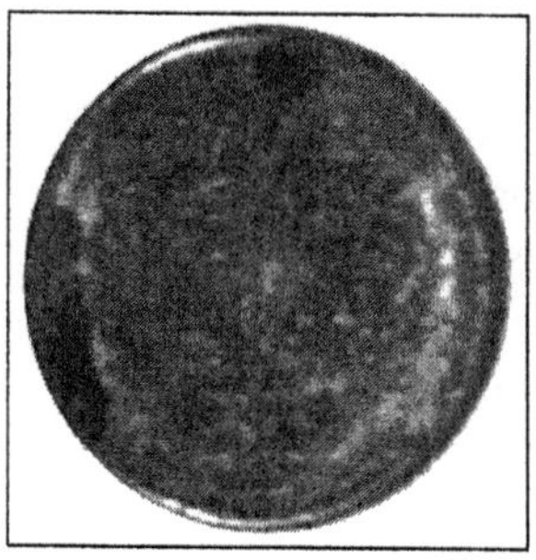

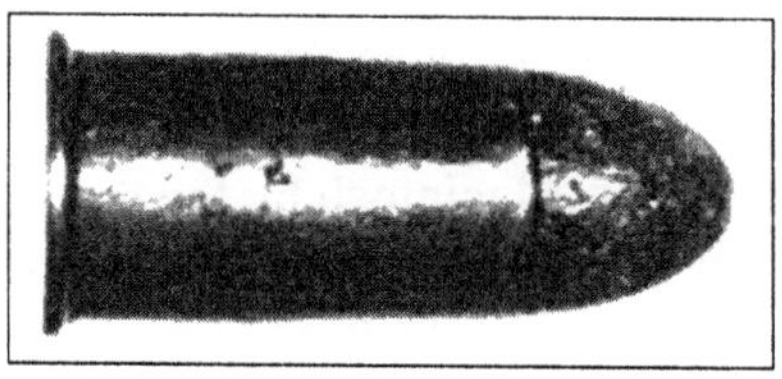

MACINTOSH TUBULAR BULLET: Caliber .577

During the mid 1860s inventors tried to improve upon existing ignition systems, invent new systems, improve and/or change the shape of the projectile and attempt to increase the range.

The Macintosh Tubular Bullet attempted to improve all of these attributes.

The cartridge was the diameter of the Enfield bullet and weighed 600 grains. The bullet was 1.77" long and designed with a hollow the entire length of the projectile. The principal of this design was that instead of a blunt or pointed ball, which had to push air aside as it, traveled through space, the Macintosh, with its' hollow design, would permit the air to pass through the open front and rear of the cartridge. This considerably reduced the resistance of air pressure when in flight. Ribs spiraled through the hole in the bullet acting much like rifling in a barrel; imparting a spin to the projectile as it left the barrel. Macintosh contended that his bullet would travel 2,000 yards with the identical sight setting that would give a range of only 900 yards with the Enfield Rifle.

The following comments are quoted from the Evening Meeting of January 21, 1867 and the March 16th, 1868 Evening Meeting at the Royal United Service Institution.

> "The next metallic cartridge to which I shall draw your attention is a novel and very ingenious invention of Mr. Macintosh's. Here the case is of lead, tubular in shape, open at both ends, having the weight and diameter of an Enfield bullet, but 1.77 inch in length. A wad of wood or other material carrying the means of ignition is secured by a covering of thin felt behind the tube, it is then filled with powder and the cartridge case has become also the bullet and a very efficient bullet, too. Having a point blank range of 400 yards, and an extreme range much greater than the ordinary projectiles owing to the dimension of the area of resistance. By reference to the diagram (Plate 1) you will see that these projectiles are fired by first shaking out the powder from one of them and placing it in the gun with its wad; next, another is inserted with the powder, and this pushes the first up the barrel. The breech being now closed and the gun being fired as usual, the first inserted bullet is driven out, and the second remains ready to be in its turn pushed forward, from having been a cartridge case to become a projectile. Some experiments have been made and others are in progress, and I shall be happy to communicate the results on some future occasion. You will see that although I hoped I had made as good, if not a better, metallic cartridge than any other, yet if this be a true system, no other metallic cartridge can compete with it, and mine must become obsolete with the others. A still farther advance has been made by Mr. Macintosh by the discovery of a new explosive compound, which he states is perfectly manageable, and so reduced in bulk as that full charge can be contained in the base of an ordinary Enfield bullet. I have now I believe given sufficient material for a discussion on the best form of cartridge and though I am aware that the subject is far from being exhausted, I think it is time to turn the arms."

In a question and answer period later on in that meeting the following was discussed:

> Mr. Macintosh: "The reason for getting a superior range with my new projectile, with the same driving charge over the Enfield bullet, exists in its being a hollow tube. After the projectile leaves the gun, the wad, from its lightness recedes before the column of air which passes through the cylinder; the air passing through, obviates in great the resistance in front of the projectile, and also the vacuum formed in the rear, which takes place in other bullets. I attained a range of 300 yards more than that of the Enfield bullet. I believe it was 600 or 700 yards. This was with the same charge and the same elevation."
>
> Captain Horton: "Another of my questions was whether it reached its destination without turning."
>
> Mr. Macintosh: "It did. The bullets which are represented on the board are smooth in the inside, and also on the outside. They are intended to be fired through a rifled barrel. You perceive that when the charge is exploded, the bullet expands and forces the metal into grooves, consequently it is then dependent for its accuracy of flight upon the spin which the groove gives it. There is another advantage which I have found in using a bullet of that sort, that is, you can have a large exploding chamber in which your gas is thoroughly ignited. This causes the greatest amount of useful effect to be obtained from the quantity of powder discharged. With regard to the sabot that has been proposed, you use a small projectile and a large bore, and make up the difference with the sabot. You leave there a difficulty which in using my bullet is overcome. In using a smaller projectile it has been proposed to wind some light material round the exterior of the projectile so as to fill up the bore; and it is supposed that in passing through the air that will come off. However, I think this would be very ineffectual. In this case, you use a large projectile with the same useful effect, in consequence of the interior of the bullet being hollow. The advantage gained is, that you give a large amount of work or a high velocity to the projectile with but comparatively little pressure of the powder gas whilst lessening the atmospheric resistance. In reference to the means of ignition, the fulminate is put in a recess of the wooden plug, or wooden wad, the same as in any other central firing."

Although much enthusiasm was shown and much praise given to Macintosh for his invention at that March 1868 meeting, little ever became of it. The members present felt that the tubular bullet would revolutionize firearms but nothing materialized out of his idea and the tubular bullet faded into obscurity.

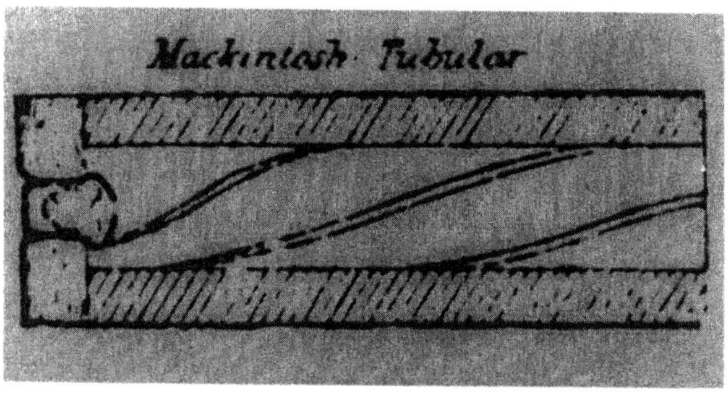

Paper III

.69 BUCK AND BALL

The .69 caliber Buck and Ball is the granddaddy of these rounds used in our military rifles and muskets. The earliest record of the use of buck and ball ammunition was dated October 6, 1777, from General Headquarters at Perkiomy which was an order that "Buck shot and ball are to be put into all cartridges which shall hereafter be made". A tactics manual circa 1814 stated that the standard musket cartridge was the buck and ball.

The specimen pictured here is 2.4" long and weighs a total of 622 grains. It contains three 43 grain .30 caliber buck shot on top of a 433 grain spherical ball .65" diameter. Powder charges varied greatly in these early rounds. The early rounds contained much greater powder charges (as much as 50%) due to the poor quality of the powder that was available. As years passed and the quality of the powder improved, the charge could be reduced and still maintain the same or increased velocity.

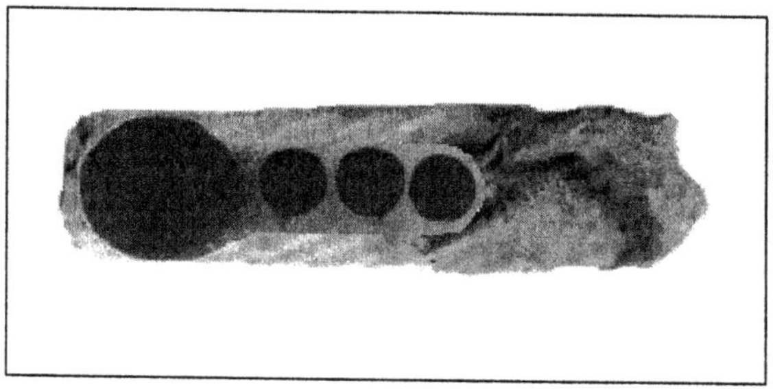

.58 BUCK AND BALL

This specimen differs in construction in several ways from the .69 caliber Buck and Ball. Notice that no string is used to hold the projectiles and paper case together. The front of the cartridge is flat caused by the three .30 caliber balls that rest on the one spherical round which is 2.27" long and weighs a total of 469 grains. The outside diameter is approximately .618"

The .58 caliber Buck and Ball is much scarcer than the .69 caliber even though the .58s were of later manufacture-circa 1866. One must remember that over six million .69 Buck and Ball cartridges were purchased during the Civil War explaining why there are still a fair sized number available in collections. Below is another example of a multi ball loading with a slightly different method of construction.

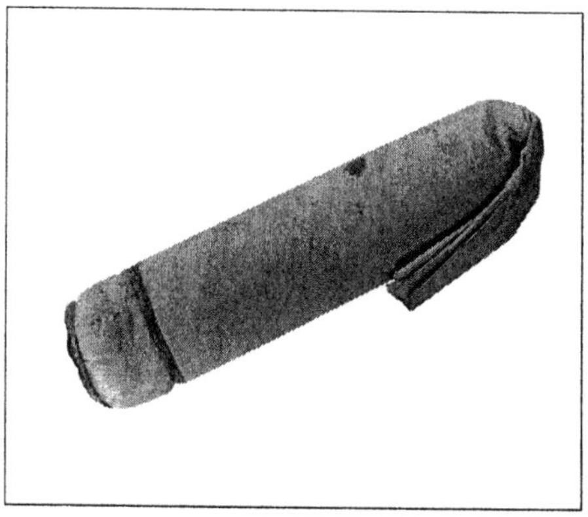

CONFEDERATE BUCK AND BALL

The coarse brown paper case identifies this round to be of Confederate manufacture. It was most likely a product of the New Orleans Depot or Richmond Arsenal.

The round is 2.1" long and weighs a total of 614 grains.

It contains the usual three buck shot which rests on a .65 caliber round ball.

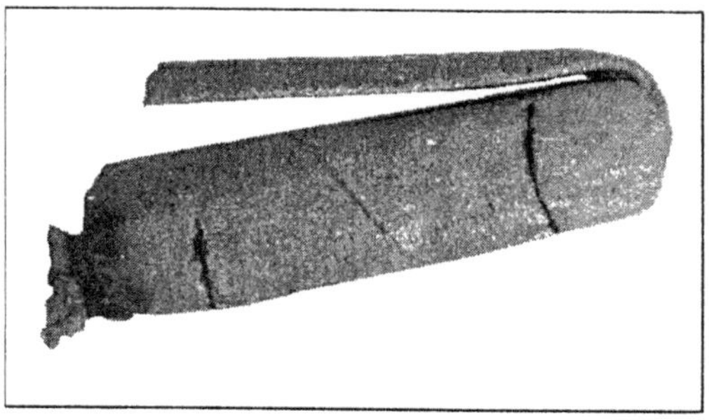

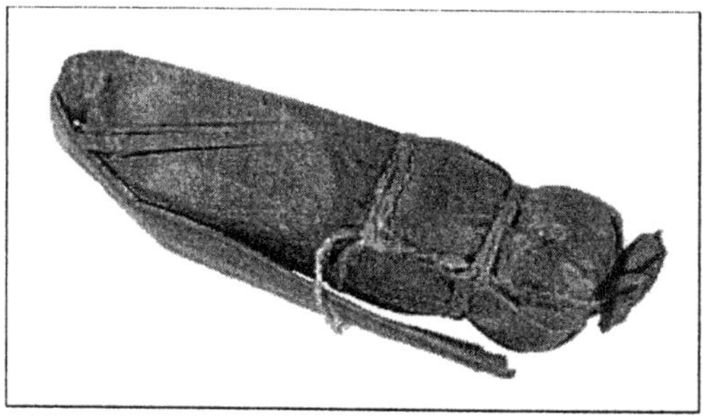

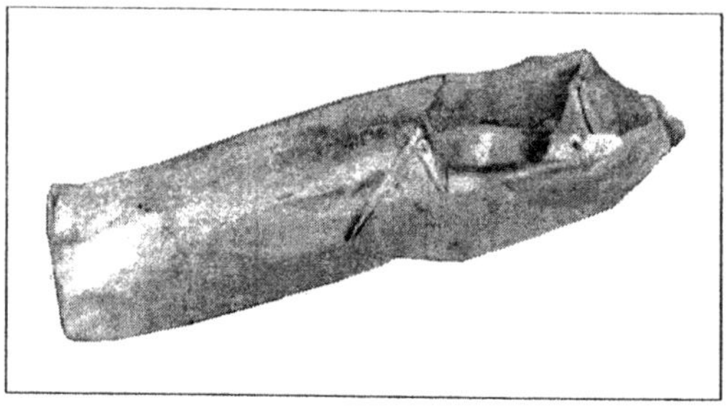

TERRY-CALISHER AND WILSON RIFLES AND CARTRIDGE

Webster's dictionary defines rare as being marked by unusual quality, merit appeal or extreme of its kind. Depending upon where your interests are in the cartridge-collecting field, it is the author's opinion that the Terry cartridge (and its close associates, the Calisher and Wilson) is truly at the top of the rare cartridge list. In thirty-five years of collecting the author has been hard pressed to find more than a handful of collectors who were able to identify a specimen much less have one in their own collection. The rifle and ammunition saw limited service by the Confederate Army

The following is a description taken from Mr. Terry's Patent #427267 dated April 12, 1855:

"The cartridge is made of strong brown paper, and is secured to the end of the ball by some adhesive substance, being glued to the back of it a wad, well saturated with tallow for preventing the gun from fouling after repeated use, the action of which will be hereafter explained.

By referring to the illustration, it will be seen that the cartridge is placed directly under the channel or bore which leads from the nipple in such a position that when the cap is discharged by the fall of the hammer, the explosion of powder takes place from the centre of the cartridge, and not from the rear, which is usually the case. The object of this arrangement is for the purpose of detaining the tallowed wad in the barrel at the time the ball is ejected from it by the force of the exploded powder; it there remains ready to be forced forward by the next ball and cartridge inserted, and leaves the barrel when the discharge again takes place. It will therefore be understood that there is always a wad left behind every discharge, ready to be pushed forward by the following charge."

The text and pictures go into the operation of the system in detail. Briefly stated, the bolt is lifted and drawn back, a cartridge is inserted in the loading ramp, the bolt is pushed forward, the hammer drawn back, a cap is placed on the nipple, the trigger is pulled releasing the hammer. The flame strikes the center of the nitrated paper case firing the round. There is a felt wad .613" diameter and .228" thick attached to the cartridge upon firing, the wad remains in the chamber. When the next round is fired, the wad is pushed out ahead of the bullet thereby cleaning the barrel as it goes. Not only does the wad act as a barrel cleaner but it effectively seals the breech against escaping gas.

The bullet design is unique because it is the first one made with its diameter bigger than the bore from land to land, thus introducing a principle that has been retained to the present day.

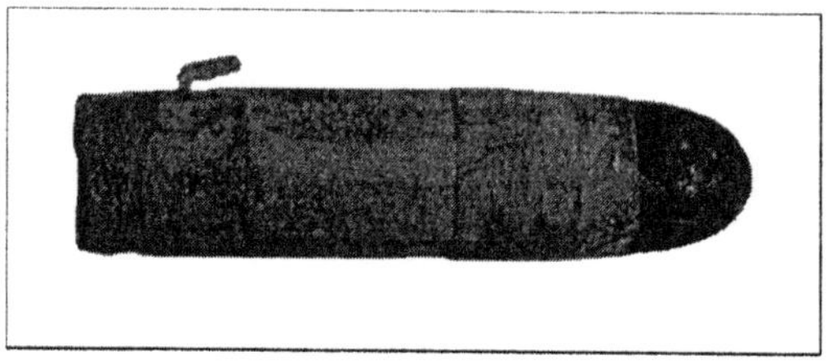

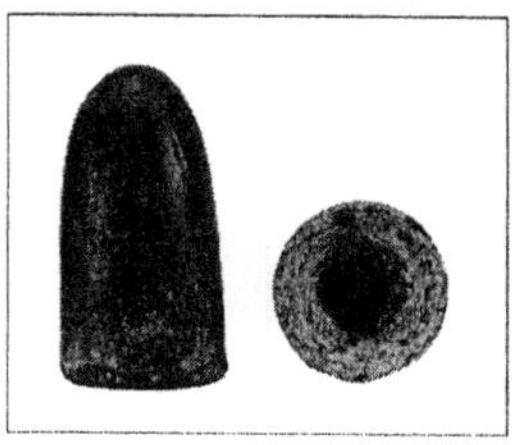

A contemporary description is as follows:

"The chamber in which the cartridge lies is of the diameter of the circle described by the bottom of the rifling grooves, and the ball used exceeds the Enfield ball in diameter by the depth of these grooves. The barrel has a gradual bore from the larger to the lesser diameter, so that when the discharge takes place the ball is gradually driven into the rifling grooves, and presents precisely the same appearance as an Enfield ball does after it is discharged. In this way the patentee states he obtains greater range, force and accuracy, with the same weight of ball and powder, than can be obtained by any other rifle."

The cartridge is 2.35" over-all and has a total weight of 622 grains. The base contains a felt wad .613" in diameter and approximately 1/4" thick. The case is 1.8" long and made of nitrated paper and glued half way up the bullet. The round nose bullet is 1.1" in length and weighs 528 grains and has a cone shaped base.

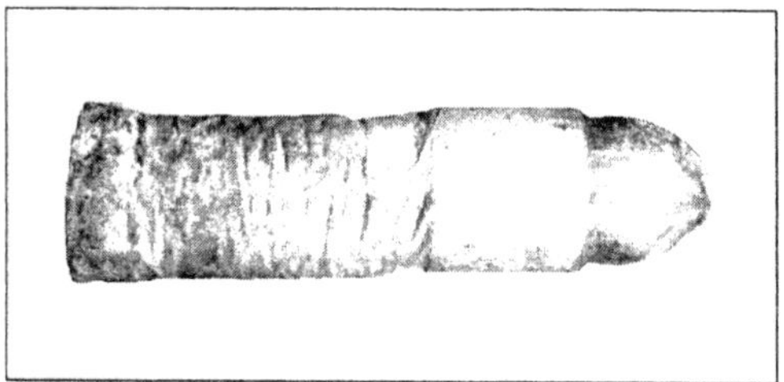

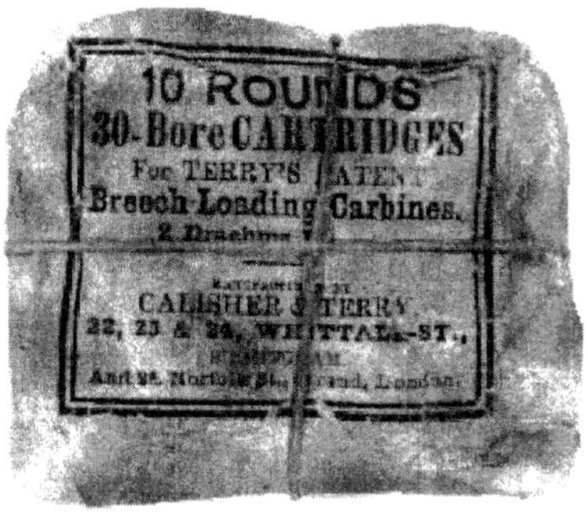

In 1862 the Terry rifle was entered in the shooting contest at Hythe, England to determine the rifle to be used in the final shooting match stage at the Queens Prize. The Terry performed poorly and perhaps with no further interest in the weapon, the inventor found a ready market for his rifle with the Confederate Army who was desperate at that time for guns.

According to the Milwaukee Public Museum Publication #2, the Terry was issued to the Confederate Army and was nicknamed the "Door Bolt Gun" because of its likeness to the closing devices on doors during those times.

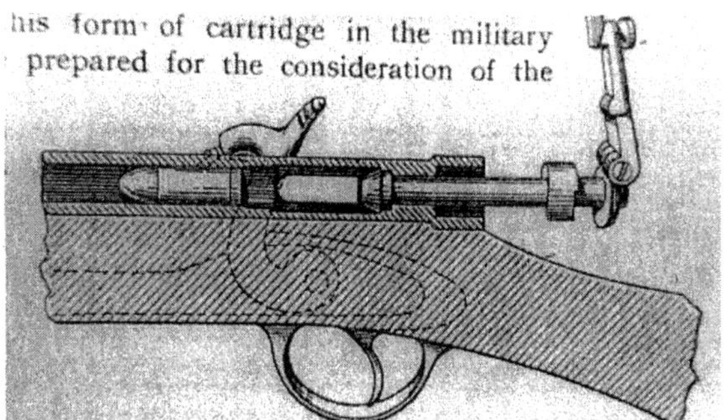

Fig. 18.—Terry's Rifle showing the cartridge with its base wad in position. No wad is shown in front of the bullet

The Terry, Calisher and Wilson Carbines all utilized the same basic cartridge construction.

Measurements of the Terry/Calisher and Wilson rounds are as follows:

	TERRY (Commercial) #1	TERRY/CALISHER (More Common) #2	WILSON (Rifle) #3
Case Length	1.09"	1.53"	1.81"
Wad	.23	.23	.29
Combined	1.32	1.76	2.10
Bullet	.58	.58	.57
Over-all	2.10	2.27	2.58
Total Grains	622	620	627

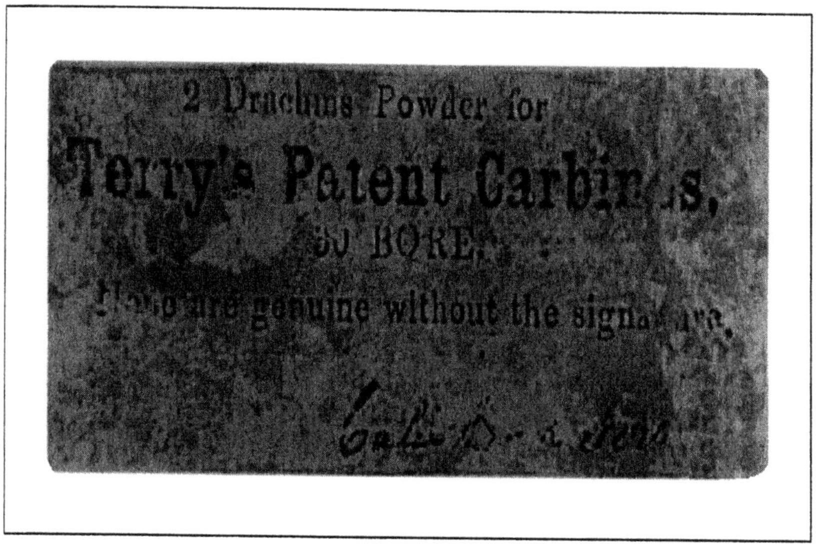

COMBUSTIBLE CARTRIDGE Johnson & Dow

Johnson & Dow Cartridge - by Elam O. Potter. Used in the U.S. Model 1855, cal. 58 rifle. Cartridge dimensions: Weight - 548 grains - diameter .575 – length 1.82" – grains of powder 60.

I have made excerpts from the patent papers, which are as follows:

> "In July 22, 1862, Elam O. Potter of the City, County and State of New York applied for a new and useful improvement in making cartridges. The cartridge comprises of two basic principals. #1. A cartridge made of pressed powder covered with collodion, and the_cake of powder further secured to the ball by a belt of collodion. #2. The pressed powder not encased but is secured to the ball by the collodion belt.
>
> My invention consists in uniting press powder to the ball running a collodion belt around the joint between the powder and the ball and extending over into each sufficiently to firmly protect them against accidents or mishandling"

The patent goes further into the virtues of the invention and these have been born out by the numerous written excellent reports from company commanders on the relative merits of the superiority of the Johnson & Dow/Elam O. Potter cartridge.

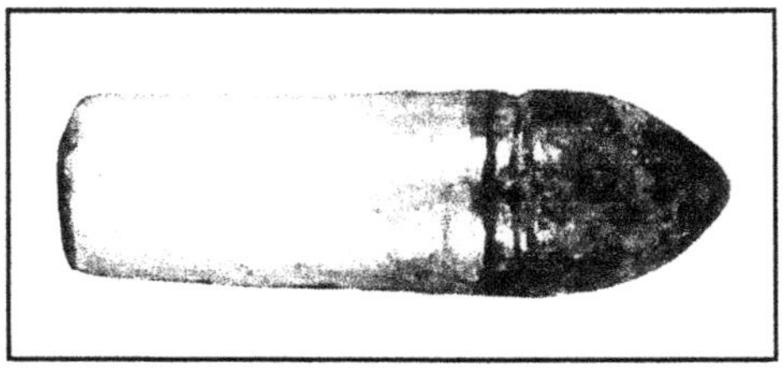

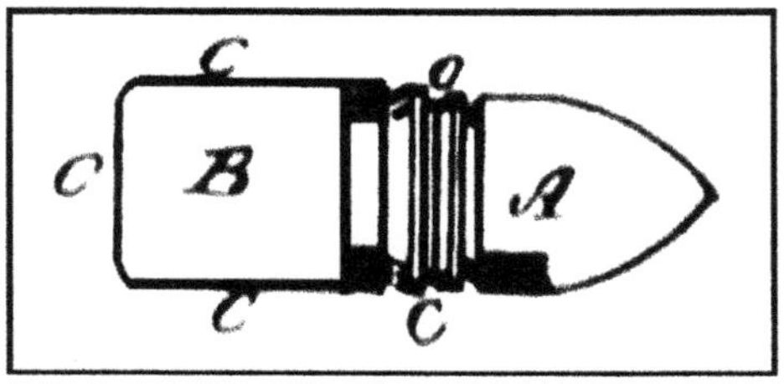

JAMES CURTIS

The letters of patent # 90732 dated June 1, 1869 granted to James Curtis of Chicago, Illinois. Shooters have always been interested in finding a better way to hit their target. Multiple projectiles appeared to be the answer. To this day, cartridges are still being designed to fire multi-ball projectiles

Here we have an extremely well made projectile submitted with the inventor's model to support his claim. The writer has observed this design in modern day rounds. Mainly in .45 pistol rounds.

United States Patent Office.

JAMES CURTIS, OF CHICAGO, ILLINOIS.

Letters Patent No. 90,732, dated June 1, 1869.

IMPROVEMENT IN PROJECTILES.

The Schedule referred to in these Letters Patent and making part of the same.

To all whom it may concern:

Be it known that I, JAMES CURTIS, of Chicago, in the county of Cook, and State of Illinois, have invented a new and useful Improvement in Slug-Balls for Fire-Arms; and I do hereby declare that the following is a full, clear, and exact description thereof, which will enable others skilled in the art to make and use the same, reference being had to the accompanying drawings, forming part of this specification.

This invention relates to an improvement in musket-balls, whereby they are rendered more effective, in battle at close quarters, than balls of ordinary construction; and

It consists in forming the ball with two (more or less) longitudinal slots, and with two (more or less) horizontal slots, said slots, or cuts in the ball being sufficiently deep to cause the ball to fly into twelve (more or less) parts or slugs when it leaves the muzzle of the gun, as will be hereinafter more fully described.

In the drawing—

Figure 1 is a longitudinal view of an elongated conical ball, divided by longitudinal and horizontal slots, or cuts, which reach nearly to the core of the ball, only sufficient metal being left to preserve the ball in its proper form in handling, and while in the gun.

Figure 2 is a view of the top of the ball, showing the slots radiating from the centre or core.

Similar letters of reference indicate corresponding parts.

A is the conical ball.

B represents the longitudinal slots, or cuts, which extend inward to near the centre or core.

C represents the transverse slots, or cuts, which likewise extend to near the centre.

It is designed that sufficient metal shall be left to preserve the form of the ball when placed in the gun or in the cartridge-case, but not sufficient to hold the parts together, when propelled by the force of the powder, against the resisting atmosphere.

After leaving the muzzle of the gun, the projectile will fly to pieces, and scatter, doing execution in every direction, thus rendering it a formidable destructive agent at close quarters.

Having thus described my invention,

I claim as new, and desire to secure by Letters Patent—

The slots, or cuts B and C, (more or less in number,) in a musket-ball or elongated conical shot, arranged and formed substantially as and for the purposes described.

The above specification of my invention signed by me this [illegible] day of November, 1868.

JAMES CURTIS.

Witnesses:

CLARENCE E. BENNETT,

C. E. [illegible]

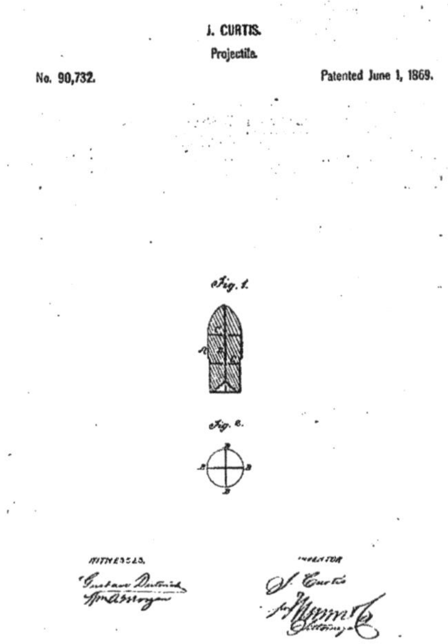

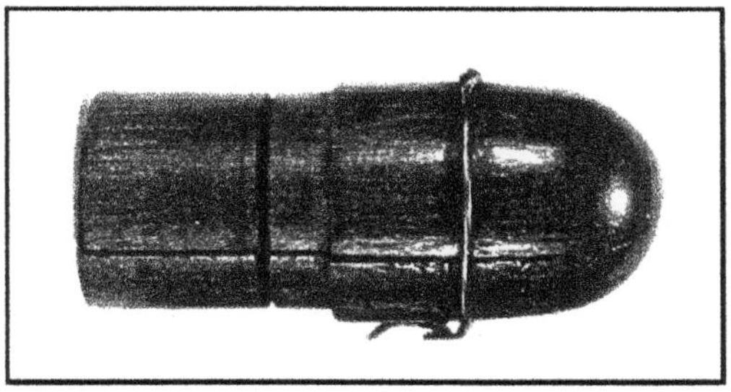

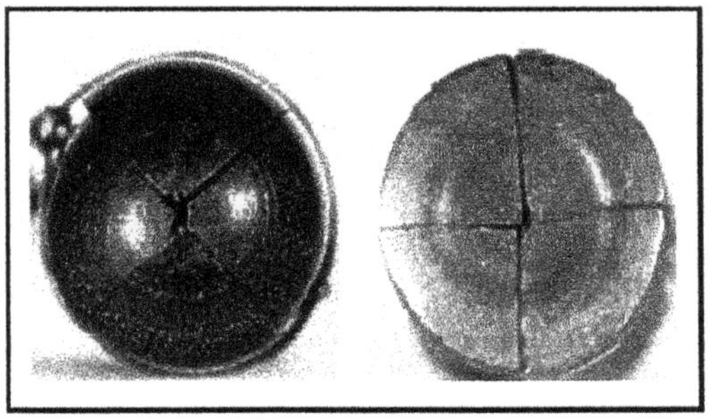

EXPLOSIVE BULLETS

History Of The Explosive Bullet

Civil War buffs are well aware of the explosive bullets used by both the Northern and Confederate Armies during some of the most intensive battles in the struggle between the States. The best known explosive small arms ammunition was the Gardiner; designed and patented on November 23, 1863 by Samuel Gardiner, Jr. of New York. Although Gardiner is generally credited as being the originator of this specially constructed cartridge and condemned by Southern sympathizers for the creation of such an inhuman contrivance. Col. B. R. Lewis points out in his book "Small Arms & Ammunition" that it is possible that the C.S.A. had an explosive round that predated the Gardiner.

According to Col. Lewis, the Confederate round lacked the copper insert and the projecting tube that was incorporated in the Gardiner. Lacking these two elements, it was more than likely the round would burst either in the barrel or at the muzzle. A Confederate packet of the .577 caliber explosive rounds and individual specimens, some of C.S.A. manufacture, are pictured in his book, along with the statement that a number of these rounds were captured at New Orleans. The rounds were made in sizes .69, .577/.58 and in caliber .54 for the Mississippi Rifle.

To strengthen the argument that the South was the first to have the explosive ammunition is that the more complex cartridge of Gardiner which contained the copper insert and projecting tube resulted from Gardiner being aware of the imperfections in the original Confederate rounds; hence profiting from their mistakes and incorporating the improvements in his own ammunition. It would also be remembered that the South purchased or traded for thousands of weapons from England and more than likely learned of the early experiments conducted with explosive projectiles in European countries and in particular England for use in their hunting rifles. Undoubtedly, the South applied these ideas for their own use.

During the 1870's there were four varieties of explosive cartridges, namely, the Jacob, Forsyth, Pertuiset and the Mead-Meigs.

> "The invention of hollow bullets charged with fulminates is credited by the English to Capt. Norton in 1824, by the French, to Capt. Delvigne in 1829. The former placed a projecting wooden plug over a charge of fulminate contained in the cylindrical cavity of the bullet. The latter used similar projectiles against the Arabs in 1830. Explosive bullets have been improved by a long list of inventors, including Minie, Devisme and Pertuiset, in France; Podewitz and Von Dreyse (inventor of the leaky "Zudnadel-Gewehr") in Germany; Gibbs, Ludlow and O'Callaghan in England; Jacob and Forsyth in India; Zimmer-haus in Russia; Gardiner, Meigs and Mead of the United States."

A series of articles appeared during the 1870's in the American Sportsman on explosive projectiles. A summary of the history of the development of the explosive cartridge is quoted from one of these articles:

"Four varieties of small arms shells are now in extensive use, two English, one French, and one American. 1. The Jacob shell much used by British sportsmen contains a pointed copper tube, charged with fulminate and black powder. 2. The Forsyth shell, a leaden cup and cover swedged together containing antimony sulphide, and potassium chlorate, equal parts; finely powdered separate, and mixed with a bone knife. 3. The Pertuist shell is charged with finely divided sulphur and potassium chlorate, mixed while moist, pressed into a very thin cup of lead, and the latter swedged into the bullet and sealed by a ring of varnish. The base contains a hemispherical cavity. This shell is powerful, but quite inaccurate, owing to its peculiar form. Dr. Gatling says that at Vienna, the inventor threw them about with impunity, but that they exploded on being fired through a loaf of bread. 4. The Mead-Meigs shell is a hollow bullet, containing either a rim-fire copper cartridge alone or filled with fine gun-powder, and having an ordinary rim-fire cartridge shell or blank inserted as fuse."

It should be pointed out at the beginning that none of the explosive bullets were intended to be used against human beings. The English sportsman desired a weapon and ammunition that could be counted upon to kill large, dangerous animals. Those shells designed for military purposes were originally intended to be used on caissons and wagons. However, good intentions not withstanding, the Gardiner was used by both the North and South against the infantryman, inflicting terrible wounds on the men.

One of the fringe benefits arising from the design of a good explosive bullet was discovered by Col. Jacobs of England, during his experiments. He found that in order to insert his nose-fuse tube in the short Pritchett-Enfield "ball", greater length was needed. The weight lost by the hole gave poor accuracy, but after creating a longer bullet, the center of gravity had shifted, thereby, getting better stabilization of the bullet. Greater distance was obtained and considerably improved the accuracy of the Weapon. Col. Boxer used this design for his rounds. The first type had an exposed nose plug while the later rounds had the hole but the lead was "spun" over the hole in the nose of the bullet.

The term "Express" bullet must not to be confused with the "Explosive" bullet. The two were similar in construction but differed in purpose. The "Express" cartridge is a hollow point bullet with an empty copper tube inserted in the hole. Comparable in size to a .22 caliber shell. The caliber of the round was approximately .450 but contained a heavy charge of black powder. The effects of a hollow-pointed bullet are well known and will not be elaborated upon here. The "Explosive" bullet is simply that! An explosive charge is incorporated within the bullet. In later years, a .22 rim-fire blank served the purpose very well. This was what the Mead-Patent called for. The advertising comments extolled the "complete safety of such ammunition" but there remains serious doubt in the author's mind as to just how safe they could be.

Types of Explosive Bullets:

There are two basic types of explosive small arms rounds, first the "impact" type and second, the "fuse" type. One of the first successful impact cartridges was the Forsyth. The bullet is cast in two segments, the detonating compound put in, the base is then joined to the first part of the body, and the whole is passed through a screw swedge, thus forming the bullet into one piece. The Jacob Shell was an improvement over some of the first exploding cartridges, evidenced by its ability to hit and destroy caissons at a distance of two thousand yards. Instead of the old system of placing a percussion cap on the bullet tip with a small charge of black powder inside, which usually failed to explode, the Jacob had in its nose a copper tube, open at the end, which contained both the detonating charge as well as the black powder. It would appear to be an extremely hazardous task to have to "ram home" a bullet having a projecting tube containing an explosive compound in its tip. The Jacob's round was still an unreliable bullet, experiencing many failures to explode.

It wasn't until Forsyth developed his previously mentioned round that hunters could depend on the exploding cartridge as a reliable hunting round.

The fuse type of exploding bullet, while more complex in design, was certainly a great deal safer to handle. Although assurances were made by their superiors that the rounds were perfectly safe to carry and handle, many thousands of rounds were left behind on the battlefields during the Civil War by soldiers who were concerned over their own safety to keep them on their person or transport them from one battle site to another.

The best known round employing a fuse type was that invented by Samuel Gardiner, Jr. of Washington, D. C. and New York City. A copy of his patent dated November 23, 1863 is shown here. A brief description of the projectile is as follows:

> "The Gardiner shell is a cylindro-conical projectile of lead. Within the interior is placed an accurately fitting acorn-shaped copper chamber, filled with fulminate and communicating with a cone and one-quarter second time fuze which is exposed to the charge at the rear of the missile; the fuze is ignited by the discharge of the piece. The bursting charge is sufficient to rend the bullet and transform it into a jagged missile".

The bullet is attached to a paper envelope containing the propelling charge and is similar to the general conformation of other paper cartridges of that era. It is assumed that Gardiner furnished the bullets only; the paper case and powder charge being attached by the purchaser.

Another interesting feature of the Gardiner cartridge is that cast onto the base of each bullet are the words:

S. GARDINER JR. SHELL PATENT SECURED

Another explosive type round was patented by a Joseph Nottingham Smith on August 18, 1863. In the rebuttal by Reverend Horace Edwin Hayden to Benjamin I. Lossing's accusations that only the Confederates purposely used explosive bullets against the Northern soldiers, mentions the use of a round similar to that patented by Smith. The description of the cartridge is quoted below along with facts stated in the Smith Patent #339593:

> "Recently it was my privilege to examine in the hands of a man just from Fortress Monroe an explosive bullet such as was used by the rebels in the six days battle. It is conical in shape, about one inch long, made of lead and consists of two parts, viz. A solid head piece, and a cylindrical chamber, which are united together by a screw. From the point of the bullet, projects a little rod, which passes down through a small hole in the

head piece into the chamber below where it was connected with a percussion cap. The chamber contains about a tablespoon full of powder. You can readily perceive that if the bullet should encounter a bone or other hard substance, when entering a man's body, it will explode and thereby produce a fatal wound." F.J.C.

"No. 39593, Joseph Nottingham Smith, New York, N. Y. Improvement in elongated projectile for firearms Patent dated August 18, 1863. It consists of an elongated cylinder having a charge chamber in its rear portion, which contains powder for propulsion. The point is a pointed axial bolt whose rear is furnished with a percussion cap to be exploded by the forward motion by the projectile."

Rev. Hayden continues:

"Not having seen this ball I cannot certainly identify it with the ball mentioned by F.J.C. but is evidently the same". "The inference is very natural that if these several projectiles patented by the United States Patent Office, as the inventions of Northern men, and used by the United States armies were ever used by the Confederates it was only as captured ammunition. It was hardly possible, at any reasonable cost to run them through the blockade to the South."

Obviously, Rev. Hayden was unaware of the manufacture of explosives by the South or was unwilling to believe they would use them except as captured of battlefield pickups.

The author was unable to find any illustrations or other mention of the Smith explosive bullet.

Use Of Explosive Small Arms Ammunition During the Civil War.

There is no question of the authenticity of the fact that both the Northern troops and the Confederate Army employed the use of explosive small arms fire. It is also an accomplished fact that both sides used the cartridges against soldiers. Explosive ammunition in the small calibers such as the Mississippi Rifle .54, the .577 Enfield and the .58 caliber Springfield Rifle were available. A limited amount of explosive ammunition was also made in caliber .69 and is pictured in "Small Arms & Ammunition" by Col. Lewis. Undoubtedly, the primary purpose of the cartridges was for use against ammunition wagons and was not intended to be used against the infantry, as their destructiveness served no additional purpose. A rifle bullet or musket ball could just as easily kill or maim. The wound caused by one of these miniature exploding artillery shells was terrible and only caused needless suffering to the wounded.

That the North used explosive cartridges is evidenced by Military Records that indicated an order was placed with Samuel Gardiner, Jr. for thirty-three thousand, three hundred and fifty rounds at a price of $35.00 a thousand. Ten thousand of these were abandoned on the battlefield for want of transportation and were undoubtedly seized and used by the South. As previously mentioned, there is a good possibility that the Confederates had explosive rounds as early as 1862. The fact that the South possessed these is quoted from Col. Lewis' book "Small Arms & Ammunition":

"Sworn testimony of several Confederate officers gives certain details:

LT. BEVERLY KENNON, CSN-"I also had shells made for the muskets and Mississippi rifles-September 19, 1862"

COMMANDER MINOR-"Lieutenant Kennon ordered the manufacture of 1,000,000 rounds of musket shells at 15 cents apiece, of which 39,000 rounds are now at the Naval Laboratory at Atlanta, Ga. Many were lost in New Orleans. They were of no use in the Navy and a dangerous projectile, and many exploded in the operation of ramming. Their sole value lay in the components of lead and fulminate of mercury. As Ordnance Officer I would not authorize their use at any time".

COMMODORE PREBLE-"C. A. sharpshooters used explosive balls. Dr. Burton has one. A conical ball, pointed and charged with fulminate."

When Reverend Horace E. Hayden wrote his rebuttal to the newspaper publication by Benjamin I. Lossing, regarding the use of explosive cartridges by the Confederates, he addressed a letter to Jefferson Davis questioning Davis as to the use of explosive ammunition. Jefferson Davis explicitly denied the use of such ammunition.

His reply to Reverend Hayden was as follows:

"My Dear Sir:

In reply to your inquiries as to the use of explosive or poisoned balls by the troops of the Confederate States, I state as positively as one may in such a case that the charge has no foundation in truth. Our government certainly did not manufacture or import such balls, and if any were captured from the enemy, it could probably only have been used in the captured arms for which it was suited. I heard occasionally that the enemy did use explosive balls, and others prepared so as to leave a copper ring in the wound but it was always spoken of as an atrocity beneath knighthood abhorent to Civilization. The slander is only one of the many instances in which our enemy have committed or attempted aims of which our people and their government were incapable, and then magnified the guilt by accusing us of the offences they had committed.

Believe me ever faithfully yours,

Jefferson Davis"

After the close of the Civil War, a great hue and cry arose against the use of explosive ammunition. Both the North and the South charging the other with employing these rounds, while denying they ever used such a "barbaric and inhuman contrivance" themselves.

An international military conference was held during October 1868 in St. Petersburg with the principal nations of Europe attending and reaching an agreement to abstain from the use of explosive projectiles weighing less than four hundred grams. Interestingly enough, the United States did not choose to join the convention or take part in the agreement. The agreement was obligatory upon the contracting parties when at war between themselves. Apparently there were serious doubts if the agreement would be adhered to in the future.

During the middle 1870's various writers, literary experts, if you will, wrote articles and commented upon the use of explosive cartridges. Two writers were particularly outspoken and expressed their opinions in no uncertain terms to the public.

One Benjamin I. Lossing, L.L.D. wrote vehemently against the South's use of these inhuman articles of war. I quote:

> "Many, mostly young men were maimed in every conceivable way by every kind of weapon and missile, the most fiendish of which was an explosive and poisoned bullet."

Lossing proceeds to illustrate a Williams "Clean-Out" and a Shaler bullet (both rounds being illustrated and discussed elsewhere in this book) and proceeds to show his lack of knowledge by describing in great detail how the Williams contained poison on the copper "scrapper" washer and that upon entering the body, the stem used to rivet the unit together, would set off an explosive charge in the projectile's nose. Of course, we know the Williams was designed as a means of cleaning out the powder residue in the barrel and was never intended to contain an explosive compound, much less poison. The Shaler was supposed (according to Lossing) to be an explosive creating "terrible, inhuman destruction". In actuality, the Shalers were a "multi-ball" projectile.

Reverend Hayden, in continuing his rebuttal of Lossing's charges against the Confederate Army, pointed out that in April 1862, the Commissioner of Public Buildings at Washington, D. C. (by the name of B. B. French) brought to the attention of Assistant Secretary of War, John Tucker, the explosive musket shell invented by Samuel Gardiner, Jr. (Mr. Gardiner was employed as an electrician in Washington D. C. at the time). The Assistant Secretary immediately referred to the matter to General James W. Ripley, who was then Chief of Ordnance. In the meantime, during May 1862, the Chief of Ordnance at West Point Military Academy made a report to the Government of trial tests they had been conducting of the Gardiner Musket Shell. President Lincoln wrote to General Ripley stating: "will General Ripley consider whether this explosive shell will be a valuable missile in battle?" Ripley replied: "It has no value as a service projectile." A request by Brig. General Rufus King, June 1862, at Fredericksberg, Va. made up a requisition for some of the Gardiner musket shells. Again, a negative reply from General Ripley: "It was not advisable to furnish any such missile to the troops at present in service."

The Chief of Ordnance of the 11th Corp. U.S. recommended in September 1862, the shell to the Assistant Secretary of War who (in spite of Ripley's comments) ordered 10,000 rounds to be purchased and made into cartridges. It would appear that Gardiner sold only the bullet and the paper case and propelling charge was assembled elsewhere by the purchaser. In October 1862, the Chief of Ordnance of the 11th Corp. sent in a requisition for two thousand Gardiner Musket shells and cartridges, which was endorsed by the General commanding the Corp. Again, the Assistant Secretary of War, as before, referred the matter to General Ripley and again for the third time recorded his disapproval of the issue of the Gardiner shells. In spite of Ripley's negative reply, the Assistant Secretary ordered the remaining nine thousand eight hundred shells to be issued to the 11th Corp. The order was obeyed.

Reverend Hayden mentions Mr. Gardiner's offer in April 1862, to sell his explosive musket shells and cartridges for thirty-five dollars ($35.00) a thousand. These were to be in caliber .58. Immediately, seventy-five thousand caliber .58 and twenty-five thousand in .54 caliber for the cavalry were ordered by the Assistant Secretary of War. During June 1863, thirty-five thousand rounds were requisitioned by the Second New Hampshire Volunteers U.S. and by order of Assistant Secretary of War, they received twenty-four thousand rounds. It is recorded that ten thousand were abandoned at Spring Vale, Virginia and thirteen thousand, nine hundred forty were distributed to the regiment. The regimental report indicated that during the period from July 1 to October 1, 1863, four thousand rounds were tested and used in rifle practice and fourteen thousand were expended in battles.

It must be assumed here that in spite of the Chief or Ordnance, General Ripley's aversion to approve the purchase and/or issuance to the troops of the Gardiner bullet, the Assistant Secretary of War simply ignored Ripley's wishes and eventually purchased one hundred ten thousand cartridges and issued thirty five thousand to U. S. Troops then in actual service, leaving an inventory of seventy-five thousand on hand at the end of the War. Military records indicate that no further issuance of explosive rounds was made after 1863.

Reverend Hayden continued to defend the charges against the South in never using explosive shells and based his conclusion that the only shells of this nature were patented in the North (the Smith and Gardiner) and that IF the South did use them, they would have been battlefield pick-ups which were used in retaliation against the Northern troops who were originally issued them.

An interesting anecdote at the end of Reverend Hayden's paper was the following:

Brownville, Pa., September 1, 1879

> "The facts recorded herein regarding the Gardiner Shell were obtained by in person at the U. S. Ordnance Office. In November 1878, after failing to obtain a reply to my written inquiries, I visited the Office and was permitted to read a paper on the subject but was refused a copy. In August 1879 I again visited the Ordnance Office when.
> Dr. V. McNally, Chief Clerk allowed me to read as often as I chose, the same paper and refusing a copy told me that I could read it and remember all I could of it and so to get the substance of it memorized which I did. He also procurred for me the box of shells from the Arsenal."

H. E. Hayden

The controversy over the use of explosive shells continued for some years after the Civil War. The afore mentioned, Reverend Hayden discussed at great lengths the inhuman use of such projectiles. Copies of Army Medical Records were processed and those injuries supposedly caused by explosive bullets-pure or imaginary-were discussed and described in minute detail. There is great doubt that many of these wounds were actually caused by explosive projectiles as the regulation musket bullet or "buck and ball" rounds, if ricocheted off a hard substance before hitting the soldier, could become distorted and render similar wounds.

The battles where these wounds were inflicted were varied. To name a few: Chancellorsville, Cold Harbor, Resaca, Gettysberg, Ringgold, Ga., Wilderness,

Missionary Ridge and Spotsylvania. All these occurred during the period from May 3, 1863 to the last reported case on June 8, 1864.

Before leaving the subject of explosive bullets, it should be pointed out, that the most glaring oversight of the inventor, Army Ordnance Officers, and the average soldier, was their failure to recognize that the Gardiner was in reality the first tracer cartridge.

Everyone ignored the advantages to which this tracer effect could be put.

The picture below is a cutaway of the Gardiner Musket shell patented on November 23, 1863. This photo clearly shows the inside chamber that contained the explosive and the groove which protrudes to the base of the bullet through which the flame from the main propelling charge set the fuse afire. Burning time was adjustable to one and one-quarter seconds.

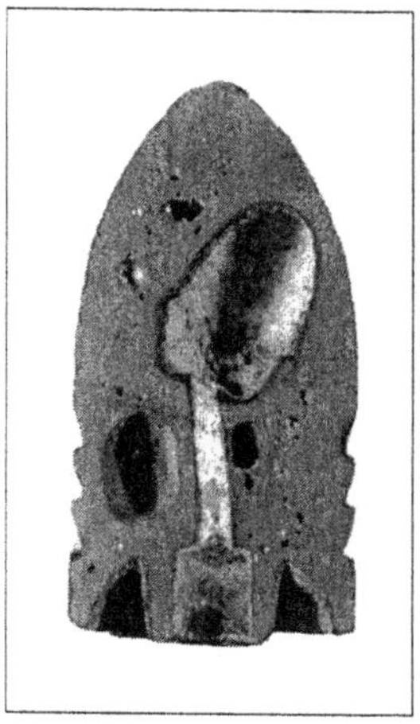

These are dug bullets which were either abandoned on the battlefield or dropped by the soldier either accidentally in the heat of battle, or intentionally, probable due to fear of having them explode while being handled or carried.

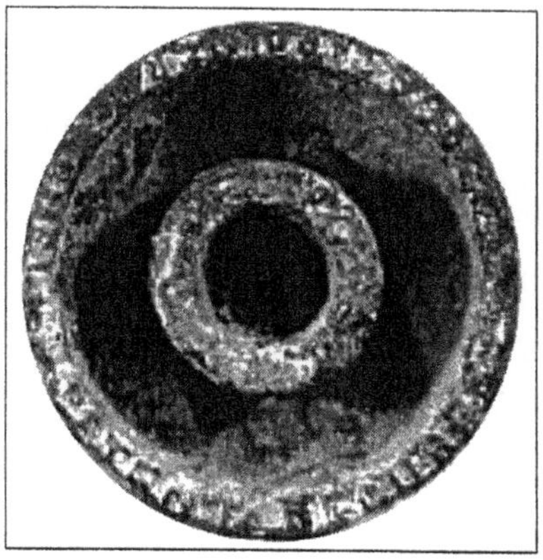

This photo shows the Gardiner in an upright position along with the base of the bullet. The projectile weighs four hundred ten grains, has an over-all length of 1.12" and measures .575" in diameter. Embossed on the base: S. GARDINER, JR. SHELL PATENT SECURED

Photo of the Gardiner Shell Patent showing an outside view and the components which set off the fuse, and the means of constructing the whole bullet. It is difficult to determine if Gardiner attached the paper case holding the propelling charge of sixty grains of black powder. Perhaps he furnished the bullets only, letting the purchaser attach the paper case.

UNITED STATES PATENT OFFICE.

SAMUEL GARDINER, JR., OF NEW YORK, N. Y.

IMPROVEMENT IN CONSTRUCTING HOLLOW PROJECTILES.

Specification forming part of Letters Patent No. 40,468, dated November 23, 1863.

To all whom it may concern:

Be it known that I, SAMUEL GARDINER, Jr., of the city, county, and State of New York, have invented a new and useful Improvement in Explosive Projectiles for Muskets and other Small-Arms, as also for Cannon of Small or Large Caliber; and I do hereby declare that the following is a full, clear, and exact description of the same, reference being had to the accompanying drawings, forming part of this specification, in which—

Figure 1 is a side view of an explosive shell for muskets and other small-arms. Fig. 2 is an end view of the same, showing the fuse-chamber. Fig. 3 is a longitudinal sectional view, representing the shell, ready for filling. Figs. 4 and 5 are perspective views of the two parts of a cup made in two halves, which are to be soldered together to form the chamber which holds the explosive substance or compound. Fig. 6 is a sectional view of the said cup, representing a small hole in each end, through which a mandrel is inserted to support the cup in the center of the mold. The said mandrel also forms the small hole leading from the fuse to the chamber which holds the explosive substance or compound.

Similar letters of reference in each of the several figures indicate corresponding parts.

The nature of my invention consists in manufacturing musket and other small-arm shells, forming and casting the shell complete and at one operation, which is done successfully by forcing the heated metal into the mold by a force-pump placed in connection with the reservoir of heated metal. The machine is similar to a squirt used by type-manufacturers. The material used in manufacturing the shells is lead or any other material that will expand enough to fill the grooves of the gun, and at the same time, when the explosion takes place, to break into several parts. I recommend using a fuse which is perfectly safe in the hands of the parties who may fire it, and is more certain in its explosion, as the fuse-mixture can be so arranged as to explode in one, two, three, or more seconds. Explosive caps may also be used to set fire to the explosive material. They are not safe to handle or transport. Therefore I recommend fuse instead.

To enable others skilled in the art to make and use my invention, I will proceed to describe its construction and operation.

A is the body of the shell, made of lead or any other material which will flow freely into the mold.

B is the inside chamber for holding the explosive substance or compound. It is made of copper or any other material.

C is the cavity which holds the fuse material.

D is the small hole leading from the fuse C to the inner chamber, B. A cartridge is attached in the usual manner to the shell. When the musket or gun is fired, the powder ignites the fuse, which will burn one or more seconds, as the time of burning the fuse may have been arranged. When the explosive material in the cup is ignited, the shell immediately explodes.

Having thus described my invention, what I claim as new, and desire to secure by Letters Patent, is—

Constructing shells for fire-arms by forcing the metal into the mold around an internal shell, B, supported on a mandrel, F, all as hereinbefore described.

In testimony of which invention I hereunto set my hand.

SAML. GARDINER, JR.

Attest:

M. M. LIVINGSTON,
J. W. COOMBS.

S. Gardiner,

Shell.

No 40,468. Patented Nov. 3, 1863.

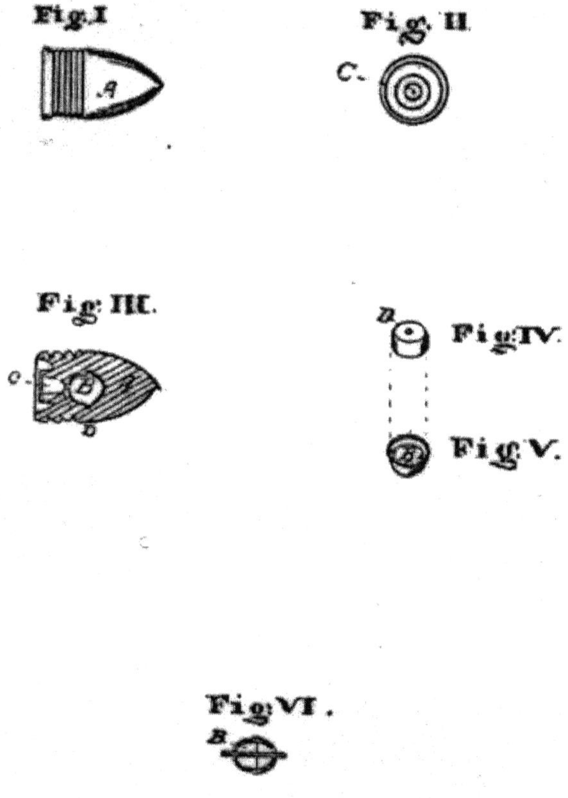

Witnesses

S. Gardiner Jr

The photo below illustrates the entire Gardiner Explosive Bullet with the paper still intact. Due to the chemical reaction of the explosive compound, lead in the bullet and the black powder propelling charge, the base of the projectile splits and eventually tears the paper container.

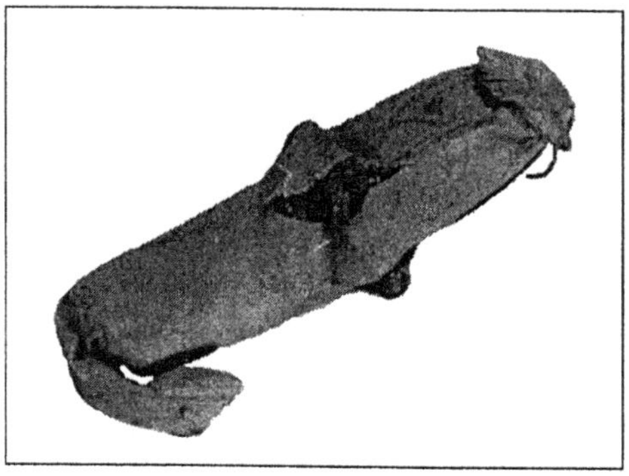

Seldom are these cartridges found in perfect condition.

The dimensions of the completely loaded round are:

Over-all	2.1"
Total Grains	530
Diameter	.574
Powder	60 Grains

There also exists a Confederate manufactured round. The method of rolling the paper case is a good means of identifying the cartridge. There is a very pronounced diagonal seam that is indicative of C.S.A. paper ammunition.

The projectile is made differently from the Northern Gardiner. The Confederate round lacked the protruding tube in the base and lacked the copper tube chamber of the Gardiner. The cavity simply had two passages protruding to either side down to the base of the bullet. When the propelling charge (60 grains) of black powder was exploded, the flame entered these two openings which communicated into the chamber holding the explosive compound. A good cutaway of the C.S.A. bullet can be found on Plate 29-K of "Small Arms and Ammunition"

Dimensions are:

Over-all	2.230"
Bullet diameter	.586"
Total grains	493

14.5mm BAVARIAN LINDNER CARBINE

The Bavarian cartridge was designed for the Bavarian Lindner percussion breech-loader circa 1868. The gun utilized a bolt action type locking system. The face of the bolt had an obturating device to prevent the rearward escape of gas. It would appear that the design was "borrowed" from the Terry, Calisher and Wilson weapons developed in the late 1850s, early 1860s in England.

Similar to these guns, the Bavarian system depended upon ignition in the center and side of the cartridge.

The cartridge has a paper case 1.525" long, a body diameter of .575", the approximately 450 grain tapered round nose bullet measures .557". The complete round weighs 535 grains and has an over-all length of 2.06".

A number of interesting features of the round are:

1. Small pinholes (7) appear on the side of the case.
   This is to insure that the flame from the percussion cap will penetrate through the powder charge. It is assumed that it is not necessary to line up the perforations with the orifice of the nipple cap, obviously to insure proper ignition.

2. A single thread is tied below the bullet to further secure the paper case.

3. A papier-mâché wad is placed at the base of the cartridge to act as an obturating devise. This wad fits snugly into the face of the bolt, remains behind after the piece is fired, and is pushed forward when another cartridge is chambered. In its forward position the wad is pushed out of the barrel, it serves to wipe the bore clean of fouling.

   The same principle was used in a number of earlier cartridges, namely, Terry, Marston, Green, Wilson and Calisher.

4. At the base of the round, a percussion cap is pressed into place. A most convenient method to permit the rifleman a quick means to "cap his piece" rather than having to fumble into a cap container.

All these "door bolt" type weapons were short lived as they were quickly replaced by breech-loaders using self-contained ammunition.

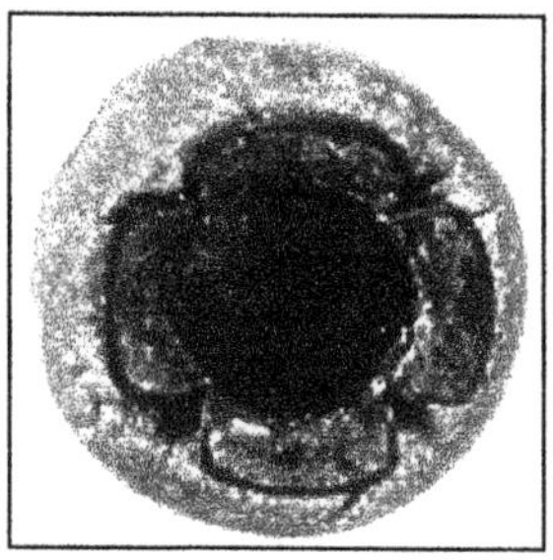

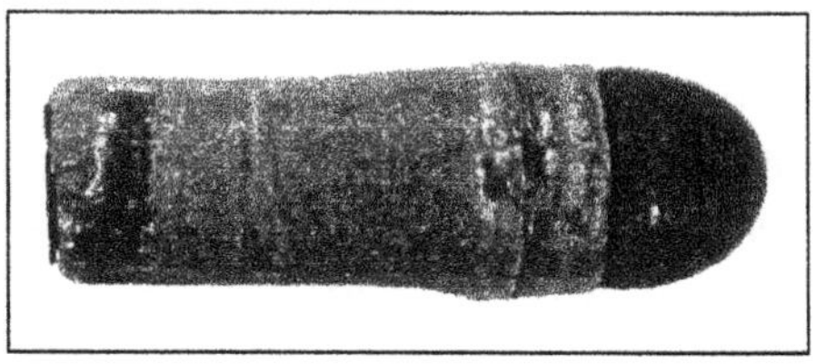

MARSTON AND GOODELL CARTRIDGE

Great strides were made in firearms when the conversion was made from smooth bore barrels to rifled barrels with the resulting increase in accuracy which was not possible before. Now hits were assured at ranges over three hundred yards instead of being limited to one hundred yards, as was the case with the old .69 caliber smooth bores.

A serious problem that did develop with the rifle barrel was the black powder residue deposited in the bore after each shot. After ten shots, it became an absolute necessity to clean the piece; other accuracy was greatly affected and if enough fouling was allowed to accumulate, loading became nearly impossible. Repeated ramming to seat the pointed bullet would quickly deform its pointed nose and cause erratic bullet flight. In order to correct the loading problem, undersized bullets were cast, and while the bullet could now be easily rammed home, it was found that the smaller diameter round became very inaccurate. We now had created a complete turnaround back to the inaccuracy of the old smooth bore weapon. Terry improved it by utilizing a felt wad at the rear of his cartridge and now William W. Marston and Frederick N. Goodell in 1851 invented the Marston cartridge with a leather wad attached to the base of his paper case round. All rounds had one thing in common; a means to wipe the bore clean after each shot.

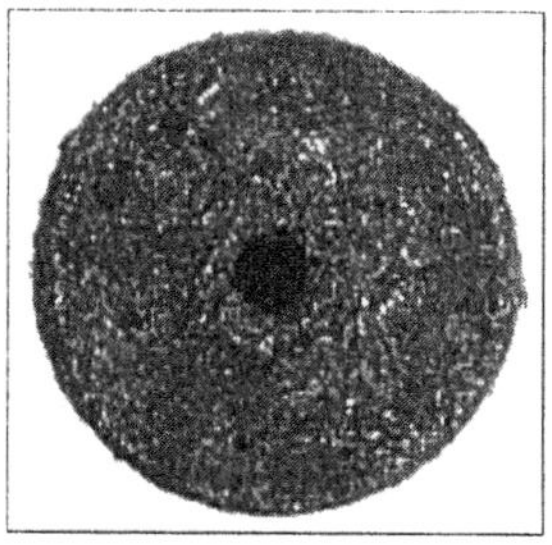

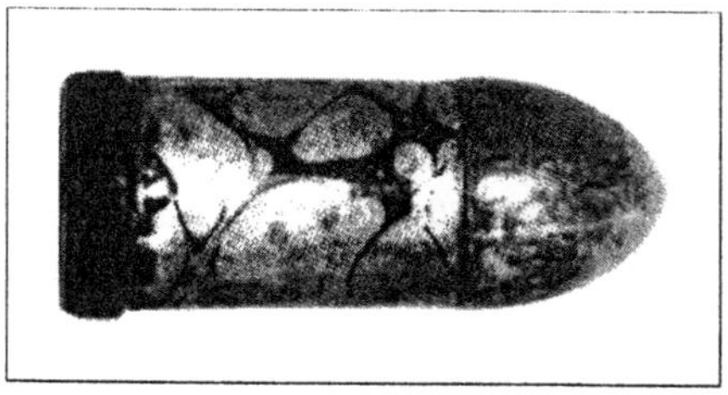

The principle of the Marston was similar to Terry's ammunition, in that after firing a shot the wad would be left behind in the chamber and when a fresh round was inserted and fired, the wad would precede the bullet, thereby, cleaning the bore as it passed through the barrel.

As stated in the text of the patent, it was Marston's and Goodell's intention that the cartridge could be made for use in any weapon from handgun to field piece. The only requirement was that the weapon must be capable of loading from the back of the breech and that the explosion was caused to be in line of the axis of the bore. Shot loads would also work and some arms were manufactured in .70 caliber.

For some unknown reason, little or no progress was made to secure their patent rights and it was not until the summer of 1851 that another request was made by Marston to have the Patent Office test the weapon. In August trials were scheduled by James M. Ward, of Orange, New Jersey, a patent tester who had his patent office on Broadway in New York City. The test was to take place in Brooklyn, New York with the following witnesses: Marston, Zebulon, Sprague and William Selleck of New York City and Messrs. Jacob Rimer, J. Groshoug, Stephen Hart and James Keene, all of New Jersey.

According to the report of the trial, different type materials besides leather were tried on the cartridges. Namely: wood, paper, India rubber, felt and cork. Leather proved to be the most satisfactory.

On September 9, 1851, James Keene and Stephen Hart reported their satisfactory findings to the Patent Office. Marston received the drawings and application back from William Serrell with the recommendation that all should be resubmitted, using the leather sole idea as their only claim. An official letter of rejection was based on the fact that Marston had only utilized the leather part as being a new idea and not that he had invented the idea of putting together with glue, a cartridge with a leather sole as he originally claimed in his application.

Marston felt thoroughly defeated by the rejection, but Anson Taylor-who was an attorney and was financing the manufacturing of the Marston Gun and ammunition now became actively involved. Taylor immediately employed Lemuel Serrell who had recently established his own Patent Attorney Office at 289 Broadway. On April 22, 1852 Mr. Serrell had James Ward, the original tester write to the Patent Office explaining in no uncertain terms that all of those in attendance at the original trials were well satisfied with the claim.

On May 5, 1852, Mr. Serrell resubmitted the original application, drawings and specifications to the Patent Office for reconsideration.

It was on May 18, 1852 that the U. S. Patent Office issued Marston & Goodell Patent #8956 much to everyone's satisfaction.

The following is an excerpt from Patent #8956 that describes the workings of the cartridge:

> "The new effects produced in cartridges thus made and fitted are that, as each cartridge is exploded, the breech piece, in front, literally swabs the barrel of the piece, by taking out, before it, the soilage caused by the previous explosion of the cartridge it had been attached to; so that after almost any number of explosions the barrel of the gun is nearly as clean as when it has been fired only once, with the additional and important advantage that the breech piece of the cartridge is not destroyed by the burning powder and serves as a shield to the breech pin of the gun and prevents any product of the explosion from causing any soilage at the end of the pin, so that no interference to the action of the pin is caused by the dirt of the soilage. All these are results which we believe have not been separately or collectively attained by any other cartridge hitherto made.
>
> We do not claim to have invented any of the separate parts described herein; but we do claim as new and of our own invention-the application of the leather breech piece to cartridges used with breech loading guns, such leather breech piece serving the purposes of a foundation for its own cartridge, a protection on to the breech pin, a wad for the next cartridge in succession, and of a swab, to clean out the soilage caused in the barrel by antecedent explosion producing a safe cartridge for pieces that load at the back of the breech, and in which explosion is also caused in the line of axis of the barrel......etc., etc.

While Marston was attempting to settle his problems here in Washington with the Patent Office, his partner, Frederick Goodell was demonstrating the rifle and ammunition in Europe in order to obtain European patent papers. He was quickly accepted in France and Shortly after his demonstrations in Paris was granted patent #8273, 1852.

In January 1853 issue of Scientific American gave a report on Marston's Rifle and its peculiar looking ammunition. Comments were made on the efficient manner in which the greased leather wad cleaned the bore of the weapon. Stating: "Barrel shines bright after 1,000 shots".

In the following issue of Scientific American (February 19, 1853) a review of the trials of the Marston Rifle and ammunition held at the Woolwich Royal Arsenal in England. Sir Henry Harding, Commander in Chief of the British Forces was greatly impressed with the gun. The weapon was fired 100 times in fifteen minutes (approximately one shot every ten seconds) and while unimpressive by modern day standards, it was a truly great accomplishment in those times. Undoubtedly the greased leather wad succeeded in keeping the bore clean enough so that there was no interruption to clean the barrel during the firing of that number of rounds. Unfortunately, no mention of the caliber of the rifle was given, but it would seem to be reasonable to assume that the British Army would be only interested in a large size projectile similar to their own .577 then currently in use.

While we are concerned here with .58 caliber ammunition, Marston did produce cartridges, rifles and pistols in caliber .31, .36, .46, .54, .58 and a shot cartridge in .70 caliber.

Very few of the cartridges have survived.

A .58 caliber Marston military musket was in the Henry M. Stewart collection. It would appear reasonable to expect that if the weapon was designed as a military musket it must have been intended to be offered for consideration at some Ordnance Board weapons testing trial. Since a weapon without ammunition would be totally useless, an ample supply of .58 caliber ammunition would have to have been made available for testing the weapon.

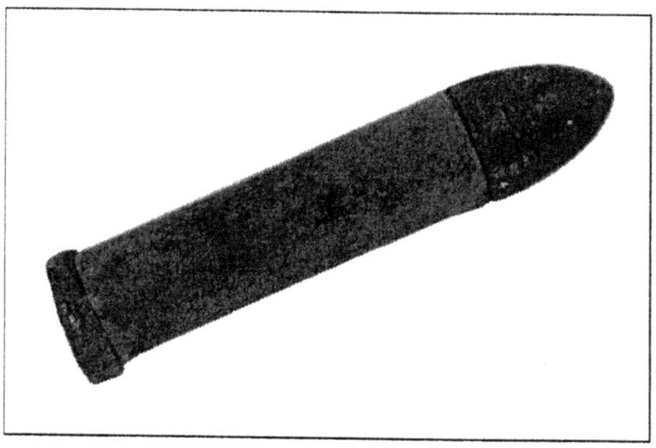

MERRILL PATENT FIREARMS COMPANY

James H. Merrill, an arms maker, inventor and patentee since 1840 seemed to have known early on the need for breech-loading firearms. He knew the outmoded muzzle loading system was on its way out and the era of the breech-loading firearms would soon take over.

From 1855 through 1861, Mr. Merrill was associated with Messrs. Thomas Latrobe and Thomas & Co. Later on the firm was known as Merrill, Thomas & Co., 239 Baltimore Street, Baltimore, Maryland. Sometime in 1864 the firm was organized as Merrill Patent FireArms Co., Baltimore, Maryland.

On July 20, 1858, a patent was issued to Merrill describing his method of creating breech-loading rifles and carbines utilizing either paper cartridges or nitre treated paper cartridges. The following describes the loading procedure (from U. S. Muskets, Rifles and Carbines, by Arcadi Gluckman, Col. USA):

> "To load, the percussion hammer was cocked and the arm was operated by a lever latched under the rear sight base. Back pressure on two flat, knurled side releases, unlatched the lever and permitted it to be raised backward, withdrawing the hinged, sliding, copper faced breech bolt and exposing the breech for the insertion of a nitre treated paper cartridge. The copper face on the bolt acted as a gas check. The lever was then lowered forward and latched under the rear sight, closing the breech, and percussion could be fired by the hammer and cone."

Merrill also converted Model 1841, 1861 and 1863 rifle muskets and rifles much on the same idea as previously mentioned. After drawing the lever to the rear and exposing the breech, the paper cartridge was chambered, a sliding breech-bolt then moved forward to seal the breech when the lever was pressed forward.

Patent #14077 was issued January 8, 1856 utilizing many of the features of the top hinged locklever as previously discussed. The text describes the "loading trough" upon which the .58 caliber cartridge is placed and chambered by pushing forward a small rammer sliding along the top of the small of the stock. When the locklever is lowered, the cartridge is seated and the excess paper at the base of the round is cut off thereby exposing the powder charge. An opening well (through the trigger guard plate) permits the excess paper to fall out of harms way.

One hundred seventy of these carbines marked MERRILL, LATROBE AND THOMAS were purchased by the government on July 26, 1855.

Opinions rendered by the Military were generally favorable to Merrill's weapons. The system was relatively inexpensive, simple to operate, reliable and not apt to get out of order. Another advantage was that the conversion could be applied to a variety of firearms.

Major R. E. Colston made the following remarks in his report to the commissioners of the Virginia Military Institute dated June 28, 1860 regarding the ammunition for the Merrill:

> "The great advantage of these arms over the others is, that they can be used with paper cartridges or with loose powder and ball, without any metallic or other cartridge case. This secures one of the indispensable conditions of an arm of war."

No firearm manufacturer's catalog "worth its salt" would be printed without many pages devoted to testimonials from satisfied customers.

Merrill's catalog of 1864 was no exception. Many of these "good words" were from the military as well as from sportsmen who had depended upon Merrill's guns to hunt and protect themselves. One letter was from Mr. J. U. Shaler, the inventor of the Shaler Sectional Bullet. The letter read as follows:

> GENTLEMEN:
>
> In testing the qualities of the "Merrill Breech-Loading Rifle", by firing my Sectional Bullets, my expectations are more than realized. Your rifle makes better hitting, and at a longer range, and with less powder than any arm I have ever used, either muzzle or Breech-Loading. My sectional Bullet weighs 700 grains, and your carbine fires them with 40 grains of powder, of the regulation grade, a result I have never been able to accomplish with any other arm. The objection I have heretofore had to breech-loading arms is overcome by the use of yours, for there is no escape of gas at the breech, and the adjustment is simple and perfect, which is a valuable consideration to the soldier and also sportsman. It is not so much trouble to keep your breech-loader in order as is the muzzle-loaders.
>
> Being satisfied by actual demonstration that your breech-loading rifle possesses all the qualities to make it good, and efficient for the different purposes, for which arms are used, I can cheerfully recommend it to the public.
>
> I am Respectfully Yours,
> J. U. Shaler

The Shaler Sectional Bullet, as discussed elsewhere in the book, consisted of three separate pointed nose projectiles nestled within one another. The purpose of its construction was to increase the chances of a hit as the three bullets diverged upon firing. Merrill's method of tightly sealing the breech thereby preventing the escape of gas, obviously enabled the gun to "shoot harder" than other weapons where the breech was not tight.

The following specimens would be representative of those Merrill combustible rounds generally encountered.

The dimensions are:

	BROWN PAPER	SALMON PAPER	TAN PAPER
Case Length:	1.61"	1.57"	1.68"
Bullet Diameter:	.580"	.560"	.590"
Total Grains	475	442	449

Merrill ammunition came in "match box" type containers holding ten cartridges. The box was approximately 1 3/4" by 2 3/4" and had an attractive pink/salmon label

When I first started on this .58 caliber ammunition book I began to realize that I lacked an ample supply of paper cased specimens that I could illustrate and identify. I did not wish to keep using the term of "unknown maker, arsenal unknown" etc. I purchased about a dozen specimens from the late James Tillinghast who bought the entire collection from a gentleman in Texas, but Jim faced the same lack of information problem, as do I.

1. .58 Chadwick

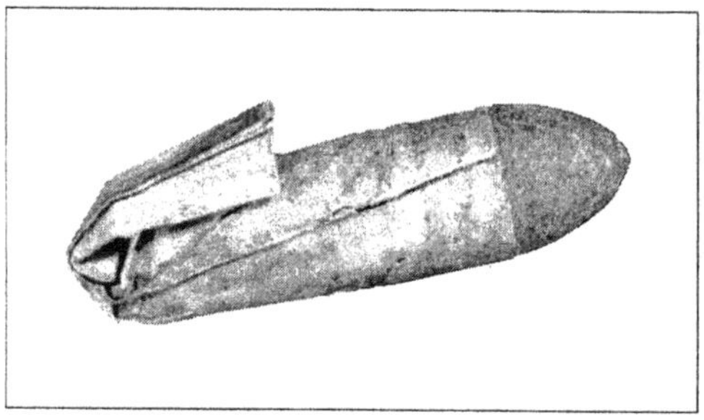

Bullet .600 Overall 2.21" Total grains 560

2. .58 Gardner Inserted

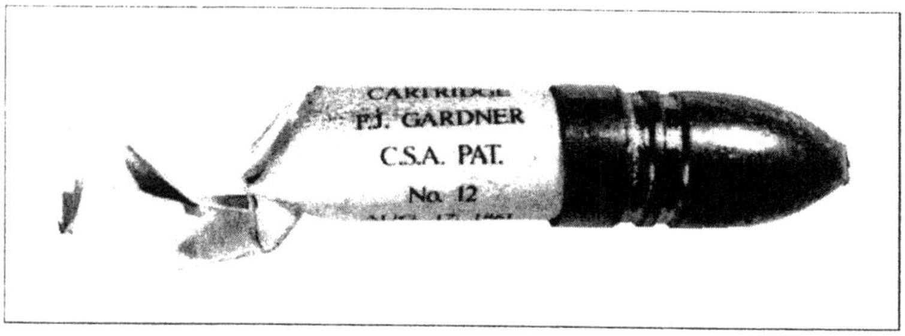

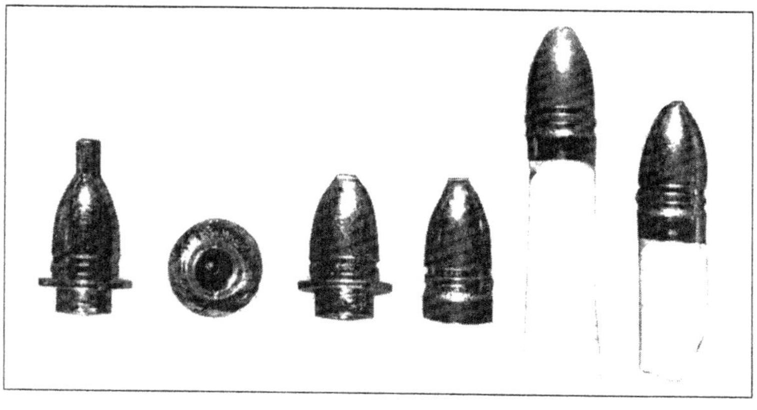

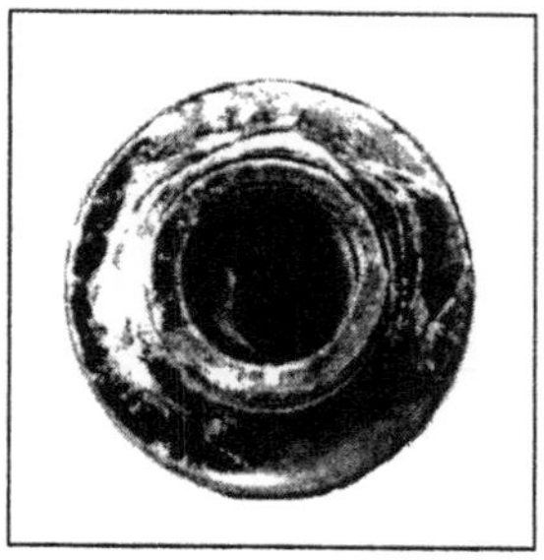

3. .69 caliber paper cartridge (predecessor of the .58 cal.)

4. .58 Rifle musket (blue paper over regulation buff)

5. .58 Rifle musket (orange paper / Williams 3 piece bullet)

6. .58 Rifle musket (buff paper with grey tail)

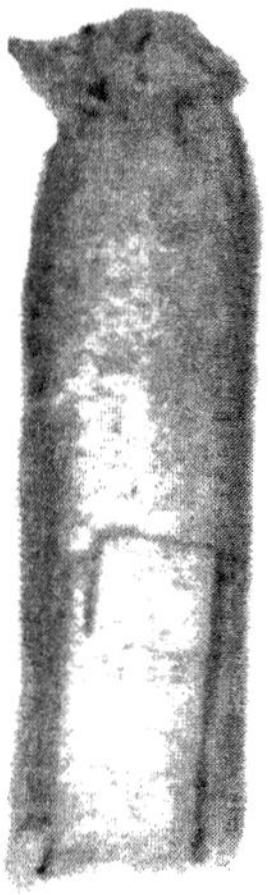

7. .58 Rifle musket (buff paper tied with yellow string)

8. .58 Rifle musket (buff paper tied with buff string)

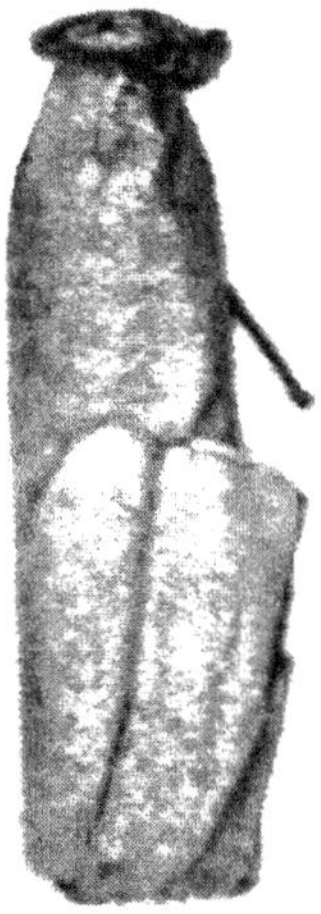

9. .58 Rifle musket (buff paper tied with red string)

10. Battlefield pickups (.58 Federal)

10 b. .57 Sharps

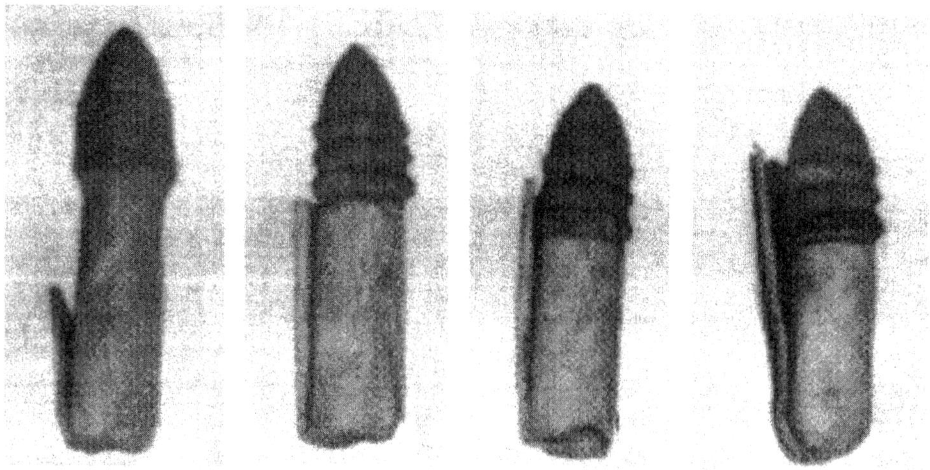

11. Sharps variants

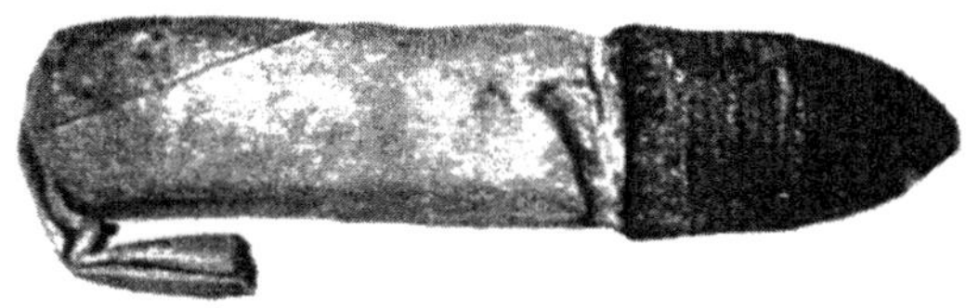

12. .58 Sharps paper

13. .577 Enfield musket

14. .58 Mount Pleasant Combustible

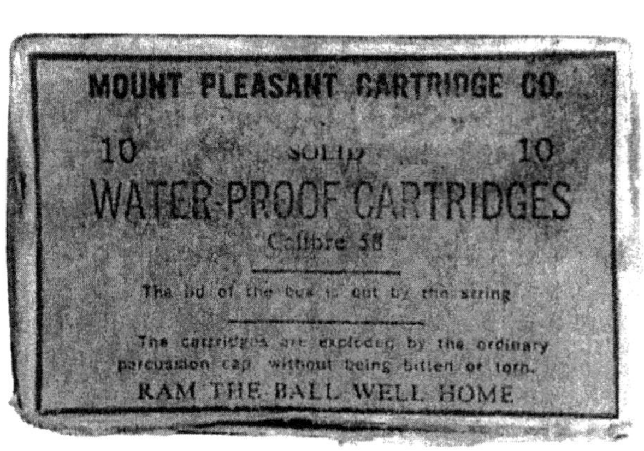

14 b. Mount Pleasant box

15. .58 Mexican musket

16. .69 Mexican musket

17. .58 Texas arsenal musket

18. .58 manufacturer unknown

19. .58 Texas arsenal musket

20. .58 Kennebunk

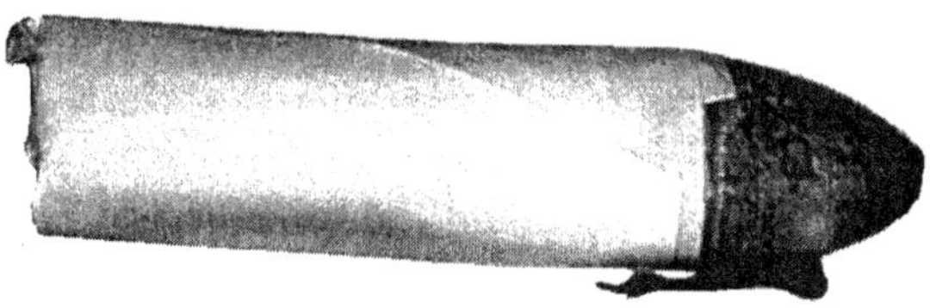

21. .577 unknown pull tab

22. .577 British

23. .58 ball unknown

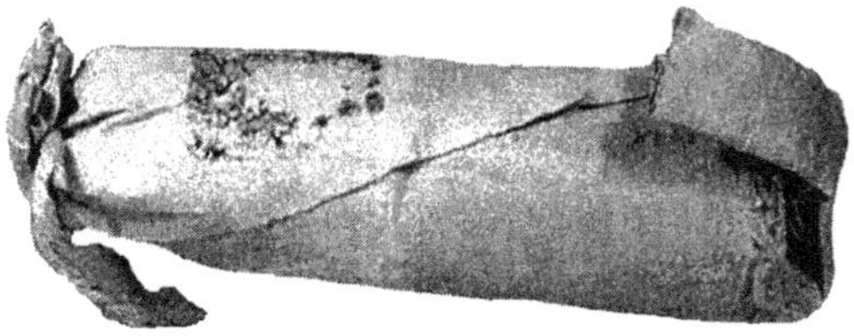

24. .577 unknown musket

25. .58 musket

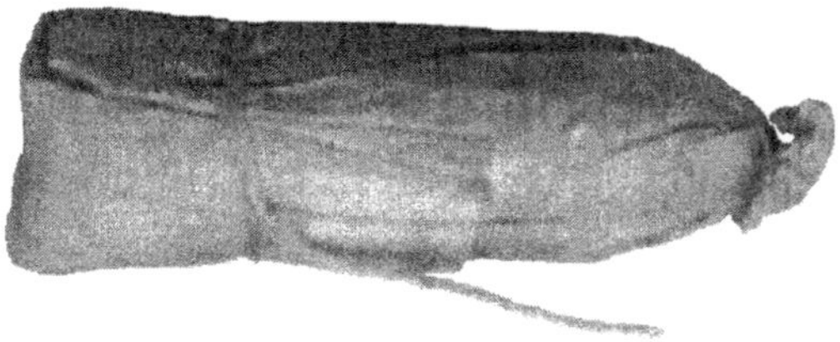

26. .58 U.S. musket

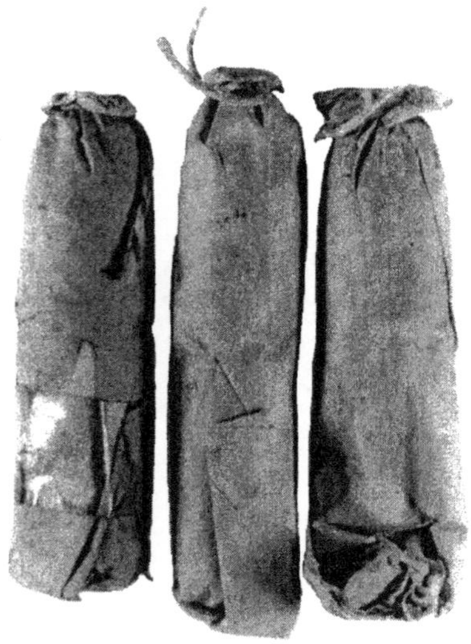

27. .58 unknown manufacturer

28. .69 Frankford arsenal box

29. .58 skin U.S. rifle unknown manufacturer

30. .58 skin unknown manufacturer

31. .577 combustable enfield rifle.

32. .58 skin

33. .58 Colt combustable

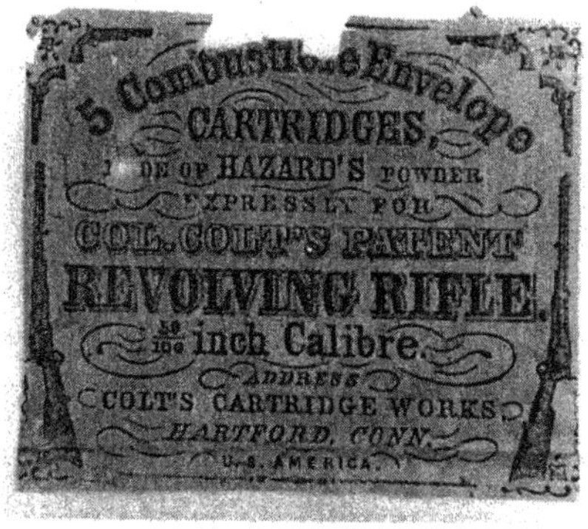

33 b. .58 Colt combustable box

34. .58 thread wound skin.

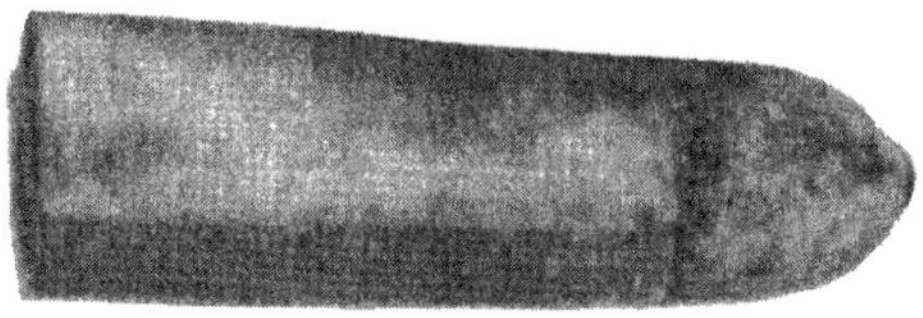

35. .54 Starr

TITLE: .58 SPRINGFIELD MODEL 55 INSERTED CHARGE

HEADSTAMP:

DIMENSIONS:

Rim:
Head:
Neck:
Bullet: .582"
Case:

OTHER DETAILS:

Bullet Grains: 497
Powder Grains:
Bullet:
Case Material: Paper
Total Grains: 570

OVERALL: 1.961"

With the adoption of the new Model 1855 came a new type of paper cartridge. This new cartridge was contained in a separate cylindrical case made of heavy rocket paper. This was rolled in a piece of cartridge paper and filled with powder; then the whole thing was placed in another piece of paper with a bullet and rolled again. The heavier paper construction resulted in much less damage in the field once issued to the troops at the approximate start of the Civil War - the exact date is not known. The old system was reinstated because it was much simpler to manufacture.

Completely sealed packages are extremely rare.

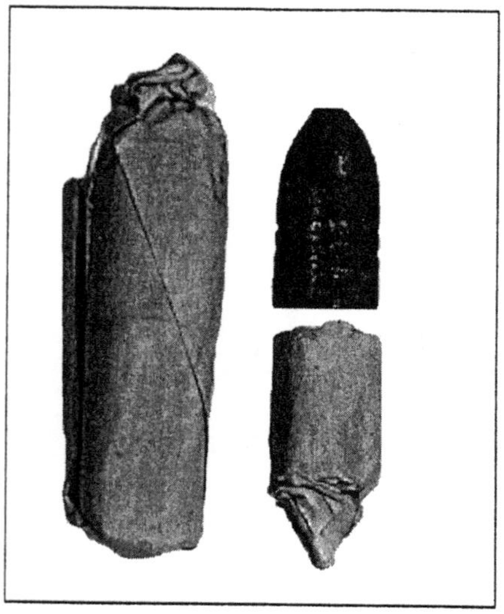

PRINCE RIFLE AND CARBINE CARTRIDGE

There was slow acceptance by the military to the idea that in order to have a successful breech-loading rifle, a self-contained cartridge was a necessity.

Inventors would not relinquish the idea that the best form of ignition was provided by a nipple and percussion cap despite the fact they had designed a breech-loading weapon.

The Prince rifle and carbine was a case in point.

The gun was of the "door bolt" design, i.e. a knob projected below the barrel and in front of the trigger guard. When pushed forward the entire barrel slid forward exposing an opening which permitted the insertion of the nitrated paper cartridge. When fired, the flame from the percussion cap had to penetrate the base of the cartridge to ignite the charge.

Mr. Prince, a well-known English gunsmith developed his carbine in 1859. He enjoyed considerable success with this weapon among sportsmen. Because of this success, he later designed a rifle for submission to the army trials. In spite of the capping apparatus, the inventor claimed to be able to fire the gun sixteen times a minute. The gun was not accepted.

The cartridges below are designed with a brown pull-tab to remove the outer paper wrap. This round has a blue wad on its base marked "ELEY LONDON". It is 2.04" in length but only weighs 555 grains, which might indicate it is for the carbine and the heavier round below that was meant to be used in the rifle model.

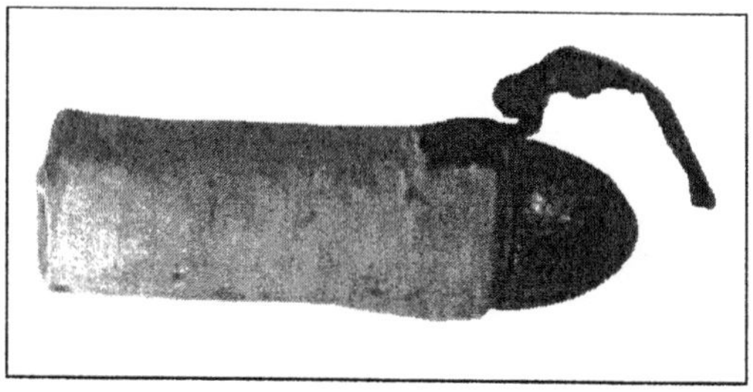

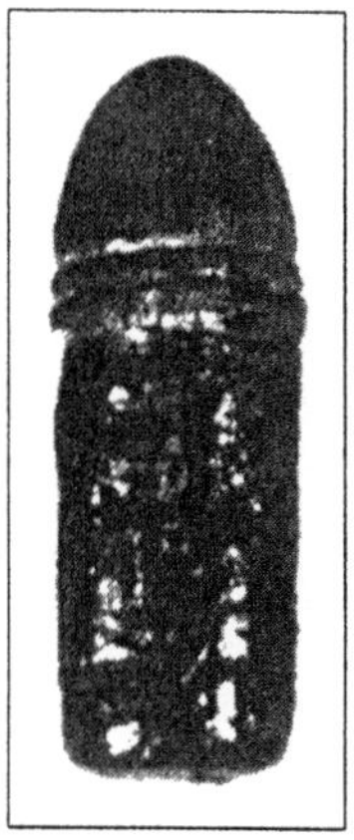

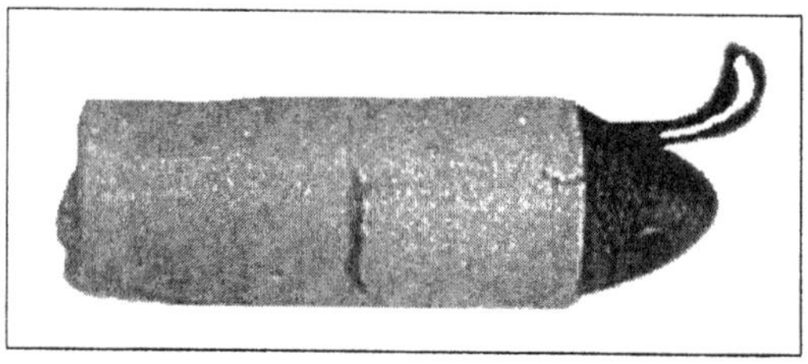

TITLE: .57 SEAMLESS SKIN CARTRIDGE

HEADSTAMP: None

DIMENSIONS:

Rim:	----
Head:	----
Neck:	----
Bullet:	.571"
Case:	1.412"

OTHER DETAILS:

Bullet Grains:	485
Powder Grains:	60
Bullet:	Conical
Case Material:	Skin
Total Grains:	550

OVERALL: 2.002"

An attempt to improve over the paper container type cartridge, hence the manufacturer resorted to animal intestines to create a self consuming round.

TEN PATENT SEAMLESS SKIN
CARTRIDGES
FOR U S GOVERNMENT RIFLE
60 GRS 574 CALIBRE
IMPERVIOUS TO DAMPNESS
MANUFACTURED BY D.C. SAGE & CO.
MIDDLETOWN, CONN. U S A
JULIUS HOTCHKISS PATENTEE FEB 11, 1862
JW.J. SYMS & BROS. SOLE AGENT BROADWAY, N.Y.

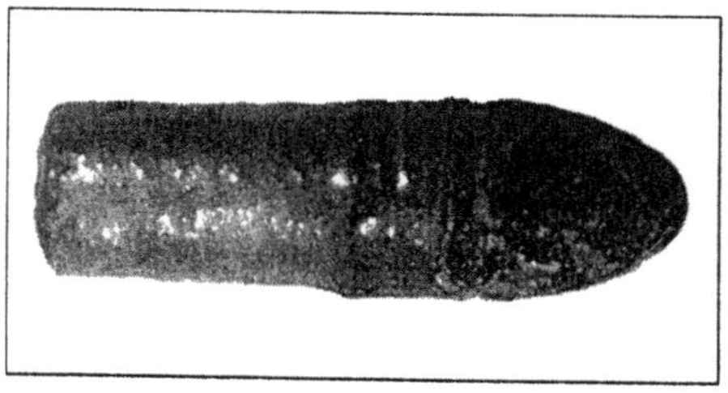

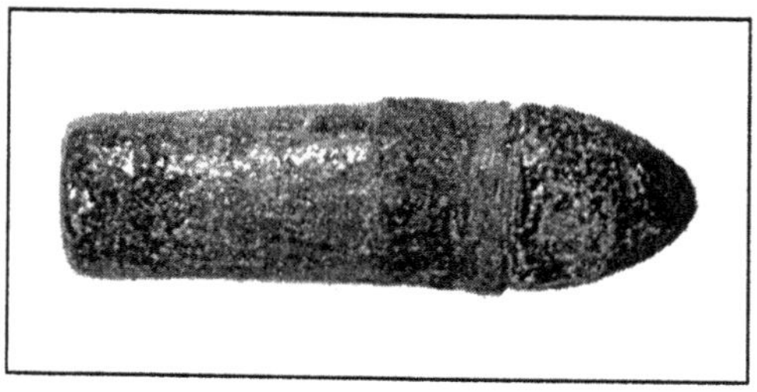

SHALER

On August 15, 1861, Reuben Shaler, 625 Pacific Street, Brooklyn, NY wrote to the War Department requesting an opportunity to demonstrate his new cartridge for the US Musket Rifle. He claimed accuracy out to 300 yds. With the 3 projectiles separating to an area of three feet. He maintained it would possess a three-fold advantage over the Minie-ball and great improvement of the much touted Multi-ball round presently in use. Approval was given for a trial of the round.

Captain S. V. Benet submitted a favorable report to Brigadier General J. W. Ripley, Chief of Ordnance who further instructed Captain Thomas J. Rodman at the Watertown Arsenal to try Shaler's bullet. The report was somewhat favorable but Shaler was upset that the projectiles didn't always enter the target point first. Unspecified adjustments were made in its construction and after more testing it was felt Shaler's cartridge was superior to the current buckshot cartridge presently issued to the troops.

Another trial on December 29, 1861 in Washington D.C. at 150-200 yards was made and General Alexander Shaler (no relationship) praised the sectional bullet highly. Powder charges varied from 50-60 grains each.

Becoming impatient with not having heard from General Ripley, Shaler traveled to Washington in an attempt to meet with him to get his bullet adopted by the Ordnance Department. Shaler pointed out in his letter to Ripley that he strongly felt his invention enabled the soldier to make three hits at each fire, and at the same time maintain an accuracy of one hit in each fire to equal that of the firing of one bullet.

During a trial – circa 1860s – the new Chief of Ordnance General George D. Ramsay, somewhat in favor of the cartridge, wrote a report from General Marsden, Ordnance pointing out some unfavorable aspects namely that "the pieces clung in flight so that never more than two hits on the target and as seldom but one." The criticisms were strongly defended by Shaler. (For a complete review see Dean Thomas' "Round Ball to Rimfire" Vol. #1 page 201)

An impassioned plea to General Ramsay contained the following comments "If this cartridge will as we allege and believe and will evidently affirm to those who examine the subject make one man equal to two or in other words as General Alexander Shaler says give our Army 100 per cent advantage over one that does not use them on the Field of battle. Then is it no small consequence to the government and the country that there should be no unnecessary delay in furnishing the Army with them."

According to J. E. Casey, agent for Ira W. Shaler, orders were very slow in being made and insufficient to supply the troops. Despite J. E. Casey's pleas to General D. Ramsey in November 1861, orders to purchase Shaler's cartridges, the reply was to the effect that a recent trial by Captain Benet at the Washington Arsenal confirmed that the result of such trial did not justify further orders.

DESCRIPTION OF THE SHALER CARTRIDGE

Overall length of the round was 2 ¾" long, the white paper case containing the sections or projectiles, and a powder charge of 55-60 grains of black powder. A pull-tab protruded approximately an inch from the base and was used to open up the paper case without the necessity of having to expose the powder as was required in the bulk of current paper cartridges then in use. The total grain weight of the round was approximately 720 grains.

The bullets were unique in configuration. The front projectile was similar to the musket round with a hollow base in which nestled another pointed bullet with a wide ring around its base. The third bullet was much smaller with a pronounced pointed nose and a solid base. As can be seen from the photos all three pieces nestled together, were placed in the paper container and set upon the 60 grains of black powder. There were four styles of projectiles types A, B and C were encountered and type D remained extremely rare.

Despite all the praise extolling the many improvements over the "buck and ball" cartridges issued to the troops. A number of complaints about the round were reported by those who were concerned with testing them. Some of which were, inaccuracy, fusing of the projectiles after leaving the gun barrel, erratic performances at certain distances including penetration in the target. Rumors were circulated that the rounds contained poison placed inside each bullet. These remarks were also made about the Williams Clean-out Cartridges. Neither of which were ever proved.

This would not be the first or the last incidence where a particular piece of military equipment did not meet with favor among the infantry and was soon dropped.

(For more detailed information see Dean S. Thomas' "Round Ball to Rim Fire – part one)

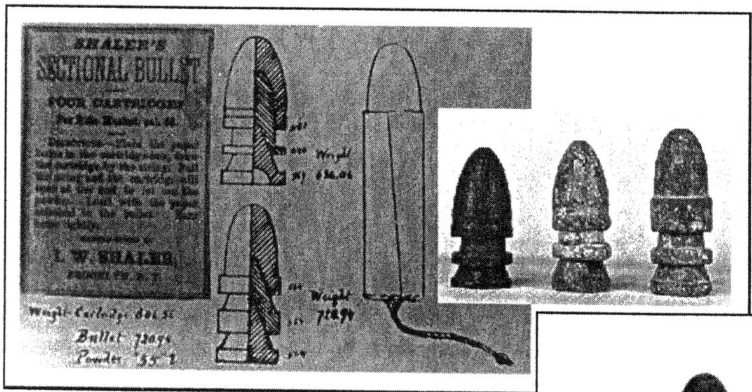

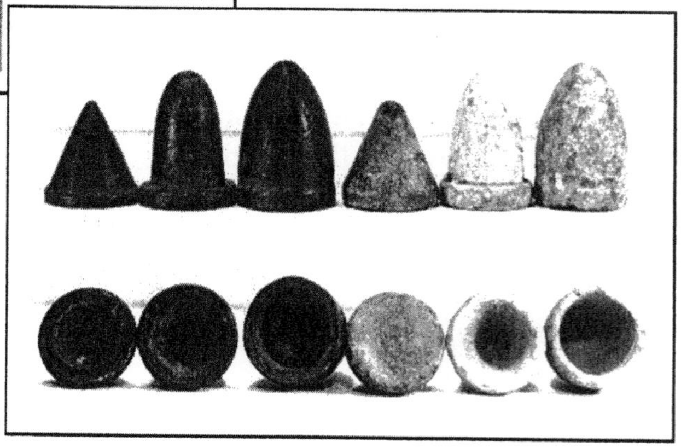

WILLIAMS CLEAN OUT CARTRIDGE

Black powder fouling deposited in the musket bores were the bane of the soldier's existence. He was constantly faced with the problem. He had no alternative but to clean the bore frequently for as bad as the accuracy of the gun was to begin with when clean, it was an untenable situation after shooting a number of rounds. In addition to poor accuracy created by the black powder residue, if enough fouling was permitted to build up in the bore, it would become next to impossible to load the weapon. Heavy pounding on the ramrod only resulted in a flattened bullet nose thus distorting its shape and making the weapon more inaccurate than what it was already. Cleaning the bore could be a nuisance which could be endured during rifle practice, but during a battle, the soldier was exposed to danger either by having a rifle with which he couldn't hit anything, or by having to stand up and expose his position while cleaning and scraping the bore.

Elijah D. Williams designed a very satisfactory cartridge to assist in cleaning the bore and reduce the number of times the infantryman had to do it manually. This was accomplished by simply firing the round once in every ten or twelve shots thereby greatly reducing the build up of powder residue. A Williams "Clean Out" cartridge was placed in every packet of ten rounds issued to the soldier. All he had to remember was to save the blue or salmon colored paper cased round to the last of each packet and he could feel assured the bore would be scraped clean.

There were three types of "Clean Out' cartridges. Type I had two separate zinc washers secured to a teat which was cast to the base of the 555 grain bullet. The three-groove rounded nose bullet was propelled by 65 grains of powder incased in a blue wrapper. In 1864, Type II was introduced with just one heavy zinc washer which was staked to the base of the three-groove round nose bullet. The 465 grain bullet was driven by five less grains of powder and was sometimes wrapped in red or salmon colored paper. Another lighter load was made to be used in the Union or "Coffee Mill" gun (discussed in detail elsewhere in the book).

Ordnance inventories also listed Williams bullets as having been loaded in Cal. .69 cartridges but to date, no specimens have been examined.

Tests conducted on the Williams bullet and reported upon by Master Armorer Allin of the Springfield Armory spoke highly of the efficiency and accuracy of the Williams bullet. During December 1861, three million rounds were ordered. Unfortunately, the troops felt it damaged the rifle bores and the prejudice led to its demise as evidenced by Circular No. 47 - Series of 1864 which reads as follows:

> Circular No. 47-Series of 1864
> Relative to Williams' Bullets
> Ordnance Office
> War Department
> Washington, September 19, 1864
>
> Circular No. 47 of 5th August 1864, which directed that six cartridges, made with Williams' bullets, should be included in every package of ten cartridges for rifle muskets, is hereby revoked.
>
> No more Williams' bullets will be made up into cartridges after the receipt of this circular. Such bullets of that kind as may be on hand will be retained until further orders, but will not be used except in cases of emergency. The cartridges of this kind on hand will be issued, but where they have not been packed, only three will be put in each bundle until those on hand are used up
>
> A. B. Dyer
> Brig. General
> Chief of Ordnance

Justified or not this would not be the first or last time that soldier's opinion of a piece of his equipment would cause it to be cast aside.

IV

Patent Ignition

TITLE: .58 BAR ANVIL

HEADSTAMP: 58 CAL

DIMENSIONS:		OTHER DETAILS:	
Rim:	.714"	Bullet Grains:	500
Head:	.632"	Powder Grains:	60
Neck:	.625"	Bullet:	Pointed
Bullet:	.612"	Case Material:	Copper
Case:	1.194	Total Grains:	673

OVERALL: 1.720"

An extremely rare cartridge utilizing the bar-anvil principal. All reference books tell us this system was first applied to .58 caliber cartridges but here we have evidence to the contrary. The bar-anvil consisted of a copper case and a rigid tinned-iron anvil recessed at the center to hold the percussion composition and grooved in the ends for crimping and securing into place.

Case construction of this nature was not successful as the crimping operation weakened the rim causing the case to burst. In assembling, the bar-anvil could be inserted upside down and if all that wasn't enough, occasionally the anvil could get lodged in the barrel. These problems coupled with a high rate of misfires, it was decided the project should be dropped in favor of the more reliable Benet Inside Primed cartridge.

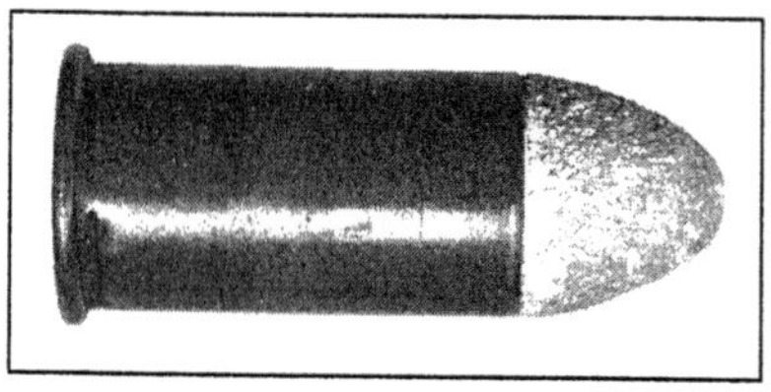

CHARLES F. BROWN'S MUSKET CONVERSION OF A .58 CALIBER RIFLE.

Charles F. Brown, of Warren, Rhode Island, if nothing more could be said for him, had an extremely fertile imagination. His patent No. 12942 was granted on May 29, 1855 to construct a "water cartridge" for the service rifle.

His idea was to construct a cartridge case with a chamber in the rear of the cartridge to be filled with water and, another area in the front of the case. Upon exploding, water in the rear compartment would leak out to put out any residual fire from the explosion and the water in the front chamber area, upon exploding would cool off the weapon if carried in an inclined position.

I would not wish to write further of this and deprive the reader of a good laugh learning of other plans he also had for his invention.

Needless to say no known cartridge specimens were ever made available.

C. F. BROWN.
Cartridge.

No. 12,942. Patented May 29, 1855.

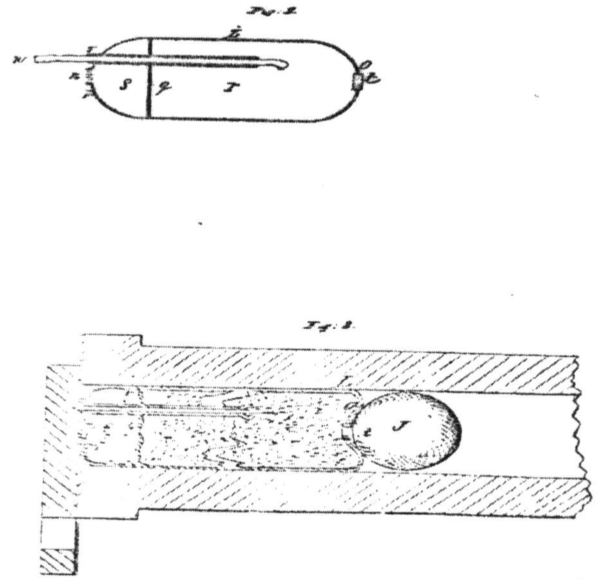

UNITED STATES PATENT OFFICE.

CHARLES F. BROWN, OF WARREN, RHODE ISLAND.

IMPROVEMENT IN CARTRIDGES.

Specification forming part of Letters Patent No. 12,942, dated May 29, 1855.

To all whom it may concern:

Be it known that I, CHARLES F. BROWN, of Warren, in the county of Bristol and State of Rhode Island, have invented an Improved Cartridge for Breech-Loading Ordnance; and I do hereby declare that the following is a full, clear, and exact description of the same, reference being had to the accompanying drawings, forming part of this specification, in which—

Figure 1 is a central longitudinal section of the cartridge; Fig. 2, a central section of the chamber of a gun, with the cartridge in the act of exploding.

Similar letters of reference indicate corresponding parts in both figures.

The case of this cartridge is intended to be made of tin-plate or other metal.

The invention consists, first, in making the ends of the case of hemispherical or other convex form. By giving them this form, when the explosion takes place they will be caused to expand in such form and direction as to fill the bore of the piece tightly, and in a great measure prevent windage both at the breech and around the ball or projectile.

The invention also consists in making a water-chamber in the rear end of the case. This chamber will be burst by the explosion of the charge, and the water will serve at the same time the purposes of cooling the breech and of preventing the escape of the fire at the breech, and the consequent loss of force.

L is the case, of which *o* is the front end, and *p* the rear end, both of nearly hemispherical form. At a short distance from the rear end there is a partition, *q*, which divides it into two parts, *r* and *s*, the former of which is the powder-chamber and the latter the water-chamber. In the front end *o* there is an opening to insert the powder in the chamber *r*, and in the rear end *p* another opening to fill the chamber *s* with water, the former opening being closed by a tight stopper, *t*, and the latter by another tight stopper, *u*. A tube, *v*, leads from the rear end *p*, through the water-chamber and the partition *q*, into the middle of the powder-chamber, to convey a fuse, *w*, to the center of the charge, the said tube being intended to be furnished at its outer end with a percussion priming, which will be ignited by the shearing off of its end by the closing of the breech; or the fuse may be ignited by any other convenient means.

The ball or other projectile, J, is put in the piece separate from the cartridge, and in order that the cartridge and projectile may be inserted easily and the ball may fit with sufficient tightness during the discharge, the chamber of the piece should be slightly enlarged toward the breech. The explosion of the cartridge drives one end toward the breech and the other toward the projectile, and the concave end *o* will be driven against the projectile with such force that before the latter has hardly started the said end will have been expanded laterally, so that it nearly forms a square corner all round the bore, and the case L, for a considerable distance from its front end, will fit very closely to the bore, so as to prevent windage around the projectile. The end *p* will be driven back against the breech, and a similar lateral expansion will take place, causing a considerable portion of the case from the rear end to fit tightly enough to the bore to prevent, in a great measure, any windage or escape of exploded powder taking place at the breech by reason of the latter not being perfectly tight. The explosion of the charge after having expanded the end of the case, as described, will burst the water-chamber, either by tearing the metal or melting the soldering of the case, and the water being discharged will not only cool the breech, but, if the breech should not be tight, it will be driven into the crevices, and thus prevent the escape of the gas or exploded powder. When the breech of the piece is opened after the discharge, the water all runs out unless the inclination of the piece is such that it runs toward the muzzle.

What I claim as my invention, and desire to secure by Letters Patent, is—

1. Making the ends of the metal cartridge-case of hemispherical or other convex form, substantially as and for the purpose herein set forth.

2. Providing a water chamber, *s*, in the rear end of the cartridge case, to be filled with water previously to the insertion of the cartridge in the gun, substantially as and for the purpose set forth.

CHARLES F. BROWN.

Witnesses:
JOS. GEO. MASON,
WILLIAM TUSCH.

V

Percussion

AGER (UNION REPEATING GUN) CHAMBER

It wasn't until 1866, five years after its development, that Wilson Alger, an American secured British patents for this weapon. During the Civil War it was known as the Union Repeating Gun and or Coffee Mill Gun. This title was given because of its likeness to the cartridge feed hopper on the top of the breech.

The ammunition was a .58 caliber Williams Clean-Out cartridge inserted in a steel chamber with a percussion cap fitted to the nipple. Lacking a supply of the clean-out cartridges ordinary paper ammunition could be used but of course they lacked the barrel cleaning potential of the Williams round. To operate the gun a supply of the loaded chambers were placed in the "hopper" mounted on the top of the piece. The operating handle was turned, and by mechanical means was aligned with the barrel, cammed forward, fired and the empty chamber fell to the ground where it was later collected and reused again.

The cycle rate was approximately 120 rounds a minute.

Evidence that the gun was used in the Civil War, observe the picture of the rusty relic found at Gettysburgh.

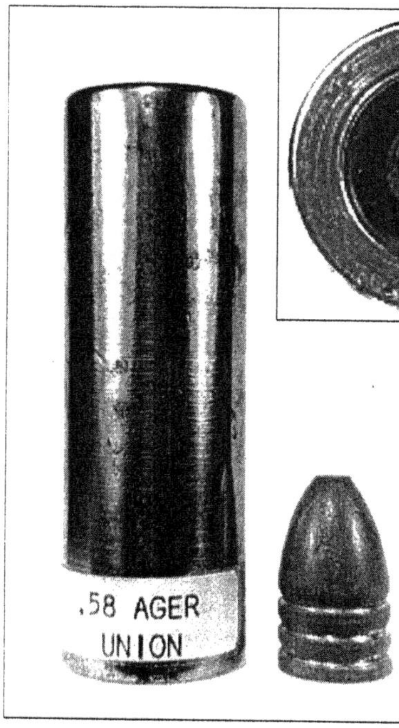

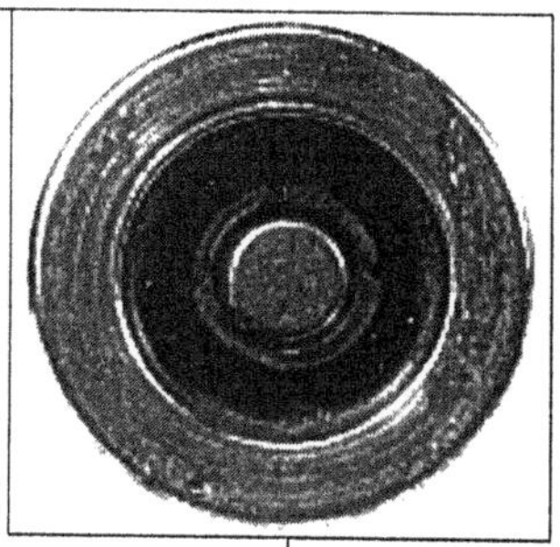

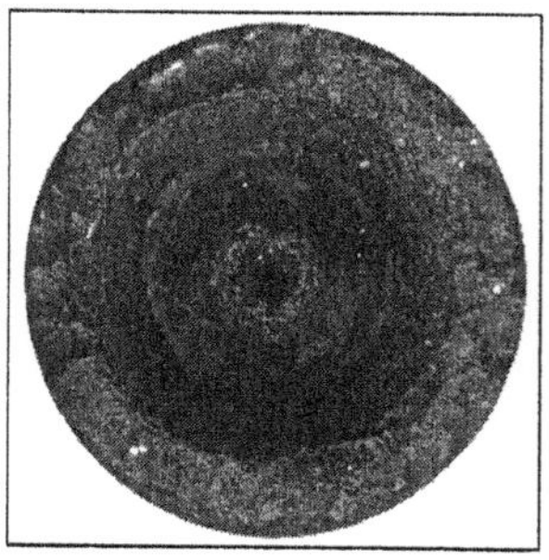

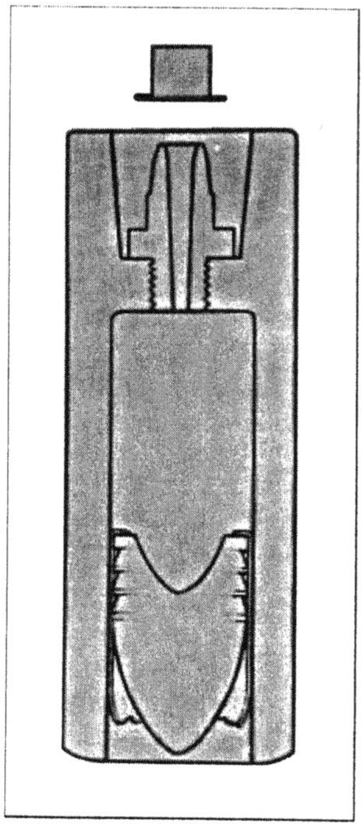

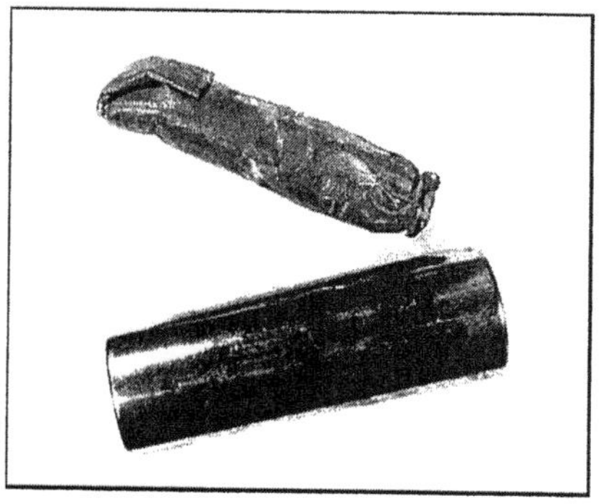

BILLINGHURST REQUA GUN

The Billinghurst Requa and Mitrailleuse battery gun is a spin off and improvement of the Danish Orgelespringol (see Hoyem Vol I pg. 27 and 93 Circa 1864). The gun consisted of 25 barrels laid across the rear of the weapon with a hinged bar at the rear of the breech so that it could be opened or closed by means of an upright lever at the rear of the breech. In order to operate the piece, a perforated strip loaded with 25 brass cartridges was used to insert the rounds are at one time into the chamber of each barrel.

A small amount of black powder was laid in a trough extending along the back of the 25 cartridges and the breech was closed. The hammer was set in a centrally located position with a percussion cap positioned on the nipple. To fire the gun a lanyard was pulled releasing the hammer to hit the cap thereby igniting the powder train. Each cartridge had a hole in its base admitting the flame from the exploding powder.

The Weapon saw limited use during the Civil War as a defense weapon on covered bridges.

Col. Berk Lewis told me one time that while the vast majority of the cases were brass, some had been made of copper. I recall seeing a copper cased round for sale many years ago.

This is sometimes referred to as a .56 caliber round.

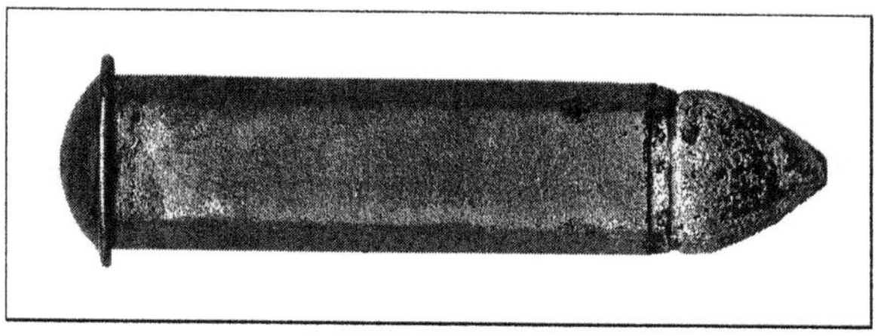

BURNSIDE
Experimental Caliber .58 Burnside

The following report was encountered by Dean Thomas while researching the National Archives Record Group 156 – entitled "Report on Poultney & Trimbles (P &T) Soft Metal Wrapped Cartridge for the Burnside Carbine." The report was made by Acting Second Lt. Of Ordnance at the Washington Arsenal, May 9th, 1864 stating the following details.

> The total weight is 597 grains, outside diameter at base .489", overall length 2.26", case length 1.91".

The report gives an outside measurement of .589" over the projectile instead of .54 cal. as it is usually referred to.

"The distinguishing characteristic of the Poultney and Trimbles cartridge however is the method of forming the metallic case. This instead of being stamped, punched or spun up from a flat disk as appears to be the case with the regular Burnside Cartridge, is formed by wrapping sheet metal upon a mandril or former of the proposed shape and closing the bases by folding down the projecting metal in four folds much as would be done with paper under similar circumstances. ' This folded base is pressed closely and neatly into shape probably by the same operation as that used in punching the central aperture to admit the flame of the cap. The joints, if any exist between the folds at the base, are apparently sealed by the wax or other matter used in closing the puncture."

"The lubricating material is applied at and around the base of the ball which is without cavity."

The conclusion reached after firing an extensive numbers of rounds was that the cases did not improve upon or remedy any defect of the Burnside cartridges, but that it is as good and cheaper.

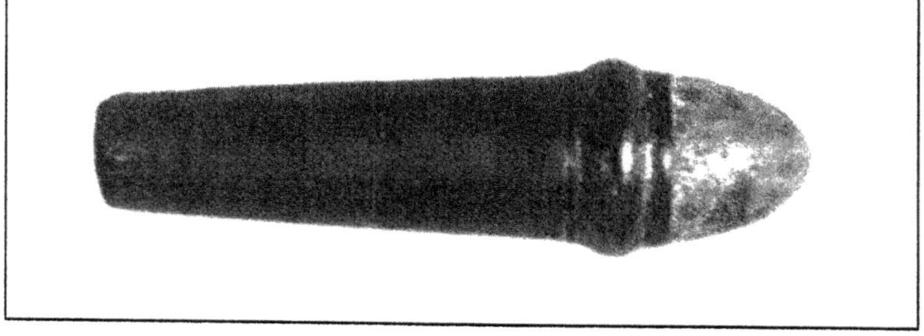

DEANE, ADAMS & DEANE REVOLVING RIFLE
(DUST BIN LID CARTRIDGE)

A novel idea for a bullet, and later on, a bullet and powder chamber was patented on October 1, 1852, by Robert Adams, King William Street, London, England for use in revolvers and a revolvers and revolving five shot rifle. The calibers were .36, .44, .50 and the one illustrated here in .60 caliber. The bullet diameter is .576".

The 1852 patent was based on an earlier idea by Adams by casting a bullet with a tang on which a wad would be attached. In order to protect his patent, he first registered under the "NONORNAMENTAL USEFUL ACT" during November 1851. Under English law, by simply registering his idea, the Act gave him three years of protection against infringement.

The text of the registration (No. 3033) stated that the purpose of the projectile was the ease of fastening a wad to the tang of the bullet and herein lay the uniqueness of the idea; the shape of a projectile with a tang.

On October 1, 1852, Robert Adams patented the idea of a "bullet with tang" attached to a metal powder chamber to be known as the "Dust Bin Lid Cartridge". This marriage of the two ideas was issued Patent No. 1-1852 for what Adams described as "IMPROVEMENTS IN BALL CARTRIDGES". A copy of the patent has been reproduced here.

The completed specimen contains a number of parts and pieces, which could easily become lost or damaged in a very short time. The necessary reloading tool for casting the bullets and a special wad cutter was essential to the system. Without any one of the parts, the entire cartridge was useless.

Briefly, the idea consists of the special cast round ball with attached tang, a leather wad, a copper powder chamber with a hole in the base, a ring riveted to the lid and a very thin piece of tissue to seal the powder in the container. This means there are six items to keep track of plus the special wad cutter and mold to cast the peculiar shaped bullet. It is no wonder the military never took any interest in this very complex system even through the pictured specimen was .577 caliber. This largest size of the "dust bin lid cartridge" was for use in a five shot revolving rifle.

The operational procedure is to cast the bullet and then attach the leather wad by forcing the wad over the tang. The bullet is then inserted into the hole in the powder chamber and riveted or peened over to attach the pieces securely together. The powder is poured into the chamber and secured by cementing a tissue (blue in color) over the end of the chamber. Over this is placed the lid to securely seal the charge.

The only identification on these chambers are the raised letters ELEY LONDON appearing on the lid.

Not satisfied with this multitude of bits and pieces, Adams also recommended the shooter obtain a metal plate perforated with holes to accommodate the rings on the lids. After inserting the rings through the holes in the plate, a wire was to be passed through each of the holes to secure a group of the fully loaded chambers to the plate. It was Adams' intention that the shooter would pull off the chambers (with the ball attached) and place them in position for firing in the rifle.

Of all the various shapes and sizes of cartridges and components of cartridges encountered and examined by the writer, nothing matches the uniqueness of the Dust Bin Lid Cartridge

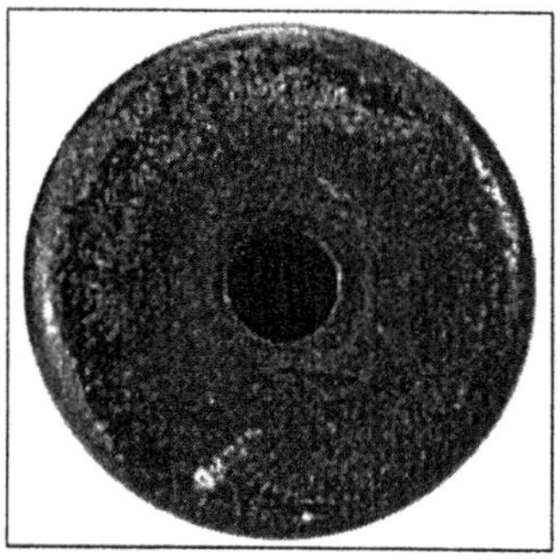

A.D. 1852 . . . . Nº 1.

Ball Cartridges.

LETTERS PATENT to Robert Adams, of King William Street, in the City of London, for the Invention of "IMPROVEMENTS IN BALL CARTRIDGES."

Sealed the 8th January 1853, and dated the 1st October 1852.

PROVISIONAL SPECIFICATION left by the said Robert Adams at the office of the Commissioners of Patents, with his petition, on the 1st October 1852.

I, ROBERT ADAMS, of King William Street, in the City of London, do hereby declare the nature of the said Invention for "IMPROVEMENTS IN BALL CARTRIDGES" to be as follows:—

This Invention consists of affixing a metal chamber to a ball and wad to contain the charge of powder, the end of the chamber is closed with paper or other material.

SPECIFICATION in pursuance of the conditions of the Letters Patent filed by the said Robert Adams, in the Great Seal Patent Office, on the 30th March 1853.

TO ALL TO WHOM THESE PRESENTS SHALL COME, I, Robert Adams, of King William Street, in the City of London, send greeting.

WHEREAS Her most Excellent Majesty Queen Victoria, by Her Letters Patent, bearing date the First day of October in the year of our Lord One thousand eight hundred and fifty-two, in the sixteenth year of Her reign, did for Herself, Her heirs and successors, give and grant unto me, the said Robert Adams, Her special licence that I, the said Robert Adams, my executors, administrators, and assigns, or such others as I, the said Robert Adams, my executors, administrators, and assigns, should at any time agree with, and no others, from time to time and at all times thereafter during the term therein expressed, should and lawfully might make, use, exercise, and vend, within the United Kingdom of Great Britain and Ireland, the Channel Islands, and Isle of Man, an Invention for "Improvements in Ball Cartridges," upon the condition (amongst others) that I, the said Robert Adams, by an instrument in writing under my hand and seal, should particularly describe and ascertain the nature of the said Invention, and in what manner the same was to be performed, and cause the same to be filed in the Great Seal Patent Office within six calendar months next and immediately after the date of the said Letters Patent.

NOW KNOW YE, that, I, the said Robert Adams, do hereby declare the nature of the said Invention, and in what manner the same is to be performed, to be particularly described and ascertained in and by the following statement thereof (that is to say):

This Invention consists of affixing a metal chamber to a ball and wad to contain a charge of powder, the end of the chamber is closed with paper or other material/and in order that my said Invention may be most fully understood and readily carried into effect I will proceed to describe the means pursued by me.

Description of the Drawing.

/Figure 1, shows a ball, a charging chamber, and a wad complete, figure 2, shows a section thereof, and figure 3, shows the three parts separately; *a*, is the ball which is formed with a projecting tang *b*, which passes through the wad *c*, and into the metal cup or chamber *d*, and is there riveted so as to combine the three parts together. The chamber *d*, which I prefer to be raised out of thin sheet copper, is charged with powder, and a disc of thin paper or other material is pasted or cemented over the end of the chamber. In order to protect the end of the chamber *d*, I employ a cap or cover *e*, into which the end of the cartridge is placed, and I form such cap or cover with a shank, as shown, by which a number of caps can have a wire passed through them, and thus I prevent their being lost; and for the convenience of stowing or packing cartridges such as above described I find it convenient to have a plate of metal perforated with holes for the passage of the shanks of the covers *e*, of the cartridges, so that by passing a wire through the shanks of several cartridges they will be attached to the plate of metal, and the cartridges may be removed out of their end covers or cases *e*, leaving such end cases or covers attached to the plate.

Having thus described the nature of my Invention, and the manner of performing the same, I would have it understood that I do not confine myself to the precise forms of the parts; but what I claim is, the affixing a metal chamber (to contain the powder) to a ball and wad.

In witness whereof, I, the said Robert Adams, have hereunto set my hand and seal, this Twenty-sixth day of March in the year of our Lord One thousand eight hundred and fifty-three.

ROBERT ADAMS. (L.S.)

Witness, J. G. Tongue.

LONDON:
Printed by George Edward Eyre and William Spottiswoode,
Printers to the Queen's most Excellent Majesty. 1853.

A.D. 1852, OCTOBER 1. N° 1. (One Sheet.)

ADAMS' SPECIFICATION

Fig. 1.

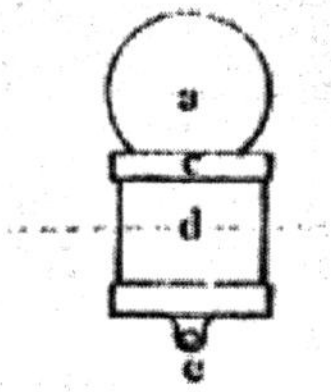

Fig. 2.

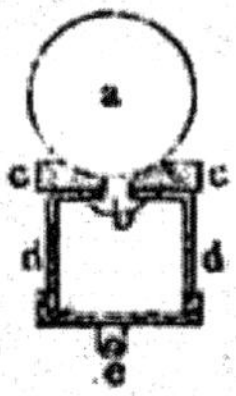

Fig. 3.

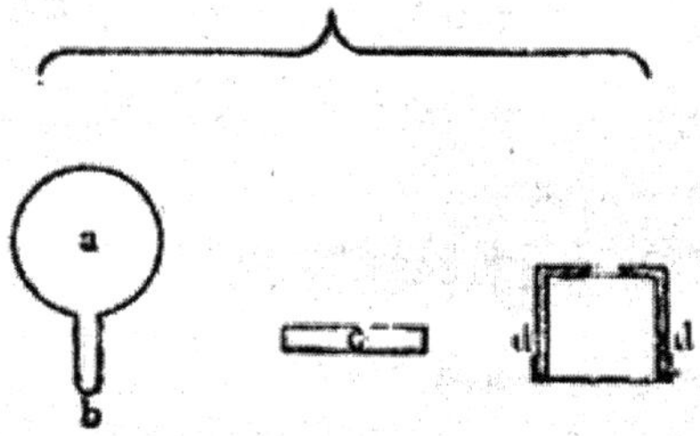

THE FILED DRAWING IS PARTLY COLORED.

R. PROSSER, LITH.

LONDON: Printed by GEORGE EDWARD EYRE and WILLIAM SPOTTISWOODE,
Printers to the Queen's most Excellent Majesty. 1853.

MAYNARD CARTRIDGES

At the mention of the name Maynard, cartridge collectors immediately think of those strange looking large rimmed cartridges of various shapes and sizes. They never fail to create interest when examined for the first time.

The creator of these strange looking cases was a dentist!

Dr. Edward Maynard first attended West Point at the age of eighteen, but due to ill health was forced to leave without finishing his Plebe year. Shortly after leaving West Point, he studied dentistry and upon obtaining his doctorate, moved to Washington, D. C. where he pursued his profession as well as his interest in firearms. Dr. Maynard was an extremely talented individual as he was responsible for many innovative techniques in dentistry-he introduced the use of anesthesia in dental surgery-as well as his many successes in the field of firearms and ammunition.

Dr. Maynard's first patent was issued to him in 1845 (Patent #4208) titled "Magazine Priming Devise" for small arms; better known as a Tape Primer. The system consists of pellets of priming composition glued between two strips of shellacked paper. Approximately forty primers are on each roll, which are wrapped in brown paper and packed in a tin container-ten rolls to a container. The tin covers will be encountered with two legends; one from Frankford Arsenal and the common one with the embossed words: "Maynard's Patent, Sep. 22, 1845. Manufactured by Mass. Arms Co., Chicopee Falls, Mass.".
The tape primer was submitted for testing to the United States Army Board that convened on January 29, 1845 at West Point, New York. On March 20, 1845 the Ordnance Department contracted with Edward Maynard for the rights to apply his lock to 4,000 muskets at one dollar each. Included in the contract was the rights reserved to the Government to put locks on a maximum of 100,000 arms at a maximum cost to the Government of .25 cents each. The tape system was applied to Sharps, Jenks and Greene rifles as well as Maynard arms.

On February 3, 1854, Secretary of War, Jefferson Davis arranged to have the War Department buy out all of the Maynard priming devises for fifty thousand dollars. By the end of 1863, the Maynard tape priming system was no longer needed and was abandoned.

Since we are interested only in the cartridge patents (and particularly those of .58 caliber) a description of those patents are hereby presented:

1. The first cartridge patent was #15141 covering a bullet in a metal case having a rear aperture closed by a waxed paper. Ignition was by an external primer (percussion cap or tape primer) placed on the breechblock nipple. A thin brass cap fitted on the base to protect the case while being handled. A Steel cartridge case of .58 caliber utilizing the service bullet was available

2. Maynard entered the Army trials in 1858 to convert arms to breech-loaders. It was at this time that Maynard offered the first of his unusual looking cases to be used in the conversions. The cases were of three types:

   a. A .69 caliber cartridge having a rounded base. Collectors now refer to this round as the "Capsule".
   b. A case similar to the previous mentioned 1856 type was offered, but this one had an extension arm added to assist in extraction of the fired case. Two sizes were available - .69 caliber for converted muskets and the other in .58 caliber for use in the converted rifles. These 1858 cases with the extension arm became known as "flop ears".
   c. A longer case length to be loaded with either buckshot, round ball or buck and ball was designed for use in smooth bore-muskets. Unfortunately none of those conversions submitted by Maynard were considered acceptable and therefore, all were of a very limited production.

3. In 1859 Patent #22565 was granted to Dr. Maynard for an "Improvement in Metallic Cartridge Cases". This patent involved the soldering of a steel disk to the outside surface of the bottom of a brass cup that has a smaller diameter than the disk. A small hole was drilled in the center through both the brass and steel bottom of the case. The purpose being to facilitate the removal of the cartridge case from the weapon's chamber.

4. The Model 1865 series cases were a modification of the 1859 patent in that a large thin rim was soldered and/or riveted to a brass tube. The rim was perforated in the center to admit the flame from the percussion cap. Ignition was by a percussion cap placed on the nipple on the breechblock of the weapon. Both the cases and arms were manufactured by the Mass. Arms Co. of Chicopee Falls, Massachusetts.

   The calibers ranged from .35 to .64. The .55 and .64 were shot cases and as it was the practice of the Mass. Arms Co. to furnish various size barrels (either rifled or smooth) there is a demand for ball and shot loads in different calibers depending upon the whim of the sportsman.

5. Patent #26364 granted on December 6, 1859 was titled "Improvement in Breech Loading Firearms". While basically a firearms patent, it included a description of a cartridge case with a flanged bottom metallic cup with a perforated head secured to the breechblock with either a chain or thong.

   These cartridges became known as "Thong Maynards". A leather thong with a hook on one end was used to connect the metallic cup to the gun. Calibers of these cases and firearms were .35, .40, .50, .55 and .64.

6. Dr. Maynard was particularly active during the year 1863 having three patents issued to him. They were:

a. Patent #39823 issued on September 8, 1863 dealt with a means of preventing the accidental explosion of the round. This was accomplished by tying a percussion cap to one end of a cord which had been secured to the base of the shell.
This round is referred to as a "cord extractor". Also as part of this patent, another drawing pictured a devise with a hollow tube going into the cartridge case that could be used as a flash tube for an external percussion cap.

b. On September 29, 1863 Patent #40000 described a system for fastening and retaining a charge of fixed ammunition for shotguns. It was accomplished by cutting slits in the top of a shot shell brass case. After loading the case, the segments were bent in and around the top wad thereby preventing the charge from going forward and out of the case. This system was copied by other manufacturers until modern times.

c. Patent #40112 dated September 29, 1863 concerned itself with a variety of extractor arms such as wires, cords and hook eyes for the cartridges.

7. The Model 1866 cartridges were described in Patent #59044 issued on October 23, 1866. This patent presented an idea for an improvement in priming metallic cartridges. It consisted of a thin metal outer cap with an extractor arm attached and having been constructed with a primer recess formed in the base of the case.

A specimen of this Flop-Ear case is pictured in "Small Arms and Ammunition in the U. S. Service" (Plate 43g) by Col. Berkley R. Lewis and in the "History and Development of Small Arms Ammunition: on page 88-PBL38, Vol. One by George A Hoyem.

Dr. Maynard submitted this self-primed cartridge along with a conversion of the Maynard Model 1865 breech-loading percussion rifle to the United States Army trials in 1866.

This system was rejected by the Board.

8. Just one year later, Patent #61225 was issued on January 15, 1867. These rounds were to be known as the Model 1867 Maynard. The text describes the detachable cup to be of two piece construction containing a fulminate primer. Only two calibers of this type were known - .40-40 and a .64 caliber shot case. The idea never got beyond the experimental stage and was soon abandoned.

9 The next series of Maynard cartridges are the 1873s. A patent for these was issued on March 23, 1869 under Patent #88202. These new cartridges

were self-contained center fire rounds using the Berdan primer supported by an iron anvil and inserted in W. F. Parker's patented reloadable shotgun case.

Dr. Maynard could now boast that with this new 1873 styled case that he offered to the public the greatest variety of calibers (fifteen) and loads of any firearms manufacturer. The sizes ranged from .35 to .64 having loads available from ball to shot, including an explosive round. The latter had the conventional conical bullet with a hollowed nose into which was inserted a .22 rim fire blank.

10 The final series of Maynard cartridges were known as the Model 1882. These utilized cases of solid one-piece construction having either Winchester or Berdan primers. This series were offered in eighteen calibers.

With the advent of repeating weapons in the 1870s, the firearms offered for sale to the public by the Mass. Arms Co. fell out of favor with the public and by the mid to late 1880s the firm went out of business.

It is not the intention of the author to offer here a total and complete discussion of the ammunition produced by Dr. Edward Maynard. The text is meant to give a thumb nail sketch of the rounds invented by a most prolific and inventive individual. Once again, as history has proven before, we see a man whose ideas and products were too far ahead of their time and their potential could not be grasped by the average layman.

During his lifetime, Dr. Edward Maynard was granted a total of twenty-three patents spanning forty-one years; from September 22, 1845 to June 18, 1886

.58 MAYNARD FLOP EAR

In 1858 the United States Army conducted trials to test various methods of converting arms to breechloaders. One of the firearms submitted by Dr. Maynard used this "flop ear" style cartridge in his converted rifle. The case on this specimen is 1.37" long and has a diameter of .621".

.58 MAYNARD MODEL 1865:

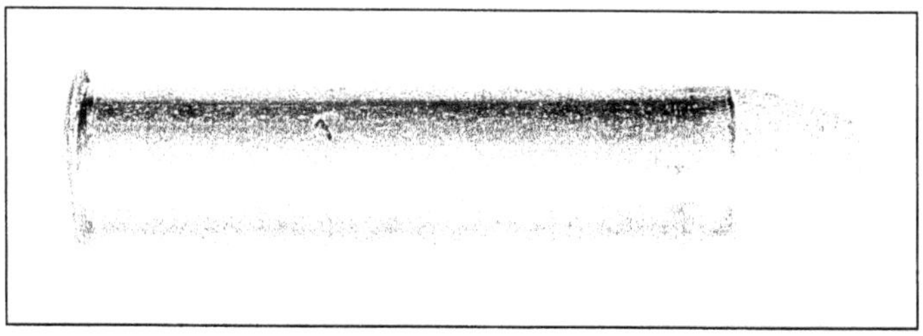

Dimensions are:

Rim	755"	Case	2.575"
Head	.640"	Material	Brass
Neck	.631"	Total Grs.	761
Bullet	.595"	O. A.	3.800"

Maynard cartridges have had a history of being faked, duplicated or "made up" so a collector can never be positive of the authenticity of many of the rare early experimental rounds. Certainly this round is no exception.

It is the author's opinion that this round was "made-up" based on a Model 1865 shot shell. The round was used in a specially made Maynard barrel for long-range shooting.

An interesting and unusual round - but then weren't all Maynards?

TITLE: .58 MAYNARD MULE EAR

HEADSTAMP: PLAIN

DIMENSIONS:

Rim: .575"
Head: .575"
Neck: .568"
Bullet:
Case: 1.394"

OTHER DETAILS:

Bullet Grains: NA
Powder Grains: NA
Bullet:
Case Material: Brass
Total Grains: NA

OVERALL: NA

This round is simular to the Maynard flop ear of 1.532 " except it is set off and beveled on one side.

The offset flop ear measures .793" in length.

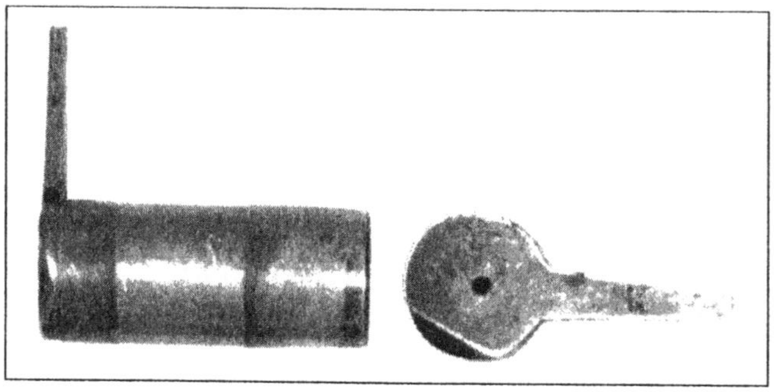

TITLE: .58 MAYNARD FLOP EAR

HEADSTAMP: PLAIN

DIMENSIONS:

Rim: .649"
Head: .593"
Neck: .575"
Bullet: .530"
Case: 1.532"

OTHER DETAILS:

Bullet Grains: NA
Powder Grains: NA
Bullet: Tapered Lead
Case Material: Brass
Total Grains: NA

OVERALL: 1.900"

This round is a rare Maynard percussion case utilizing the 1867 model primed cup. The flop ear measures .688 in length.

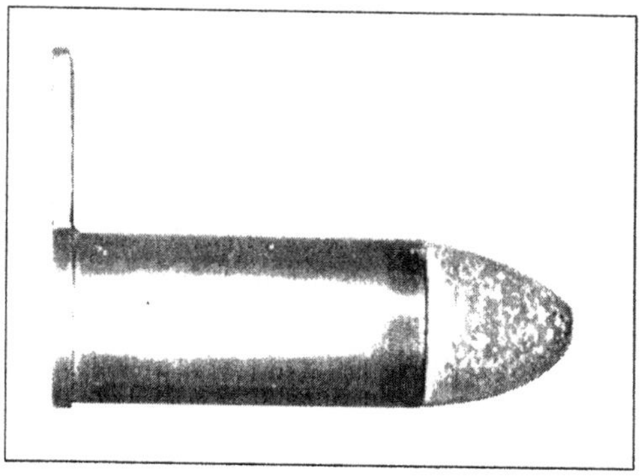

PAULY

One of the most persistent and troublesome problems of early breech-loading firearms was obturation, i.e. the ability to effectively seal the breech of the weapon to prevent the escape of gasses, which occurs at the moment the cartridge explodes.

Any looseness of fit between frame and chamber face meant leakage resulting in loss of power, flash back and extreme discomfort to the shooter.

In 1812 one of the first complete cartridges was invented by Jean Samuel Pauly, a Swiss gunsmith living in France. After moving to England he obtained additional patents in 1814 and in 1816. The 1816 patent pertained to the idea of using heat from compressed air to ignite a priming pellet. The idea being similar to the Daisy VL caseless.22 rifle.

The 1812 design of his first cartridge utilized a fulminate wafer ignition system. The rounded base of the cartridge, similar to a sink stopper, quite effectively solved the obturation problem previously mentioned.

In 1814 the cartridge case was altered to take a percussion cap nipple which was screwed into the base of the shell. The cartridge should properly be referred to as an auxiliary breech rather than a cartridge case. Obviously the container was not meant to be expendable but was intended to be reloaded and used over again.

The Milwaukee Public Museum has in its firearms collection a cased pair of Pauly Patent Pistols which are considered to be the prototype for improvement in the original wafer ignition system which was patented in 1812. The cartridges in the pistol's case are virtually identical to the wafer type except for the addition of a percussion cap nipple.

A description of the cartridge is as follows:

RIM:	.822"	HEAD:	.688"
NECK	.660"	BULLET:	.582"
C. L.	1.493"	WEIGHT:	775 GRS.

The case held thirteen grains of black powder.

The design of Pauly's ignition system was years ahead of other weapons of its time and like all things too advanced for its period was not accepted by the shooting public and soon fell into obscurity.

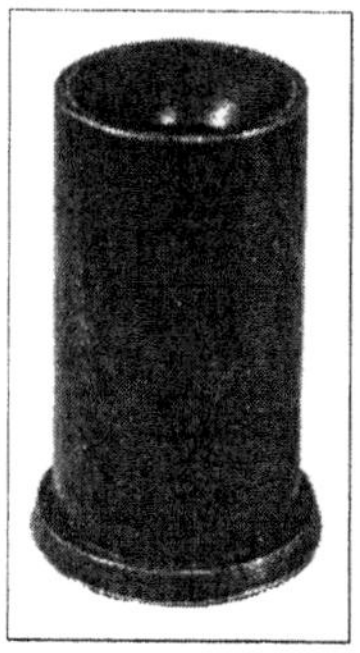

TITLE: UNKNOWN .58 PERCUSSION OF ENGLISH ORIGIN

HEADSTAMP: Plain

DIMENSIONS:

Rim:	.680"
Head:	.591"
Neck:	.591"
Bullet:	14.22 mm
Paper Case:	1.821"

OTHER DETAILS:

Bullet Grains:	N/A
Powder Grains:	N/A
Bullet:	Lead, Sharp Point
Case Material:	Light Brown
Total Grains:	N/A

OVERALL: 2.560"

A rare and little known round for a percussion rifle dating back to 1869. In Bartlett & Gallatins "Cartridge Manual" which lists patents and pictures of a similar round on plate 11 of the English section – 2071 of 1869. The description reads, "case of paper, with metal thimble at base and rubber wad in between. Wire from front wad removes paper case when fired."

This is a variation on the same idea patented in England, by Manceoux in 1859, through Rudolph Schmidt dates it 1861. In this English specimen, the wire would tighten around the unburned paper case material as the bullet moved forward. The wad apparently remained at the base of the chamber against the breech face.

The arm that used this cartridge is unknown. The English modified a few Model 1855 Sharps carbines by removing the tape priming device and situating a percussion nipple on top of the chamber so that flame from the percussion cap would enter the center of the chamber around the middle of the paper case. The wood plug, a variation from the patent description, would be forced tightly against the chamber wall and lay against the face of the sliding breech block when the powder charge exploded, thus providing a gas seal. When the gun was opened, the tapered plug would fall out or was picked out of the chamber. Although not apparent, a rubber was described in the patent summary, which may be inside the base in front of the wooden plug. See Sharps Firearms, Frank Sellers, 1978 pg. 57 illustration of a modified Sharps 1855 carbine.

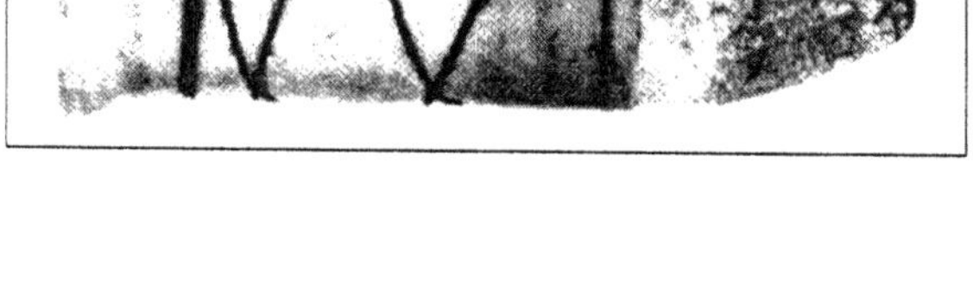

TITLE: .58 GALLAGHER

HEADSTAMP: Plain

DIMENSIONS:		OTHER DETAILS:	
Rim:	NA	Bullet Grains:	NA
Head:	.565"	Powder Grains:	NA
Neck:	NA	Bullet:	Round Nose
Bullet:	.584"	Case Material:	NA
Case:	1.937"	Total Grains:	NA

OVERALL: 2.350"

A questionable tittle for this round which resembles the common .50 Gallagher round. The round is percussion with a deep dish base.

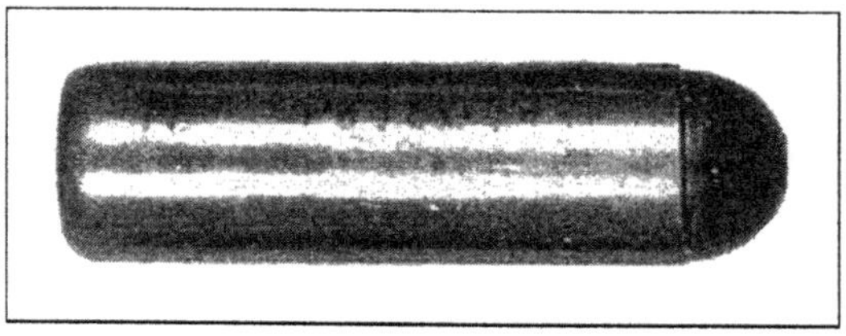

250
Waterproof Primers
No. 1
WARRANTED
J. GOLDMARK'S PERC. CAPS.
U.S. MUSKET CAPS.
PAT. SEP. 24.
1867
100
Warranted Waterproof
THEO. WEHLE, NEW-YOR
100
No.
250
Water Proof
MUSKET CAPS
HICKS
U.S MUSKET CAPS
WATERBURY
U. S.
MUSKET
AMORCES DE CHASSE.
CANNELÉES
AU FULMINATE
DE MERCURE
1859
100
U. S. MILITARY MUSKET CAPS.
WATERPROOF AND INFALLIBLE.

VI

Pinfire

PIN FIRE AMMUNITION

Undoubtedly the greatest advancement in the evolution of the cartridge was the invention of LeFaucheaux, a French gunsmith. LeFaucheaux should get credit for making the first breech-loading gun, he designed a new type of ammunition for use in the weapon. He patented in 1835 the gun and a cartridge having a metal base and a paper body not unlike a modern-day shot gun shell. Inserted in its base was a primer cup for ignition and directly above the cup was a thin wire (pin) which protruded through to the outside of the case.

LeFaucheaux's pin-fire cartridges were prone to leak gas much to the dislike of the shooter. The gas escaped through the hole in the case and around the pin. Because it was necessary to extend the pin above the breech so that the outside hammer could strike the pin, the gas seeped occasionally through the breech.

Obturation problems were yet to be completely solved.

About ten years later, in 1846,Houllier, another French gunsmith patented (#1936-1846), an improved pin-fire cartridge. There was considerable doubt in the minds of historians whether Houllier designed his pin-fire without the knowledge that a similar round had already been made. It did not appear reasonable that he would not have been aware of he work done by another of his countrymen (LeFaucheaux) ten years earlier. While Houllier took credit for being the originator of the pin-fire, he really improved the original round by reinforcing the head and setting the cap in a recessed base washer. To quote from Houllier's patent:

> "I claim the use of thin metal for the entire case to provide obturation by the stretching of the case."

Regardless of whether Houllier created an entirely new cartridge or simply improved on the LeFaucheaux cartridge it provided tremendous advantages over any other ignition system. No other cartridge of the time was waterproof and gas tight. Ironically, this 1846 patent also alluded to making cartridges on the rim-fire principal, but apparently the cartridge was ignored until a few years later.

During the next fifteen years, a proliferation of pin-fire cartridges was designed in sizes varying from five millimeters to fifteen millimeters. While the ammunition was very popular in Europe, it never caught on in America-except to a limited extent during the Civil War. In 1862, Christian Sharps took out a patent for pin-fire cartridges, while E. Allen and C. D. Leet of Springfield, Massachusetts manufactured them for use during the Civil War. The U. S. Army issued seven and nine-millimeter ammunition as "Official Type" as evidenced by a note in the Ordnance Memo #1 of 1865. The LeFaucheaux twelve-millimeter revolver was a popular personal side arm carried by Officers. Ammunition for this revolver was manufactured by the C. D. Leet Co. of Springfield, Massachusetts. Specimens, while scarce, can be found for sale by cartridge collectors.

The 15mm size cartridge was never as popular in America as it was in Europe. To the Author's knowledge, no 15mm size was ever manufactured in America.

Weapons using the 15mm ammunition were single and double-barreled pistols, revolvers, and carbines. Case lengths varied from .575" to .881". Loadings included blanks, shot loads and lead bullets. As indicated on the listing, varied head-stamps appear. Some were plain, some had other lettering, and others were simply numbered "15". The later were made by S. F. M., while the "G" stood for Gevelot. Both were French manufacturers.

Union Metallic continued to make pin-fire ammunition for a few years after the Civil War, but they finally ceased production due to lack of demand. Surprisingly, European ammunition makers continued to make these cartridges available until 1962! They were in use approximately one hundred thirty years after their invention. A worthy tribute to LeFaucheaux and Houllier of Paris, France.

LISTING OF PIN FIRE CARTRIDGES:

CASE LENGTH	OVER-ALL	BASE DIAMETER.	TOTAL GRAINS	CASE MATERIAL	HEAD STAMP
.575"	1.001"	.615"	214	Copper	Plain
.765"	1.340"	.620"	448	Copper	V F M & Co 15 Liege
.795"	1.371"	.620"	559	Brass	.Eley.15M:. London
.829"	1.440"	.612"	573	Brass	.Eley.15M:. London
.850"	1.462"	.618"	92	Brass	15
.855"	1.493"	.612"	560	Brass	15
.862"	1.484"	.610"	88	Copper	15
.881"	1.407"	.618"	583	Copper	G

TITLE: 15 m/m (.590") EARLY PIN FIRE

HEADSTAMP: Brevet Le Faucheux Fabr Gevelot (Raised)

DIMENSIONS:

Rim:	.700"
Head:	.664"
Neck:	.678"
Bullet:	.567"
Case:	.710"

OTHER DETAILS:

Bullet Grains:	
Powder Grains:	
Bullet:	Round Lead Ball
Case Material	Heavy Paper
Total Grains:	430

OVERALL: .943"

A very old and possibly one of the first 15 m/m Le Faucheux pin-fire rounds. This specimen differs from later rounds. Instead of having a metallic case this one has a paper case with black ink lettering. None of the lettering is legible except for the letters heux, eron, gne, bert, relat, and a large number 4. One other similar round listed for sale some years ago describes the raised head-stamp as "Brevet Le Faucheux Fabr Gevelot". Like the writing on the side of paper case, the raised lettering is barely legible.

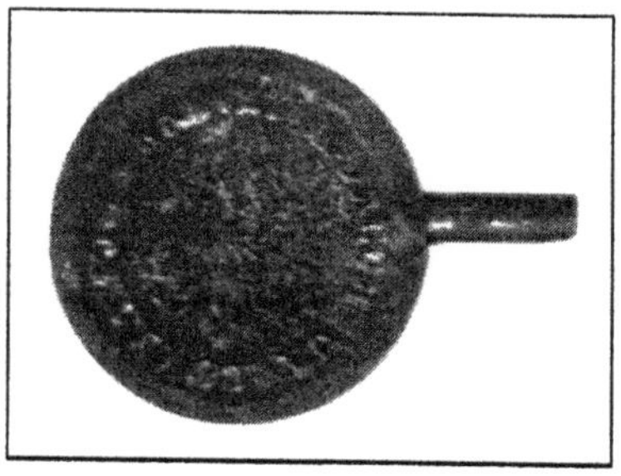

TITLE: PIN FIRE DUMMY 15 mm

HEADSTAMP: 15

DIMENSIONS:

Rim: 614"
Head .610"
Neck: .585"
Bullet: -0-
Case: .858"

OTHER DETAILS:

Bullet Grains: None
Powder Grains: None
Bullet: None
Case Material: Copper
Total Grains: NA

OVERALL: 1.476"

A Gevelot pin fire dummy round constructed for a shot load. There is a hole in the left side of the case where the ignition pin sticks into the case.

It is the author's opinion that the round is a product of Gevelot and was intended to be part of a cartridge board display.

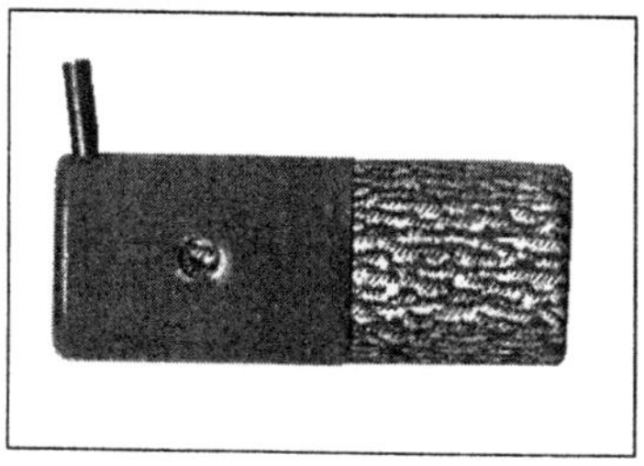

TITLE: .58 GALLAGHER & GLADDING

HEADSTAMP: Plain

DIMENSIONS:

Rim:	.815" (over pin)
Head:	.800" (body)
Neck:	.643"
Bullet:	.608"
Case:	1.160"

OTHER DETAILS:

Bullet Grains:	550
Powder Grains:	70
Bullet:	Minie
Case Material:	Brass
Total Grains:	715

OVERALL ?

The only one is in the Smithsonian Institute – exteremly rare

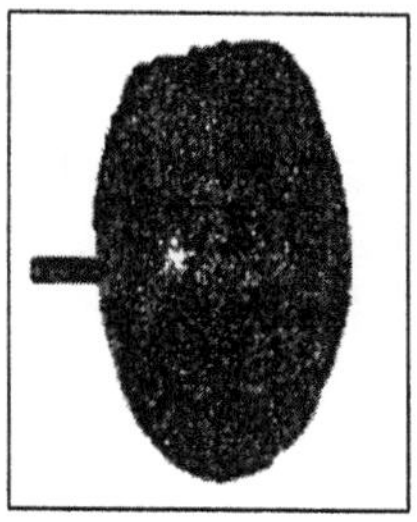

TITLE: .58 GALLAGHER & GLADDING

HEADSTAMP: Plain

DIMENSIONS:

Rim: None
Head: .622"
Neck: .615"
Bullet: .578"
Case: 1.899"

OTHER DETAILS:

Bullet Grains: ----
Powder Grains: ----
Bullet: Lead, Round Nose
Case Material: Steel
Total Grains: ----

OVERALL: 2.261

No specific .58 caliber Gallagher has been found, but this cartridge definitely follows the design of the common.50 Gallagher but with larger dimensions. Two types of rounds have been located; one has a steel case and the other of tinned copper.

Drawings below are not to scale.

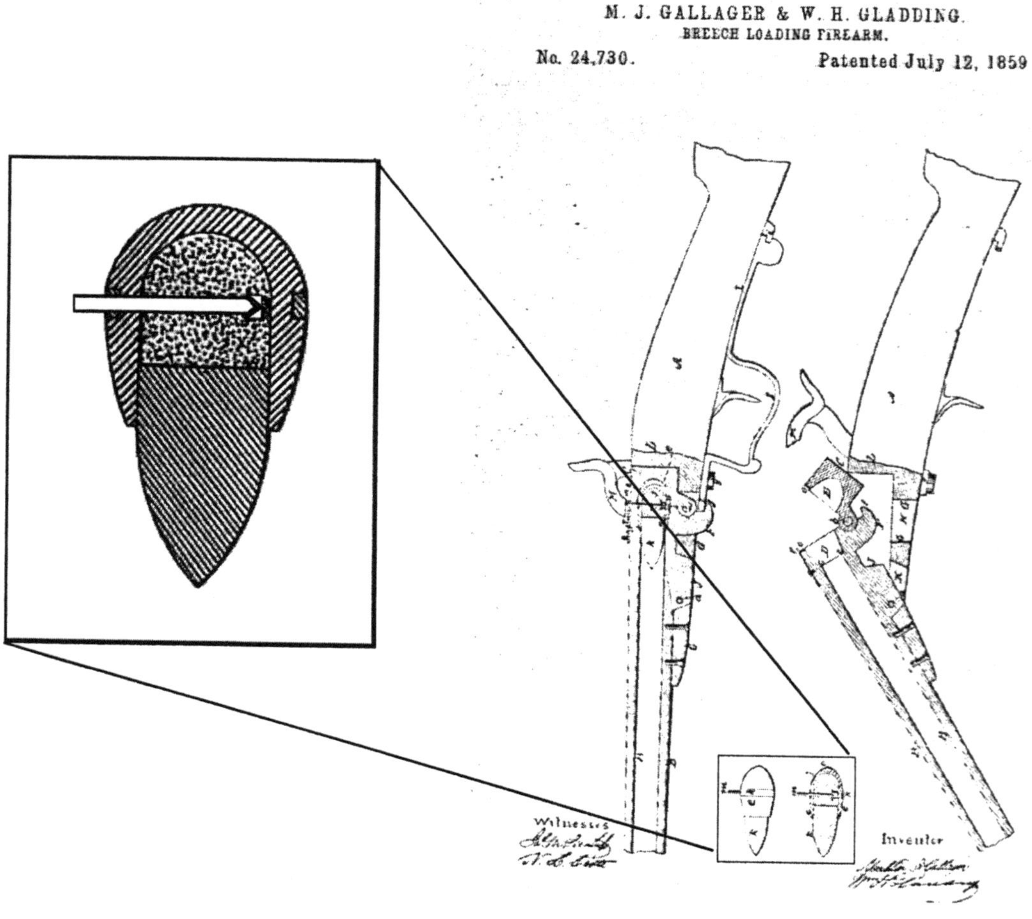

.58 SCHUBARTH

The majority of cartridge collectors, regardless of their own specialty interests, will agree the Gallagher and Gladding and the Schubarth cartridges to be one of the most desirable and rarest of collector's cartridges.

Webster's dictionary defines the word RARE as:

> Uncommon, not frequently found, scarce, unusual

No one will disagree that the Gallagher and Gladding and Schubarth qualifies on all four counts.

M. J. Gallagher and W. H. Gladding of Savannah, Georgia obtained a patent (#24730) on July 12, 1859 for an "Improvement in Breech-Loading Firearms". The patent claimed the following:

> "Forming the chamber in the barrel and breech of the gun of the shape of two frustums of cones or of a frustum of a cone, and a section of a parabolic spindle, whose brass meet at or near the line of the joint between said barrel and breech, for the purpose of containing a cartridge-case of the form substantially at that herein represented."

The cartridge case referred to in the patent is described as follows:

> "The cartridge-case may be made of paper, wood, or metal. If of paper or wood, they may have a light metallic band, h, around them: and though the case is swelled out on its outer perimeter, as shown in the drawings, its chamber is cylindrical and of the same diameter throughout.
>
> j is the charge of powder in the cartridge-case, and k the ball, which is tightly inserted in its open end, and is of the "Minie" form, or pointed. The cap n or other thing which ignites the powder is placed inside of the cartridge, and a pin or rod, m, is set in said cap, and extends through the cartridge-case, upon which rod the hammer m falls and explodes the cap or other priming and ignites the powder."

If you further read the text of the patent you will find that the principle of the system is similar to pin-fire cartridges. The purpose of the pin or wire is to fire the priming cup. The pin extends down to an inverted primer cup, and when struck by the hammer, it slides down far enough to crush the fulminate and ignite the propelling charge of seventy grains of black powder. The patent drawing depicts a piece of wire (the pin) protruding a considerable height above the case. A hole in the top of the breech permits the pin to protrude high enough to enable the hammer to strike the top of the pin, driving it down far enough to hit the primer. This system is similar to the conventional pin-fire cartridge.

The only known specimen with the protruding pin as described in the patent papers is the one chambered in the Gallagher and Gladding carbine now in the Smithsonian Institution collection.

The patent specifications mention that in addition to a metal case, the round could be constructed of wood or paper. In the event wood or paper was used, a lightweight metallic band should be placed around the case to provide a firmness to the case. It would appear logical that the inventors experimented with wood or paper cases, and, finding them flimsy placed a reinforcement band around the case. The empty case found in the chamber of the weapon in the Smithsonian Institution collection is a molded, fibrous, paper-like material with an iron band encircling the case.

The inventors also designed a means of partially withdrawing the fired case. To quote from the patent text:

> "An adjustable gage could be attached to the barrel, and the face of the hammer could be made so that when the piece had been discharged, and the lever swung out to release the barrel, and the barrel raised up, the hammer continuing in contact with the rod (m) would draw out the cartridge case in the barrel."

This describes one of the earliest devices to provide an extractor on a rifle using a metallic cartridge.

On July 23, 1861, Casper D. Schubarth of Providence, Rhode Island was granted a patent (#32895) for another "Improvement in Breech-Loading Firearms." Since this patent closely resembles the Gallagher and Gladding, it would seem reasonable to describe Schubarth's patent as "Schubarth's Improvement Of The Gallagher and Gladding System."

The cartridge for the Schubarth is described in the text of his patent as follows:

> "The cartridge employed is entirely waterproof, being formed on its upper side with a projection which, without penetrating the metallic casing, may receive the end of a wire resting in an inverted percussion-cap at bottom. The hammer strikes the casing covering the end of the said wire, either by means of a projection on its face or through the medium of a spring-pin: or the cavity in the casing may be filled with fulmination powder to be exploded by the hammer without the use of a wire."

The author cannot resist describing an incident that happened some years ago involving the Schubarth cartridges. I regret the vagueness of the details such as time, place, and person connected with the incident, but it did happen and was described in a very old issue of the American Rifleman magazine.

A collector of guns and ammunition was travelling somewhere in Vermont and became lost. He was driving on one of those rural country roads for which Vermont is famous. Having lost his way stopped to ask directions from two boys, he spotted playing on some rocks by a small stream. He walked over to them to inquire directions to get to his destination. As he approached the boys, he saw hat they were hitting some object with a rock. Much to his surprise, he realized the object they were striking was a cartridge. Yes, a cartridge and no less a cartridge than a Schubarth! The boys said they were trying to make the thing "go off", and much to their disappointment not one of the eight had exploded. All of the eight rounds were now a hopeless, twisted mess. Knowing how rare these rounds were even in those days (mid-1930's), the collector offered to buy the remaining two cartridges for ten dollars. Needless to say, they happily accepted his offer.

Upon questioning the boys, they said there used to be a gun around their house that the cartridges fitted. While the gun had long since disappeared, they had found the ten rounds wrapped up in a paper packet. What better fun than to try to make them explode by hitting them with a rock!

TITLE: .58 SCHUBARTH

HEADSTAMP: Plain

DIMENSIONS:

Rim: .815" (over pin)
Head: .800" (body)
Neck: .643"
Bullet: .608"
Case: 1.160"

OTHER DETAILS:

Bullet Grains: 550
Powder Grains: 70
Bullet: Minie
Case Material: Brass
Total Grains: 715

OVERALL: 1.968"

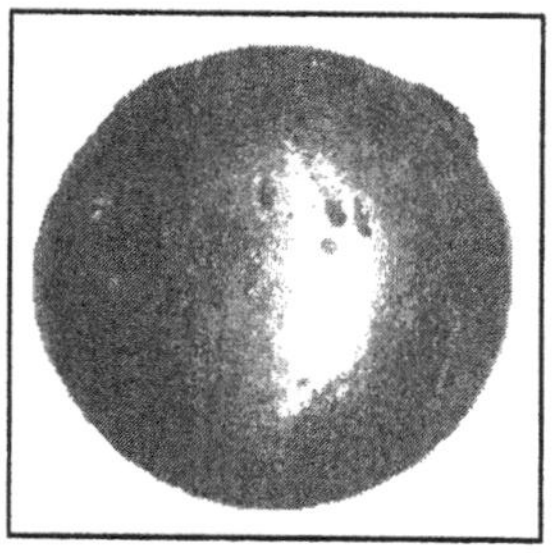
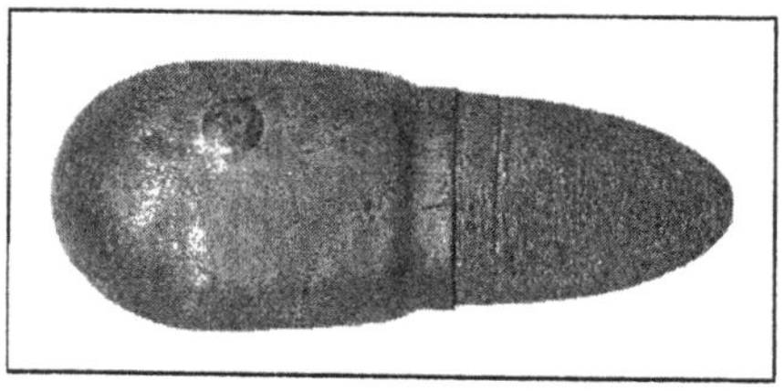
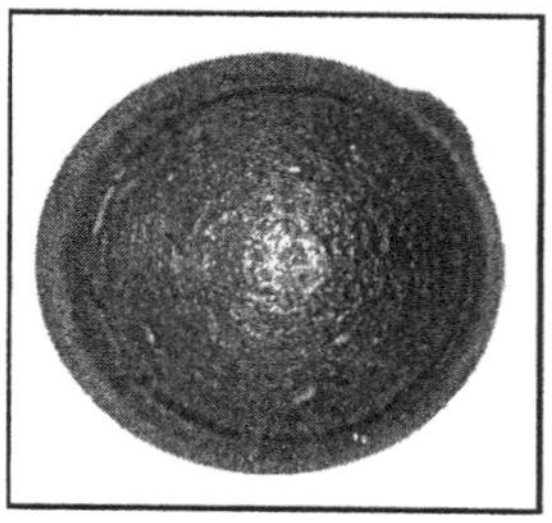
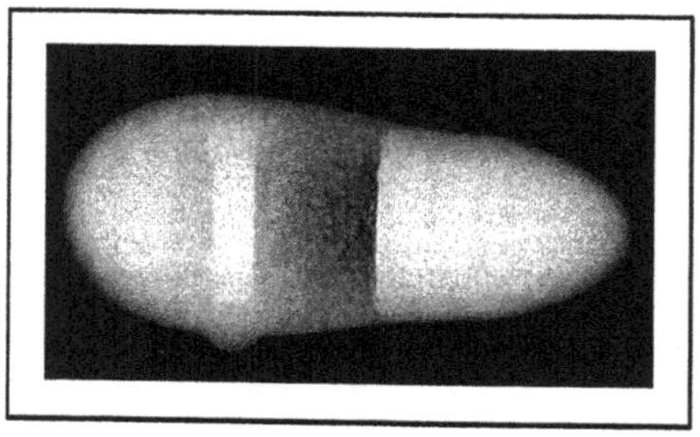

37TH CONGRESS, 2d Session. } SENATE. { REP. COM. No. 69.

IN THE SENATE OF THE UNITED STATES.

JULY 14, 1862.—Ordered to be printed.

Mr. TEN EYCK submitted the following

REPORT.

The Committee on the Judiciary, to whom was referred a resolution for the expulsion of James F. Simmons, a senator of the United States from the State of Rhode Island, have had the same under consideration, and, after full deliberation, make the following report:

That in October, 1861, Senator Simmons being in Washington on his own private business, was called upon by a Mr. Schubarth with a letter of introduction from Messrs. A. D. & J. Y. Smith, a wealthy business firm in Providence, Rhode Island, requesting him to aid Mr. Schubarth in procuring an order from the government for the manufacture of a breech-loading rifle for the use of the army and navy, of which he was the inventor, and stating that if the order was procured they (the Messrs. Smith) were to be concerned in its execution.

Mr. Schubarth was introduced by Senator Simmons to the Secretary of War, General Ripley, of the ordnance, and Captain Harwood, of the Navy Department. His rifle was tried at the navy yard, and a favorable report made, but the government declined ordering his breech-loading rifle, and, preferring the Springfield rifled muskets, gave an order on the 11th of October last for 20,000, at the rate of $20 for each arm, with the appendages.

When called upon by Mr. Schubarth with the letter of the Messrs. Smith, as above stated, Mr. Simmons determined to aid him in procuring the order for the manufacture of the rifle invented by him, by going with him to the departments, recommending his arm, and soliciting a contract in the same manner as he had done before in other cases, and has since done in behalf of constituents of his without any offer of compensation having been made to him for his aid, and without expecting or dreaming of receiving any. That after he had thus undertaken to assist Mr. Schubarth, either before or after he had introduced him at the departments, Mr. Simmons thinks it was after and some time during the stay of Mr. Schubarth in Washington, while going to or returning from the departments, Schubarth said he had a letter from the Messrs. Smith authorizing him to pay a commission for obtaining an order, which he offered to Mr. Simmons to be read, but which he declined reading, saying, playfully, "that the Messrs. Smith were rich enough to pay, and that if he got commissions for all he did for his constituents he should be as rich as some of them were;" but it appears that no suggestion was made to Mr. Simmons to pay him a commission or compensate him in any way for any aid he might render, until after he had determined to assist the parties in getting the contract; and that nothing further ever passed between Mr. Simmons and the Messrs. Smith on the subject of commissions.

That afterwards, in November, 1861, Mr. Simmons being in Washington, on his way to Hatteras Inlet, for the purpose of consulting the government about procuring a vessel load of cotton, Mr. Schubarth informed him that he could do nothing with the order he had obtained unless it was made for the same number and as long a time allowed for delivery as other parties had, and requested him to aid in getting such alteration, which he did; and on the 26th of November last an additional order was made for the manufacture of 30,000 more.

In January last, Mr. Schubarth informed Mr. Simmons that the Messrs. Smith declined pursuing the business. And seeing him afterwards in February, in New York, Schubarth told him he had made arrangements with other parties for means to carry out the contract, and that they wanted to have an understanding with him about his commissions. Mr. Simmons states that he saw these parties, and took from them an agreement, in writing, to pay at certain stipulated times (in about a year from May last) the amount agreed upon. The agreement was without qualification or condition, and had no connexion with what the government might subsequently do, only binding themselves to a due execution of the order on their part. Under and in pursuance of this agreement they subsequently gave him one note, payable in August, and one in September next, both amounting to $10,000, which he expected would be paid, as he had no doubt of their responsibility.

Mr. Simmons further states that he never expected Schubarth to pay him a cent for the assistance he rendered him; that he was in moderate circumstances; that the Smiths were rich and that he expected them to pay; and that he should have aided just as soon, and acted as earnestly in endeavoring to procure the orders, without compensation, as he had repeatedly done in other cases for his constituents. Mr. Schubarth states, on the contrary, that he considers himself bound to give Senator Simmons five per cent. commission on both orders, and he says he made the offer of commissions to Senator Simmons because he had understood that a commission was paid for obtaining contracts, and he thinks it was one of the Messrs. Smith who told him so. It may not be amiss to state that Mr. Simmons, who has appeared before the committee and with great frankness and fairness detailed all the circumstances of the transaction, and answered all questions put to him promptly and without hesitation, also stated that he is now, and was at the time, unconscious of having done anything wrong in acting as he did, and that he was not aware that it was unlawful, there being no law prohibiting it; that he considered he was not only benefiting his constituents, but was also doing the government a service by procuring for it responsible contractors for the prompt supply of arms, which were then so greatly needed, he having about that time learned in high quarters that although 500,000 men had recently been called to the field, the government had then only 200,000 stand of arms to put in their hands.

It further appears that Senator Simmons still holds the notes given him for these commissions as agreed upon, and expects to receive payment on them when they fall due.

It also appeared to the committee that both before and after the contract for these arms was made, many other contracts were made for similar arms, between July 1, 1861, and January 1, 1862, and the lowest price paid by the department for arms of this description was twenty dollars, the price agreed to be paid to Mr. Schubarth; that twenty dollars was the lowest bid that was made at the department for the manufacture of such arms during the periods aforesaid, and that the government lost nothing in consequence of the offers of Mr. Schubarth to pay Mr. Simmons a commission for his assistance in this particular, after he had voluntarily undertaken to assist in procuring the contract without reward.

The committee, on these facts, are of opinion that, although Senator Simmons may have thought, as he declares, that his act was neither unlawful or wrong, (a declaration which his age, respectability, and long, useful, and honorable life, it would seem, should favor, whilst it is true that these very considerations should make a wilful departure from the line of correct conduct more glaring and censurable,) still, that such a practice is entirely indefensible, and that it was highly improper for a senator of the United States to have acted thus, even where the government sustained no loss thereby; but especially at a time when the very existence of the government and the Union were endangered by a fierce and huge rebellion, to suppress which the patriotism of the country was everywhere engaged—a rebellion which, first by robbery and afterwards by occasioning unlimited requisitions for the purchase of supplies and arms—had subjected the public treasury to untold and frightful drains.

That Congress disapproves of such conduct is manifest from the fact that they have promptly passed a law making it a penal offence hereafter; this will prevent all such acts in future; but to visit a severe and striking penalty upon an act which at the time of its commission was not punishable or forbidden by any public law would be retroactive in its effect, and render the step liable to that objection to which all *post facto* laws are justly subject.

The committee would further state, in this connexion, that it was the common practice, until a recent period, for members of Congress to prosecute claims against the government for fees and rewards, and that for many years our most distinguished public men were in the habit of prosecuting claims under such circumstances, without censure or criticism; but this was made a misdemeanor by an act of Congress in 1853, and since that period the practice has been discontinued.

The committee unanimously report back the resolution, accompanied by the statement of facts herein contained, (about which there appears to be no dispute,) in order that the Senate may take such action in the matter as they in their wisdom and discretion may think fit.

VII

Rimfire

TITLE: .58 JOSLYN CARBINE

HEADSTAMP: Plain

DIMENSIONS:

Rim:	.691"
Head:	.612"
Neck:	.612"
Bullet:	.600"
Case:	.730"

OTHER DETAILS:

Bullet Grains:	390
Powder Grains:	44
Bullet:	Conical
Case Material:	Copper
Total Grains:	489

OVERALL: 1.242"

Benjamin F. Joslyn obtained patents on August 28, 1885 and October 8, 1861 for the Joslyn Carbine and successfully obtained a contract to sell eleven thousand carbines to the United States Government. The arms were manufactured by the Joslyn Company of Stonington, Connecticut.

This round was not used during the Civil War as was commonly thought. It was developed some years later probably during the early 1870's

As part of the contract with the United States Government, the Union Metallic Cartridge Company and the Winchester Repeating Arms Company also manufactured the rounds.

The carbine fired a short stubby .58 caliber cartridge. It had the shortest case of any of the .58 rim-fire cartridges. The case was only .730" long with an over-all length of just 1.242".

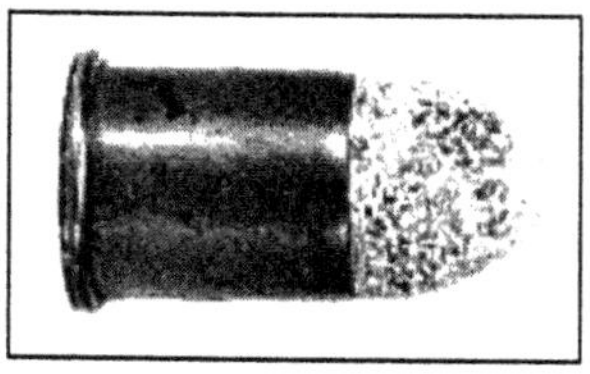

TITLE: .58 GATLING-MEDIUM CASE

HEADSTAMP: Plain

DIMENSIONS:

Rim: .677"
Head: .607"
Neck: .606"
Bullet: .602"
Case: .980"

OTHER DETAILS:

Bullet Grains: 565
Powder Grains: 55-60
Bullet: Pointed, Ribbed Grooves
Case Material: Copper
Total Grains: 692

OVERALL: 1.879"

A variant medium cased Gatling made by an unknown manufacturer.

The cartridge has the following four differences from the more common variety:

1. A greater over-all length
2. A longer and larger diameter bullet
3. In spite of the larger bullet, the total grains of this round is seventeen grains less than the short
4. The base has a deeply depressed circle with a pronounced "dimple" in the center. X-rays reveal the usual rim-fire type construction

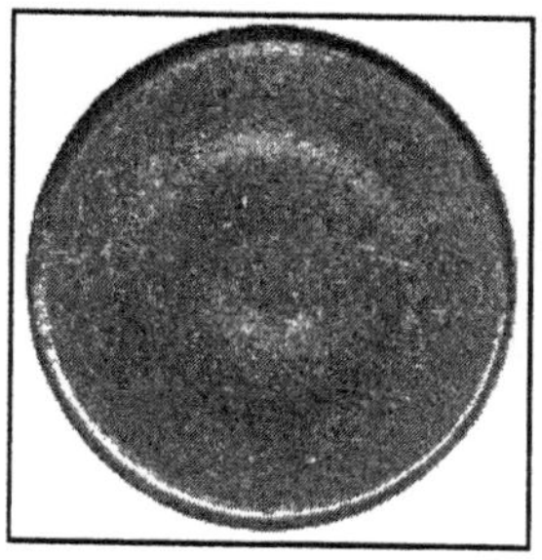

TITLE: .58 GATLING-MODEL 1862

HEADSTAMP: Plain

DIMENSIONS:

Rim: .705"
Head: .615"
Neck: 617"
Bullet: .585"
Case: 1.000"

OTHER DETAILS:

Bullet Grains:	565
Powder Grains:	60
Bullet:	Lead, Pointed
Case Material:	Copper
Total Grains:	701

OVERALL: 1.813

This round is commonly referred to as the "Mountain Gatling". There is no verification for the correctness of the title "Mountain Gatling". To date no such weapon has been located.

The first Gatling Gun utilized a steel auxiliary chamber that accommodated a .58 caliber paper cartridge. The chamber had a percussion nipple built into the case for ignition purposes. This percussion model was soon replaced by the Model 1862 Gatling that fired this short case rim-fire round. The auxiliary chamber was altered to enable this .58 rim-fire, (which approximated the powder charge of sixty grains of the paper cartridge) to be inserted and fired from the chamber.

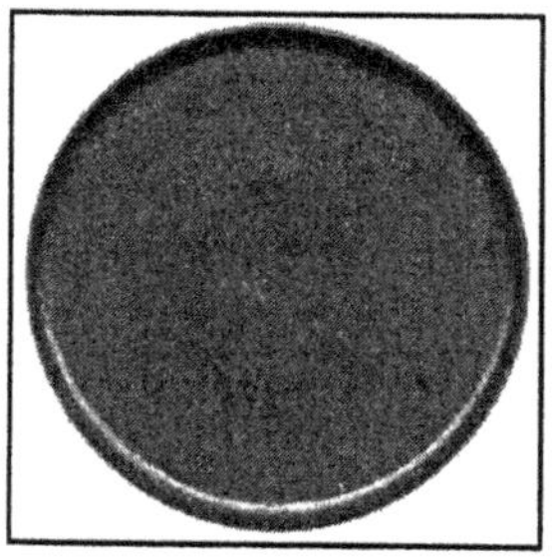

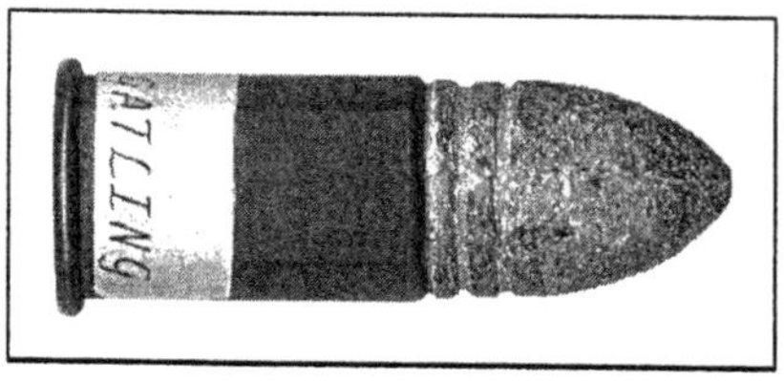

TITLE: .58 GATLING-SPECIAL

HEADSTAMP: Plain

DIMENSIONS:		OTHER DETAILS:	
Rim:	.704"	Bullet Grains:	565
Head:	.612"	Powder Grains:	60
Neck:	.614"	Bullet:	Pointed, two grooves
Bullet:	.587"	Case Material:	Copper
Case:	1.000"	Total Grains:	709

OVERALL: 1.705"

A variant Gatling cartridge by an unknown manufacturer.

The bullet and the case differ in minor ways from the common short cased Gatling round.

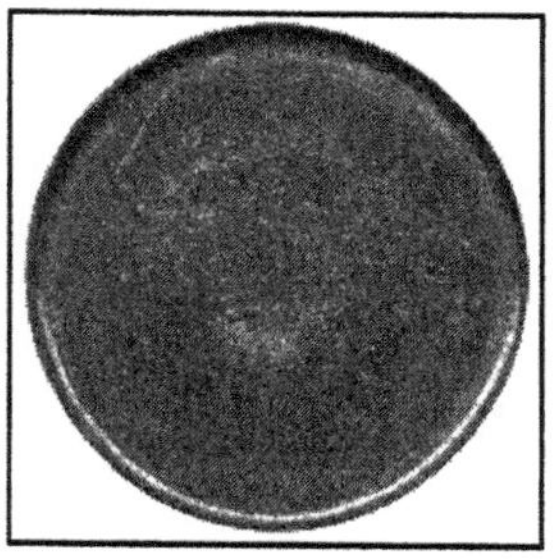

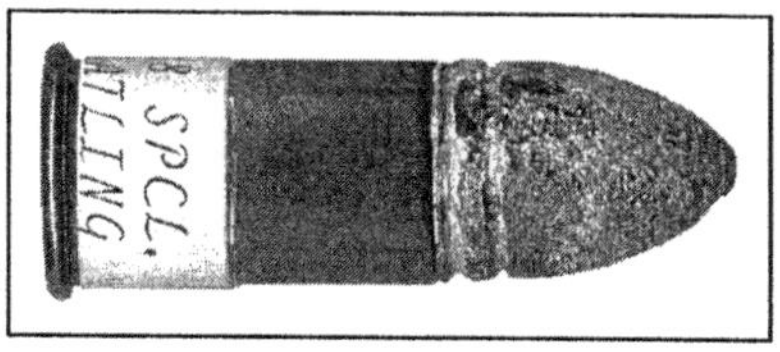

TITLE: .58 MUSKET-UNKNOWN

HEADSTAMP: Plain

DIMENSIONS:		METRICS:	OTHER DETAILS:	
Rim:	.700"	17.2 mm	Bullet Grains:	Unknown
Head:	.638"	16.0 mm	Powder Grains:	Unknown
Neck:	.621"	15.0 mm	Bullet:	Lead, Flat Nose
Bullet:	.593"	10.14 mm	Case Material:	Copper
Case :	1.088"	27.3 mm	Total Grains:	609

OVERALL: 1.541"/38.10 mm

This round is presently unidentified. The weapon that it was intended to be fired in has not been located.

The case is a shortened version of the common Miller/Allin rim-fire cartridge. The rim and body dimensions are very similar.

The bullet was molded with a pronounced flat nose.

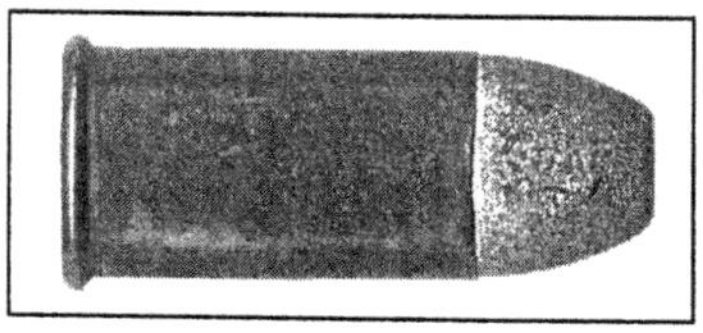

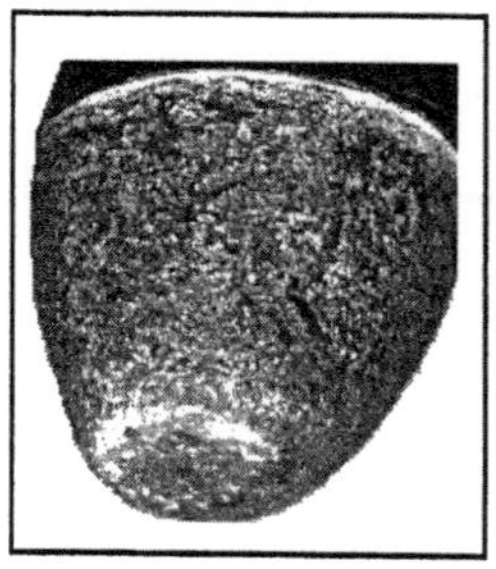

TITLE: .58 SPRINGFIELD MUSKET DUMMY

HEADSTAMP: Plain

DIMENSIONS:

Rim:	.706"
Head:	.631"
Neck:	.624"
Bullet:	.570"
Case:	1.180"

OTHER DETAILS:

Bullet Grains:	?
Powder Grains:	?
Bullet:	Flat Nose
Case Material:	Copper
Total Grains:	610

OVERALL: 1.710"

The title is pure conjecture on the part of the author because the case measurements agree with the Miller/Allin rounds and occasional reference has been made to the .58 Springfield round.

Three exceptions have been noted:

1. A very pronounced case mouth crimp.
2. The total weight of the round is sixty grains less.
3. The bullet is tapered with a slight flat nose.

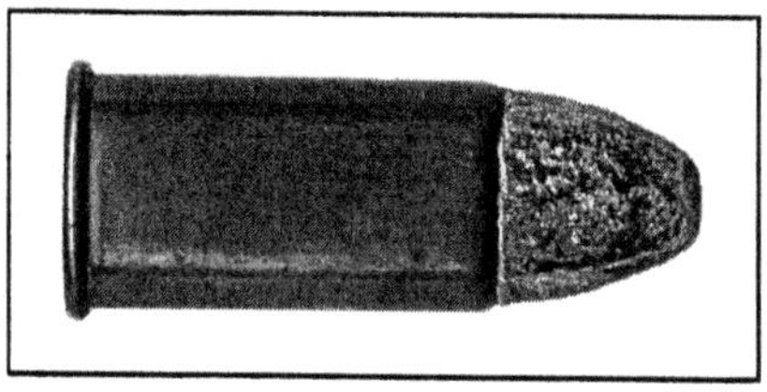

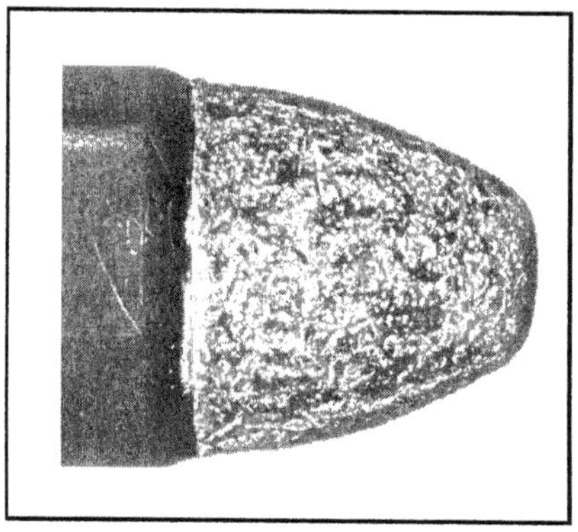

TITLE: .58 MUSKET-BOARD DUMMY

HEADSTAMP: U

DIMENSIONS:

Rim: .715"
Head: .635"
Neck: .621"
Bullet: .568"
Case: 1.182"

OTHER DETAILS:

Bullet Grains: 500
Powder Grains: 60
Bullet: Lead, Pointed
Case Material: copper
Total Grains: 1090

OVERALL: 1.627"

This is a much sought after round of the .58 caliber variety. Only a small number of these cartridges are available because they appeared only on the Union Metallic Cartridge Company cartridge display boards. Although no loaded round has ever been found with the U head-stamp, a box of twenty .58 Caliber Musket cartridges has a label depicting the round with the U marking. It would be interesting to learn the reason why the Company chose to go to the trouble of head stamping these board dummies when there was a plentiful supply of the unmarked .58 Miller/Allin cases available

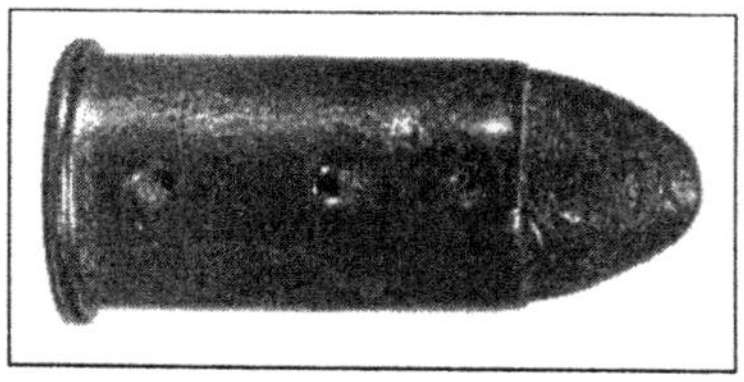

TITLE: .58 MUSKET-BLANK-NECKED

HEADSTAMP: Plain

DIMENSIONS:

Rim: .701"
Head: .633"
Neck: .418"
Bullet: Blank
Case: 1.184"

OTHER DETAILS:

Bullet Grains: Blank
Powder Grains: ?
Bullet: Blank
Case Material: Copper
Total Grains: 137

OVERALL: 1.184

This round appears to be a special purpose cartridge because of its unusual, extreme, case neck taper. The total grain weight of only 137 grains indicates a one-half charge loading.

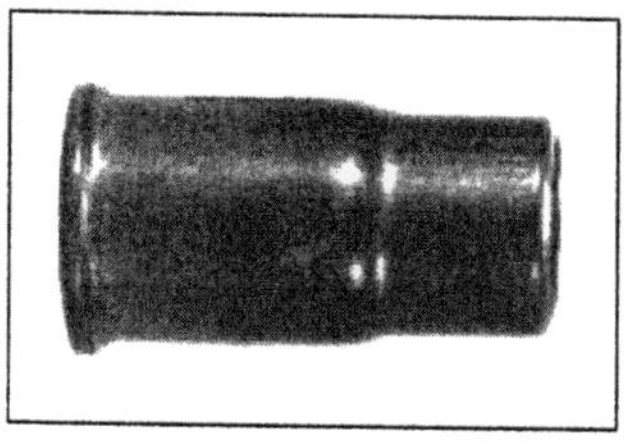

TITLE: .58 MUSKET RIM FIRE-BLANK

HEADSTAMP: Plain

DIMENSIONS:

Rim: .702"
Head: .631"
Neck: .623"
Bullet: Blank
Case: 1.188"

OTHER DETAILS:

Bullet Grains: Blank
Powder Grains: ?
Bullet: Blank
Case Material: Copper
Total Grains: 177

OVERALL: 1.188

There has always been a need for blank loads. Blanks are used for signaling or training purposes. The fifty-eight rim-fire was no exception. Many different manufacturers made blanks for the .58. Some blanks have different colored paper wads possibly to indicate different powder charges. Each manufacturer had a unique case crimp design. Generally the total grain weight varied from 165 to 177 grains.

TITLE: .58 CONVERTED – STORM/ROBERTS

HEADSTAMP: Plain

DIMENSIONS:		OTHER DETAILS:	
Rim:	.735"	Bullet Grains:	480
Head:	.648"	Powder Grains:	80
Neck:	.620"	Bullet:	Conical
Bullet:	.606"	Case Material:	Copper
Case:	1.188"	Total Grains:	802

OVERALL: 1.906"

William Mont Storm was granted patent U.S. #15307 on July 8, 1856 and U.S. patent #24414 on June 14, 1859 for the conversion of the M1842 percussion musket to metallic cartridges. The U.S. Government purchased the right to alter 2,000 muskets using a Storm front hinged tip-up breechlock action for $2,500.00.

The alterations were to be completed by the Harpers Ferry Arsenal, but very few alterations were actually completed because the arsenal fell into Confederate hands soon after the start of the Civil War.

Mr. John L. Barber, author of the Rim Fire Cartridge 1857 to 1884 is of the opinion that Smith & Wesson supplied an unknown quantity of .58 rim-fire cartridges to the Springfield Arsenal to be used in the trials of the Storm/Roberts. Col. Berk Lewis was also of the opinion that the .58 Storm/Roberts rim-fire cartridges were tried in the M1860 musket.

Contracts for manufacturing the ammunition were given to Smith & Wesson, but the actual manufacturing was done by Leet, Goff & Co..

No action was taken with the cartridge until after the Civil War. Another conversion patent (#52887-February 27, 1860) was issued and used in the B. S. Roberts muskets.

U. M. C. made the cartridges but gave it the title .58 Roberts Converted Enfield or converted Springfield.

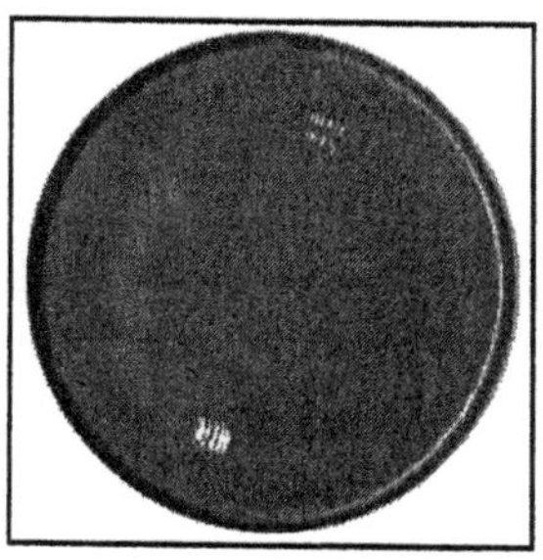

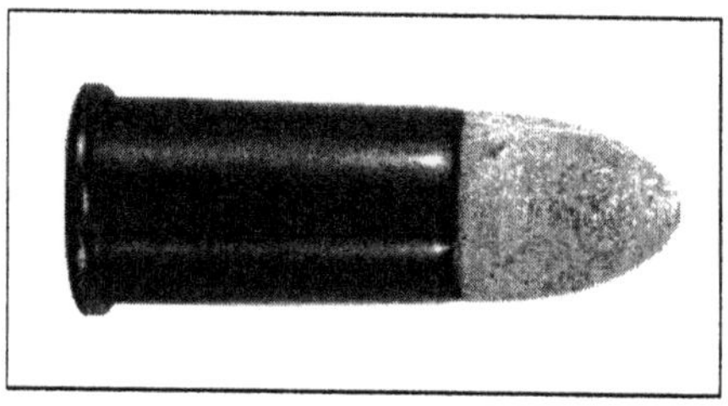

TITLE: .58 CHABOT

HEADSTAMP: Plain

DIMENSIONS:

Rim:	.695"
Head:	. 624"
Neck:	.622"
Bullet:	.598"
Case:	1.200"

OTHER DETAILS:

Bullet Grains:	?
Powder Grains:	?
Bullet:	Conical
Case Material:	Copper
Total Grains:	?

OVERALL: 1.200"

C. Chabot obtained a patent on September 5, 1865 for his tip-up action to convert muzzle-loading muskets. The rim-fire round closely resembles the Miller/Allin .58 caliber cartridge. There are very minor differences in dimensions between the two.

No picture available

TITLE: .58 MUSKET

HEADSTAMP: .58
CAL

DIMENSIONS:	
Rim:	.708"
Head:	.637"
Neck:	..622"
Bullet:	.585"
Case:	1.200"

OTHER DETAILS:	
Bullet Grains:	500
Powder Grains:	60
Bullet:	Lead, Pointed
Case Material:	Copper
Total Grains:	1050

OVERALL: 1.720"

The .58 Cal. head stamp indicates the round was a product of Frankford Arsenal. It is also thought to be experimental in nature.

In Major T. J. Treadwill's report of metallic cartridges tested an experimental front ignition round with the following dimensions and .58 CAL head stamp with the following measurements:

I quote:

Case Length	1.211"
Rim	.705"
Body:	.624"
Grains:	106 case only
Head stamp	.58 CAL

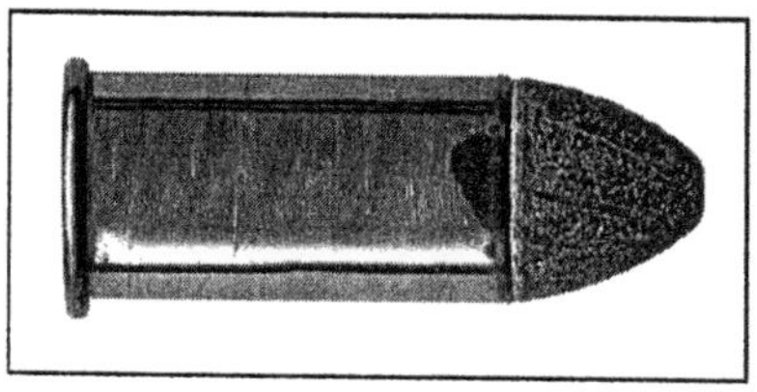

TITLE: .58 MUSKET RIM FIRE

HEADSTAMP: F.V.V. & Co.

DIMENSIONS:		OTHER DETAILS:	
Rim:	.710"	Bullet Grains:	500
Head:	.630"	Powder Grains:	60
Neck:	..617"	Bullet:	Conical
Bullet:	.590"	Case Material:	Copper
Case:	1.203"	Total Grains:	657

OVERALL: 1.782"

The head stamp, F.V.V. & Co. stands for Fitch Van Vechten and Co. another very old manufacturer of rimfire cartridges dating back to 1864. The first location was at 257 Broadway, New York City, New York. See John L. Barber's excellent book , "The Rimfire Cartridge in the United States and Canada".

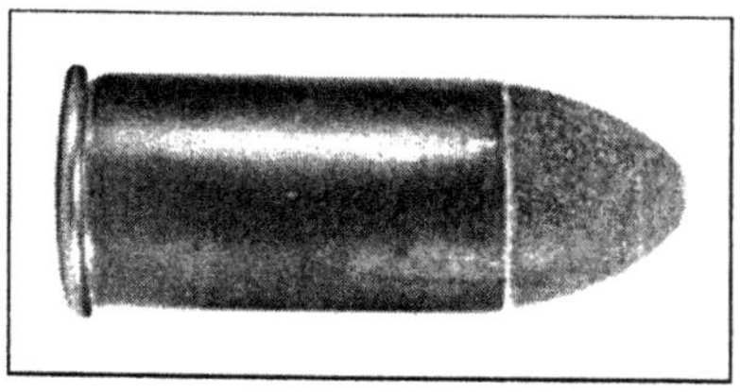

TITLE: .58 MUSKET RIM FIRE

HEADSTAMP: C. D. L.

DIMENSIONS:

Rim: .707"
Head: .635"
Neck: .630"
Bullet: .598"
Case: 1.203"

OTHER DETAILS:

Bullet Grains: 500
Powder Grains: 60
Bullet: Tapered
Case Material: Copper
Total Grains: 652

OVERALL: 1.684"

The head stamp, C.D.L. stands for C. D. Leet & Co. a manufacturing firm dating back to 1860. A complete history of the firm is discussed in John L. Barber's excellent book, "The Rimfire Cartridge in the United States and Canada", page 3.

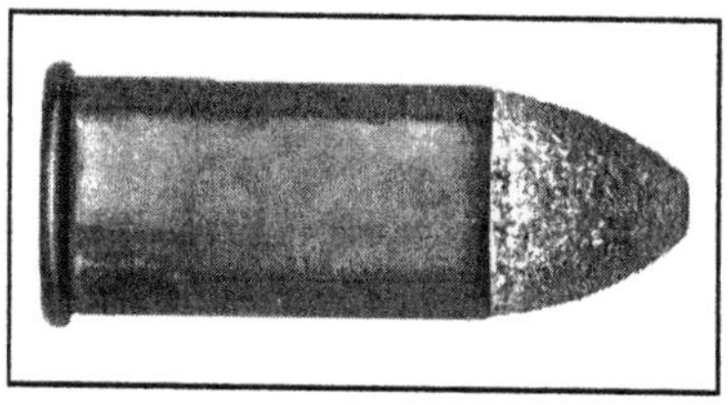

TITLE: .58 SOUTH AMERICAN ROBERTS

HEADSTAMP: Plain

DIMENSIONS:

Rim:	.739"
Head:	.650"
Neck:	.614"
Bullet:	.599"
Case:	1.218"

OTHER DETAILS:

Bullet Grains:	500
Powder Grains:	60
Bullet:	Conical
Case Material:	Copper
Total Grains:	790

OVERALL: 1.906"

For a long time, this round was referred to as the .58 Mont Storm South American rim-fire. Actually the cartridge is not a Mont Storm which it strongly resembles, it was used in a Roberts conversion breech-loader.

All of these specimens have the very distinctive neck crimps not found on any other rim-fire ammunition. Thought to have been hand loaded in South America.

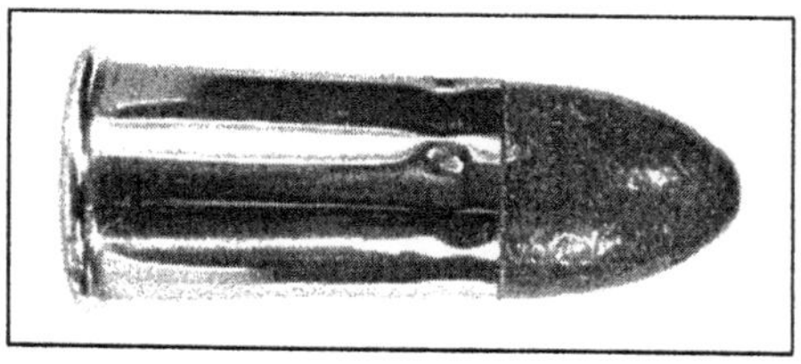

TITLE: .58 MUSKET

HEADSTAMP: 58 CAL

DIMENSIONS:

Rim: .710"
Head: .632"
Neck: .621"
Bullet: .600"
Case: 1.219"

OTHER DETAILS:

Bullet Grains: 500
Powder Grains: 60
Bullet: Conical
Case Material: Copper
Total Grains: 648

OVERALL: 1.719"

This round, with the .58 CAL. head stamp, was manufactured at the Springfield Arsenal during the late 1860's.

Experiments using this head stamped case are also found in the Bar Anvil and Corliss front ignition specimens mentioned elsewhere in this book.

Close examination of the tick marks around the .58 CAL imprint reveal slight variations from similar markings on the other musket rounds.

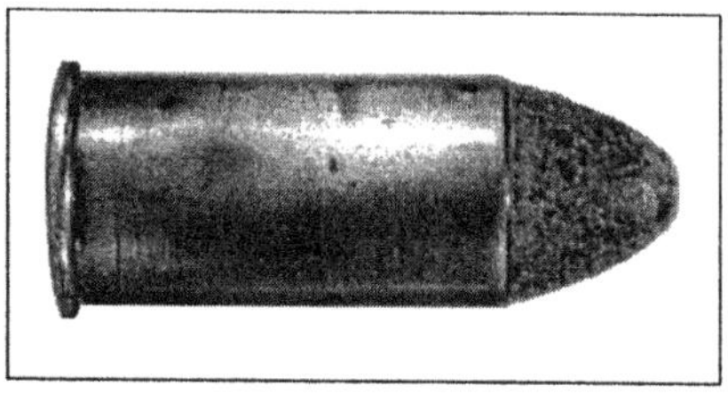

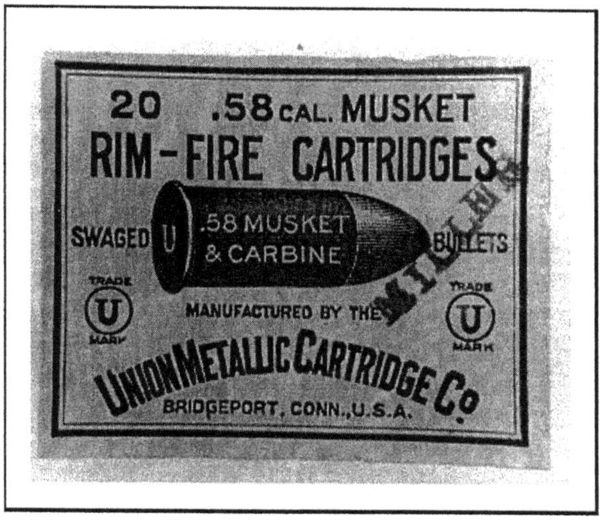

TITLE: 14.5 x 32.5 Rmm WAENZL

HEADSTAMP: R. W.

DIMENSIONS:

Rim: .685"
Head: .601"
Neck: .588"
Bullet: .575"
Case: 1.279"

OTHER DETAILS:

Bullet Grains: 458
Powder Grains: 68
Bullet: Pointed, two grooves
Case Material: Brass
Total Grains: 622

OVERALL: 1.965"

This round was designed and developed in 1865 by the Austrian gun maker J. Waenzl. The round was to be fired in the Model 1857 muzzle-loading rifle after conversion to a breech-loading system.

Due to the rifle's slow rate of fire and the low velocity of the round, the rifle and cartridge were soon discarded.

The cartridge appears with a variety of head stamps. The R W on this specimen stands for the manufacturer Robert Wolner of Vienna.

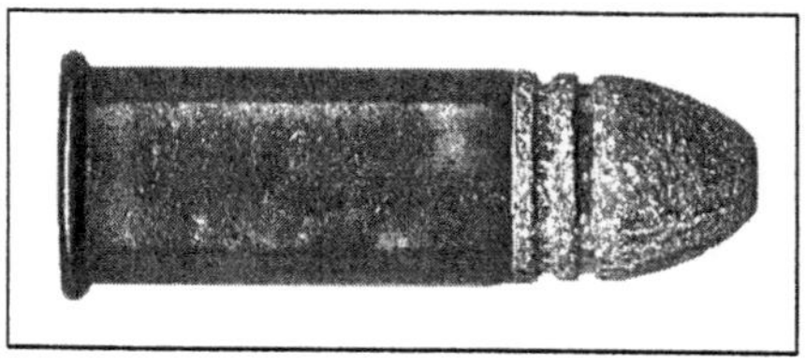

Computer Enhanced

TITLE: .58 ROBERTS

HEADSTAMP: H (r)

DIMENSIONS:		OTHER DETAILS:	
Rim:	.755"	Bullet Grains:	550
Head:	.654"	Powder Grains:	60-65
Neck:	.632"	Bullet:	Flat Nose
Bullet:	.595"	Case Material:	Copper
Case:	1.280'	Total Grains:	656

OVERALL: 1.791"

The inventor and designer of the Roberts Rifle and cartridge was Brevet Brigadier General Benjamin S. Roberts. He was granted patent #52887 on February 27, 1886 for a breech-loading system to convert muzzle loading Springfield Muskets to a rim-fire system.

In order to facilitate the extraction of the fired cartridge case, he designed the Roberts cartridge case, with a pronounced taper. No other rim-fire cartridges of the period were designed this way.

This much sought after head-stamped round was manufactured by the Winchester Repeating Arms Company.

A center-fire version was also manufactured.

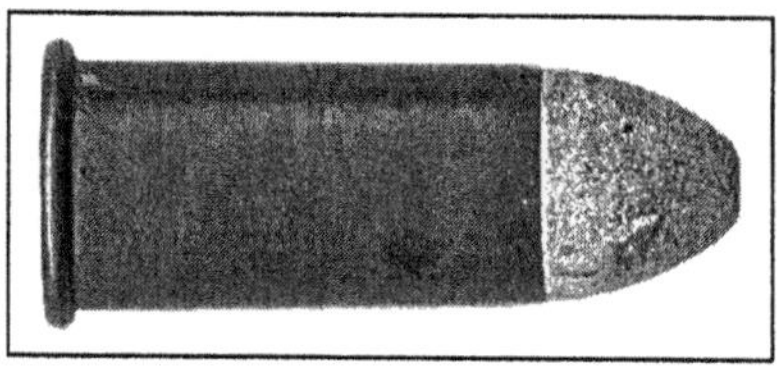

TITLE: 14.5 x 32.5Rmm WAENZL

HEADSTAMP: Plain

DIMENSIONS:

Rim: .690"
Head: .607"
Neck: .590"
Bullet: .568"
Case: 1.329"

OTHER DETAILS:

Bullet Grains: 425
Powder Grains: 70
Bullet: Knurled, Hollow Pt.
Case Material: Brass
Total Grains: 614

OVERALL: 1.870"

This round was used in an early conversion of the Austrian Waenzl military rifle to a rim-fire system.

Because it had a brass case instead of copper, it was probably commercially manufactured.

The bullet is unusual in that it has a hollow nose which appears to be filled with clay.

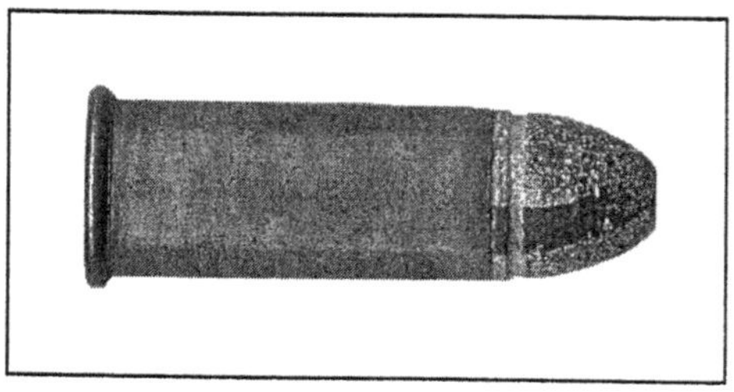

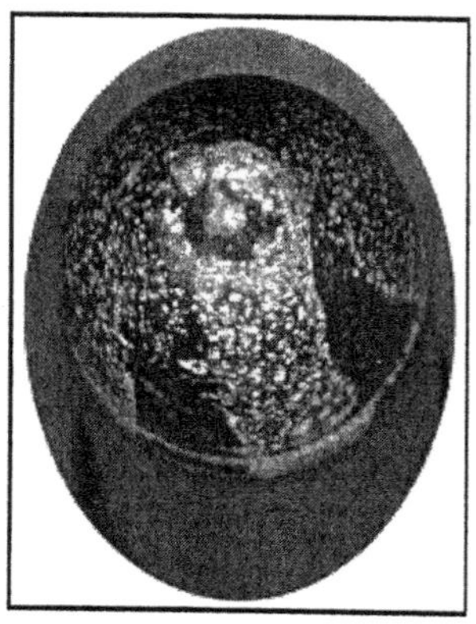

TITLE: .58 GATLING-FIELD

HEADSTAMP: Plain

DIMENSIONS:

Rim:	.706"
Head:	.611"
Neck:	.614"
Bullet:	.594"
Case:	1.344"

OTHER DETAILS:

Bullet Grains:	575
Powder Grains:	70
Bullet:	Pointed, two grooves
Case Material:	Copper
Total Grains:	757

OVERALL: 2.125"

The Model 1865 Gatling chambered this round. This was the last of the rim-fire Gatling cartridges.

There is also a center-fire cartridge that is an approximate duplicate of this round.

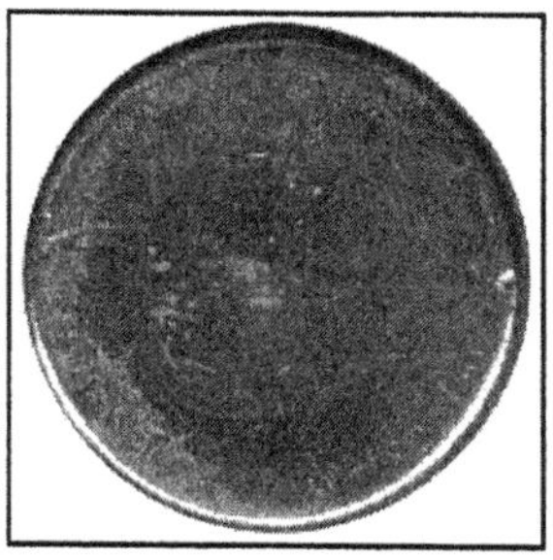

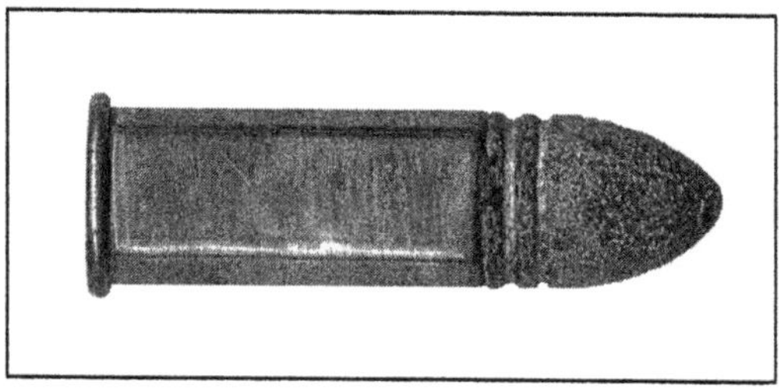

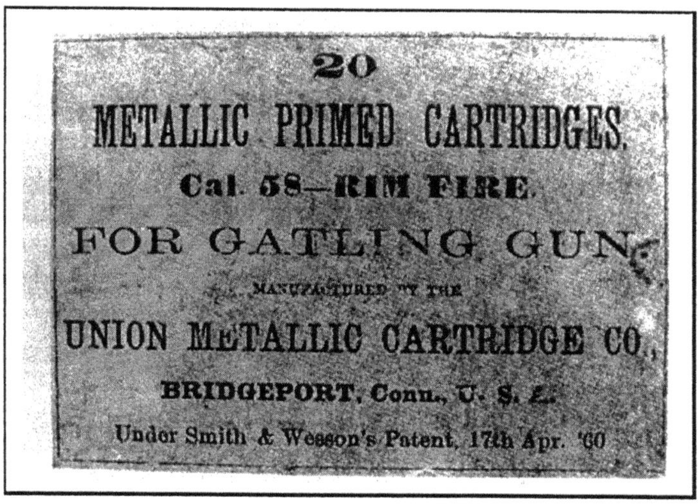

TITLE: 15 x 35 Rmm

HEADSTAMP: Plain

DIMENSIONS:

Rim: .744"
Head: .654"
Neck: .623"
Bullet: .606"
Case: 1.356"

OTHER DETAILS:

Bullet Grains: 500
Powder Grains: -----
Bullet: Smooth, tapered
Case Material: Brass
Total Grains: 689

OVERALL: 2.021"

This is another cartridge from the White & Munhall collection, as evidenced by the inscribed code numbers 8476 on the side of the case.

It is the author's opinion that this round is of European manufacture. It is quite possible that it is for the .58 Gatling gun as the general dimensions are similar.

This is a cartridge display board specimen.

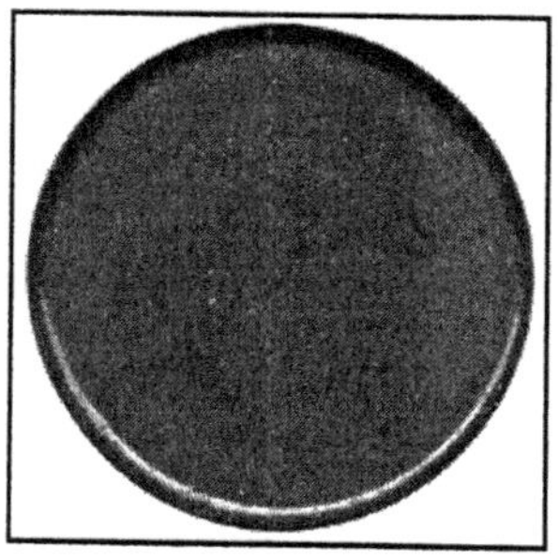

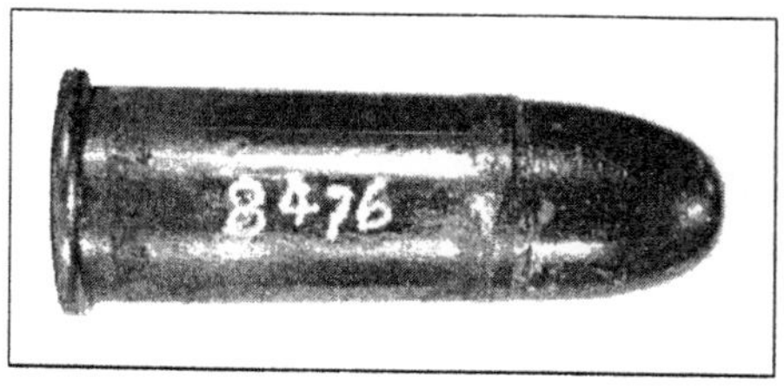

ALLEN-BRAND RIM FIRE

One of the earliest breech-loading arms submitted to the Army and Navy Ordnance Boards for testing was the Allen-Brand rifle and musket.

Christopher Brand was a manufacturer of Whaling guns and harpoons in Norwich, Connecticut. During 1862 and 1863, he obtained patents for two weapons. One was a patent for a .54 Caliber rifle, and the other was for a .58 caliber smooth bore musket.

Apparently Brand did not want to take time to promote the weapons to obtain a government contract. He entered into a business relationship in 1862 with Colonel Enos G. Allen, a Boston businessman, to promote the sale of the breech-loaders to the U. S. Government. Allen had had some previous experience in the firearms field in the mid 1800's. He unsuccessfully attempted to introduce the Lindner conversion rifle into the United States Army and the Massachusetts militia.

On June 30, 1862, Captain John A. Dahlgren of Washington Navy Yard ordered Lt. Cmdr. Robert L. May to obtain a rifle musket for testing. Although the Allen-Brand rifle performed well, no orders were forthcoming. This test was typical. The weapon failed to interest either the Navy or the Army to purchase the rifle. It was common knowledge that Chief of Ordnance, Brig. Gel. J. W. Ripley was opposed to the modern breech-loading guns and the equally new fixed or self-contained rim-fire cartridges.

The rifle musket and the smooth bore model used rim-fire ammunition. The rifle was .54 caliber, and the smooth bore used a .58 caliber shot or buck-and-ball load.

The interesting round is the .58 caliber buck-and-ball (or canister) cartridge. It was covered by patent #41590, dated February 16, 1864 granted to E. G. Allen.

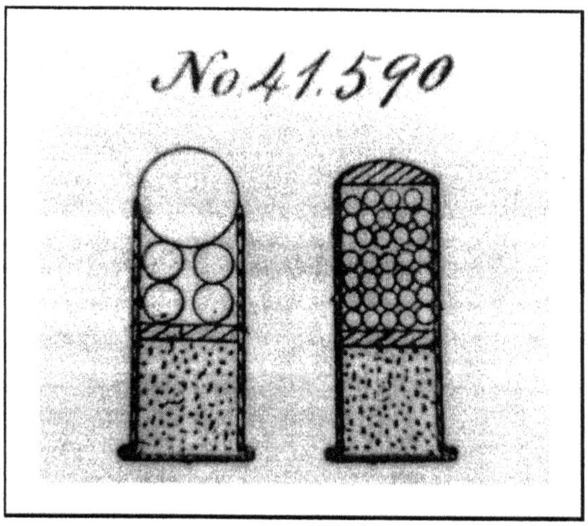

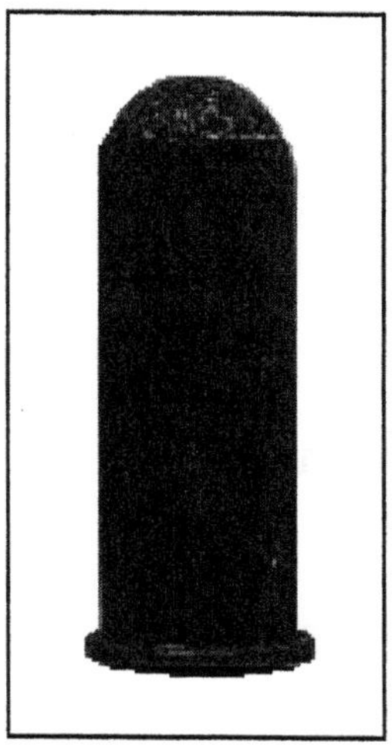

His patent further claimed:

> "Forming a metallic flanged cartridge-case, imperforate and charged with the fulminate at its flanged end and at its outer rim and swaged over the equator of the ball, or head cap, so as to hold it close, and tapering, so as to give certainty to the entry of this end into the barrel without obstruction, and secure a close fit to the chamber, to prevent the escape of the gasses therein, the said cartridge case containing buck or other shot in its anterior portion, and powder in its posterior, the two to be separated by a wad of sufficient capacity, and fitting with sufficient accuracy to prevent the escape of the explosive gas among the shot, all arranged precisely and specifically as described in the specification and drawing."

Another test of the Allen-Brand rifle and its ammunition was made on February 26, 1863, at the Washington Arsenal. A report by lst. Lt. J. C. Bradfordand and 2nd. Lt. L. S. Babbitt stated:

> "both arms and ammunition possesses unqualified merit".

In spite of this favorable report, the Allen-Brand breech-loader failed to entice General Ripley into a contract for either weapon.

Subsequent improvements in the guns failed to obtain any orders from the military. They soon became obsolete.

There is no mention in the records as to the manufacturer of the ammunition or the specific dimensions of the copper case.

The following information was obtained from the test records at the sites where the guns were inspected.

WASHINGTON NAVY YARD - February 19, 1863:

Caliber	.58
Powder Charge	48.8 Grs.
Case	97.3 "
Ball	275.5 "
Buck Shot (3)	75.9 "
Wad	3.5 "
TOTAL GRAINS	501.0 "

WASHINGTON ARSENAL - February 26, 1863:

Caliber	.58
Powder Charge	44.0 Grs.
Case	94.4 "
Ball	283.7 "
Buck Shot (3)	75.0 "
Wad	4.0 "
TOTAL GRAINS	501.1 "

RIFLE CARTRIDGE

Caliber	.58
Powder Charge	38.2 Grs.
Case	62.4 "
Ball	474.0 "
TOTAL GRAINS	574.6 "

No case length is given but it is safe to assume that the smooth bore round must be longer because the weight of the case is greater than that of the rifle cartridge.

TITLE: .58 LAMSON

HEADSTAMP: Plain

DIMENSIONS:

Rim:	.705"
Head:	.634"
Neck:	.603"
Bullet:	.567"
Case:	1.390"

OTHER DETAILS:

Bullet Grains:	475
Powder Grains:	72
Bullet:	Tapered
Case Material:	Copper
Total Grains:	?

OVERALL: 2.045"

The measurements for this specimen are nearly identical to the ones used in the Lamson Conversion. I believe it is a Lamson.

A group of firearms submitted for the New York State trials of 1867 listed a Lamson conversion. The cartridge used in the arm appears to be the same round.

The rifle was made by the Windsor Manufacturing Company, Windsor, Vermont. A similar, but unidentified cartridge appears in the American Cartridge, by Charles R. Suydam. The 1869 and 1871 tear sheets by Union Metallic Cartridge Company also refer to the .58 rim-fire as the .58 Lamson.

No picture is available.

TITLE: .58 JOSLYN CONVERSION RIMFIRE

HEADSTAMP: Plain

DIMENSIONS:		OTHER DETAILS:	
Rim:	.720"	Bullet Grains:	?
Head:	.617"	Powder Grains:	?
Neck:	.617"	Bullet:	?
Bullet:	.615"	Case Material:	Copper
Case:	1.440"	Total Grains:	?

OVERALL: 1.440"

These dimensions were obtained from a chamber cast of the actual rifle tested at the 1863 trials.

No cartridge specimen was available. No picture available.

ERSKINE S. ALLIN AND THE .58" CALIBER RIMFIRE CARTRIDGE
By George Murphy

Erskine S. Allin was born in Enfield, Conn. on February 3, 1809. His father, Diah Allin, was employed by the Springfield Armory starting in 1810 and worked as an armory foreman for over 25 years. His brother Lucius Allin also worked at the armory starting in 1826 and was a foreman in the filing department in 1860. Erskine S. Allin became an apprentice and watchman under his father in June of 1829. He spent time in a majority of the work activities of the armory and in January of 1842, he was appointed foreman of the middle shops by Col. Ripley who was the post commander at this time. On October 1, he was appointed acting master armorer. He held this position until appointed clerk to the ordnance storekeeper on May 1, 1848. He was again appointed acting master armorer on October 1, 1848 until June 14, 1853 at which time he received a Master Armorer Commission from the President. During 1854 he acted as the armory superintendent for 7 months. His memoranda of service ended on May 1, 1873.

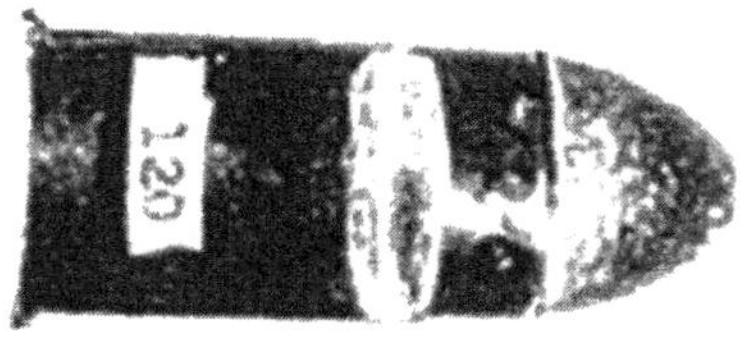

Mr. Allin died on September 11, 1879. Col. J.G. Benton, the Commander of the Springfield Armory issued Post Orders No. 27 and stated "His skill as a mechanic is attested by improvements in the arms now employed by the Government, and also in the mode of their manufacture. His fidelity to his trusts was rewarded by the confidence reposed in him by his superior officers, and the exercise of the power confided to him was ever kind and considerate to all."

"As a mark of respect to his memory, the Flag at this Post will be displayed at half-staff during Sunday, the14th inst., the date of his funeral."

The Civil War proved that the muzzle loading rifle was improper for military use and with hundreds of thousands of this style in existence a means of converting them economically to breech loaders was required. Mr. Allin on May 1, 1873 personally summarized his work activities at the armory and stated "I might perhaps also state that the breech loading gun, the leading principle of which is my own invention, I assigned to the govt. without roalty or pecuniary benefit, stands after the trial of the best for the Military Service." What Mr. Allin was referring to was patent number 49,959 issued to him on September 19, 1865 for the conversion of the .58" caliber muzzle loading to a breech loading system.

Springfield Breech Loading Rifle Model 1865

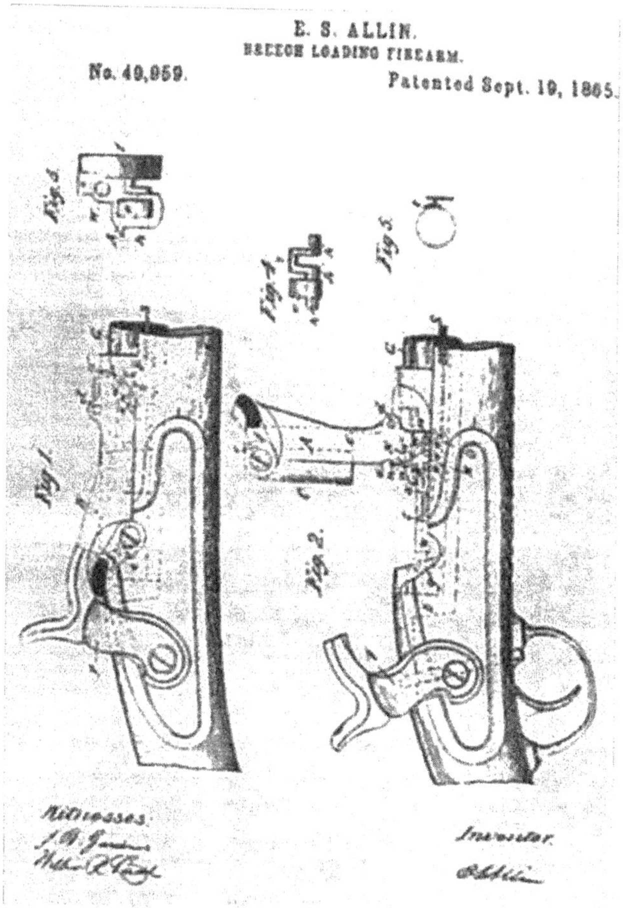

This model was produced following the patent of Mr. Allin and is commonly referred to as the "Allin alteration". This model was converted by milling open the breech section of the barrel and inserting a hinged bolt to the top of the barrel in front of the opening. This firearm was .58" caliber thus utilizing the existing barrels of the muzzle loading rifle. It used a .58" caliber rimfire cartridge.

Brig. General A.B. Dyer, Chief of Ordnance, directed the Springfield Armory to convert 5,000 muzzle loaders utilizing the Allin plan. Bvt. Col. T.T.S. Laidley, Commanding Officer of the Springfield Armory responded on January 22, 1866. He stated that 38 new pieces were required and 5 of the old parts had to be altered. The work required 95 new milling fixtures, 18 new drilling jigs and 44 new gauges. This was not a very favorable report on the feasibility of an "economical" conversion.

Springfield Breech Loading Rifle Model 1866

Model 1866 is a modification of the Model 1865. Two additional changes were requested: (a) reduce the caliber from .58" to .50" and (b) developing cartridges of both rimfire and centerfire configuration. On July 28, 1866, General A.B Dyer, Chief of Ordnance, directed Co. J.G. Benton, Commanding Officer of Springfield Armory, to convert 25,000 rifle muskets into breech loaders following the plan of Mr. Allin, with such modifications of the original plan as recent trials of the arm may have suggested, and the preparations of an adequate supply of ammunition. He further stated that "the ammunition will be made at Frankford Arsenal but you (Col. Benton) are charged with determining the proper charge of powder and weight and form of the ball and preparing model cartridges. A center primed cartridge is preferred to the rim primed, if one suitable for military service and equally sure of fire can be devised. You will direct your attention particularly to getting up a proper cartridge, but the conversion of the arms will not be delayed in order that you may determine whether center primed cartridges may be adopted.

A board of officers was convened in January of 1868 with Bvt. Brig. General P.V. Hugner as president. The board approved the Model 1866 with some alterations. This was then designated as Model 1868.

The .58 caliber musket rimfire cartridge had a very short and uneventful life as a military cartridge.

ALPHABETICAL INDEX

CASE LENGTH INDEX

Cartridge	Page No.	Case Length
Howdah .577 Carbine	125	1.315
Serbian Peabody .577 Model 70	164	1.321
14.8 mm Serbian Peabody	165	1.326
14.5x32R mm Waenzl	316	1.329
Berdan Original Frankford Arsenal	13	1.330
Serbian Austrian Carbine .58	166	1.340
Roberts Breechloader	188	1.344
Gatling Field	317	1.344
Roberts Breechloader	189	1.345
Gatling Centerfire .58	167	1.356
15x35R mm	318	1.356
15x34 Cannister Loading Serbian Peabody	169	1.360
15x34R mm G on Primer	170	1.360
Roberts Blank	190	1.375
Lamson .58	322	1.390
Snider Boxer Blank	48	1.430
Joslyn Conversion	323	1.440
Snider Mark III Blank	49	1.455
Snider Mark II Blank	50	1.456
Berdan Russian Kranka .60	14	1.470
Pauly	278	1.493
Snider Mark II Carbine & Rifle Blank	51	1.500
Snider Blank .577	52	1.524
Howdah Boxer Pistol	126	1.530
Snider Turkish Carbine	53	1.553
Snider Russian	54	1.562
Snider Turkish	55	1.565
Berdan Russian Carbine	15	1.570
Snider Blank	56	1.572
Berdan Russian Kranka	16	1.575
Berdan Musket Blanks	17	1.585 to 1.646
15.2x40R mm	171	1.586
Snider Martini Henry Blank .577	57	1.590
Snider Sporting .577	58	1.609
Snider Sporting .57	59	1.611
Snider Sporting .57	60	1.615
Snider Sporting D.C.Co .57	61	1.617
Snider Sporting Shot D.C. Co .57	62	1.620 to 1.625
Berdan Madrid Rifle Musket	18	1.621
Berdan Madrid Japanese	19	1.627 to 1.638
Berdan Musket Copper Case	20	1.638
Berdan Russian Imroved No. 1	21	1.640
Snider Shot D.C. Co. Board Dummy .57	63	1.640
Roberts .61	191	1.640
Berdan Musket Phillips Conversion	22	1.641
Berdan Musket Magnetic	23	1.642
Berdan Musket Board Dummy	23	1.644
Howdah Pistol .577 .N.A & A Co. Snider	127	1.650
Berdan Musket	25	1.652
Snider Sporting Paper Patched .57	64	1.654
Berdan Musket Iron Head	26	1.656

Cartridge	Page No.	Case Length
Berdan Musket Milbank Primer	27	1.656
Berdan Musket Patched	28	1.657
Berdan Musket	29	1.658
Berdan Assembler's Funtioning Dummy	30	1.658
Berdan Musket	31	1.661
Berdan Musket Orcut Primer	32	1.662
Snider Dummy .57	65	1.662
Berdan Musket W.R.A.	33	1.664
Snider .577	66	1.665
Snider Sporting	67	1.670
Snider Turkish Rifle .58	68	1.675
Snider Sporting Shot .57	69	1.682
Snider .577	70	1.730
Gatling Model 1862	300	1.813
Unknown Percussion	279	1.821
Unidentified .58	71	1.838
Snider Whitney Mark D	72	1.850
15x48R mm	172	1.868
Snider Whitney Mark III	73	1.875
14.7x49R mm Snider Turkish .577	74	1.892
Snider Copper Case	75	1.900
Snider *SCB* 95	76	1.900
Snider Mark IX	77	1.903
Snider Mark VI	78	1.904
Snider Transition .577 Kynoch's Patent	79	1.909
Snider Copper Case	80	1.912
Snider .577 Kynoch & Co.Birmingham	81	1.912
Snider Mark VI	82	1.928
15x49R mm Snider SK***	83	1.928
Snider Daw's Patent, 1867	84	1.930
Snider Mark IV .577	85	1.931
14.5x50 mm Snider Chinese .577	86	1.932
Snider Remington .57 REM UNC 57 S	87	1.933
Snider Boxer Sporting	88	1.933
Gallagher .58	280	1.937
Snider mark VIII	89	1.940
15x49R mm	173	1.940
15x49R mm GR	174	1.943
15x49R mm	175	1.943
Snider .577 Kynoch Patent Base	90	1.944
Howdah Pistol .577	128	1.944
Snider Belgium Coiled Case	91	1.952
Snider Canadian Mark IX	92	1.953
Snider .577	93	1.958
Snider Mark II	94	1.963
Snider .577 Kynoch Patent Base	95	1.967
Snider .577 Kynoch .577 S	96	1.983
Snider Modern Empty Case	97	2.000
Snider Sporting .577 ELEY LONDON	98	2.000
Blank .577 .577/450 Petal Crimp	99	2.065
Winchester Shot Shell	192	2.125 to 2.500

Cartridge	Page No.	Case Length
Shot Shell .58	193	2.130
Dangerfield & Lefever	176	2.135
14.7x58R Pottet	100	2.175
15x57R mm Board Dummy	177	2.217
Black Powder .577 3 1/4	101	2.250
Snider .577 Colonial Contract	102	2.324
Alexander Henry 20/577	103	2.738
Alexander Henry 20/577 ELEY BROs	104	2.741
Express .577 2 3/4	105	2.750
Boxer Coiled Case .577 2 3/4	106	2.750
Black Powder .577 2 3/4 Express Coiled Case	107	2.750
Black Powder .577 2 3/4 Coiled Case	108	2.750
Nitro .577 2 3/4 Kynoch	109	2.750
Black Powder Express .577 2 3/4	110	2.750
Police Greener Gun .577	111	2.860 & 2.483
Express .577 Kynoch L 577 C	112	3.000
Three Inch .577	113	3.000
Rewa .577	115	3.000
Bell Base N.E. .577	178	3.000
Tyrannosaur .577	179	3.000
Express .577 3"	114	3.000 & 2.980
Chinese Jingal .60	180	3.637

BIBLIOGRAPHY

TITLE	AUTHOR
THE AMERICAN CARTRIDGE	C. Suydam
CARTRIDGES	H. Logan
EARLY PERCUSSION FIREARMS	
I.C.C.A	B. R. Lewis
AMERICAN RIFLEMAN - JUNE 1955	
Meigs and U.S.C. Co. Ammo.	B. R. Lewis
NOTES ON CIVIL WAR AMMUNITION	B. R. Lewis
ORDNANCE MEMORANDA NO. 14	
MILWAUKEE PUBLIC MUSEUM PUBLICATION #2	
Terry Rifle	
WINCHESTER	Williamson
RIFLES IN AMERICA	P. Sharps
SMALL ARMS AND AMMUNITION	B. R. Lewis
1889 HARTLEY AND GRAHAM CATALOG	
THE RIFLED MUSKET	Fuller
MORSE ARMS CO. VS U.S.GOVERNMENT	
WORCESTER, MASS. CITY DIDRECTORIES	
AMERICAN ANTIQUARIAN SOCIETY	
Worcester, Mass.	
WORCESTER PUBLIC LIBRARY	
SCIENTIFIC AMERICAN	
1854 through 1870	
SMALL ARMS MAKERS	Col. Robert Gardner
AMERICAN FIREARMS	Stephen Van Renseller
AMERICAN GUNMAKERS	Satterlee
ENCYCLOPEDIA OF AMERICAN HAND ARMS	
SILAS BRONSON LIBRARY	
Waterbury, Conn.	
CHICAGO PUBLIC LIBRARY	
JOHN CRERAR LIBRARY	
Chicago, Ill.	
RIFLED FIREARMS	Colburns New Monthly 1861
Breech loading rifles	Cornhill Magazine Sept.1866
BULLETS USED IN THE CIVIL WAR	S. Phillips
BOXER CARTRIDGE IN THE BRITISH SERVICES	B. A. Temple
RIFLES AND AMMUNITION	Ommundsen & Robinson
MUZZLE FLASHES	
FIREARMS OF THE CONFEDERACY	Fuller & Stewart
I.C.C. BULLETIN-MARCH/APRIL 1979	
Evening meeting, March 16, 1868 of the Royal United Service Institution	
AMERICAN SPORTSMAN-1873	
U.S. CARTRIDGES AND THEIR HANDGUNS	Charles R. Suydam

MAYNARD'S MODEL 1866 AND 1867 PATENTED CARTRIDGES	John Belton
DR. MAYNARD-THE MAN, HIS PATENTS, HIS FIREARMS, HIS CARTRIDGES	George Murphy I.C.C.A
HISTORY AND DEVELOPMENT OF SMALL ARMS AND AMMUNITION VOLS. I, II, III	George Hoyem
RIM FIRE CARTRIDGES 1857 - 1984	John L. Barber
HIRAM BERDAN, CHIEF OF SHARPSHOOTERS	Roy M. Marcot
MILITARY BREECH-LOADING FIFLES AND BOXER AMMUNITION	Capt. Majendie S. Browne
SNIDER 577 CARTRIDGES	Charles H. Yust, Jr.
A HISTORY OF AMMUNITION MANUFACTURE IN NEW ZELAND	G. Sherriff

www.ingramcontent.com/pod-product-compliance
Ingram Content Group UK Ltd.
Pitfield, Milton Keynes, MK11 3LW, UK
UKHW051457290926
15471UKWH00008B/1249

9 780788 425929